Remediation:
Contaminant Transport, Particle Tracking, & Plumes

by D. James Benton

Copyright © 2019-2021 by D. James Benton, all rights reserved.

Preface

This is a combination of three books: *Contaminant Transport, Particle Tracking*, and *Plumes*. These three topics combine to build a system for modeling remediation strategies, which is essential to obtaining optimal results. I have used the principles and software described in this text over decades of practice and many successful environmental cleanup projects for the US DoD, EPA, ACoE, and USGS.

All of the examples contained in this book,
(as well as a lot of free programs) are available at:
https://www.dudleybenton.altervista.org/software/index.html

All of the color figures can be found here (click on cover):
https://djamesbenton.altervista.org/

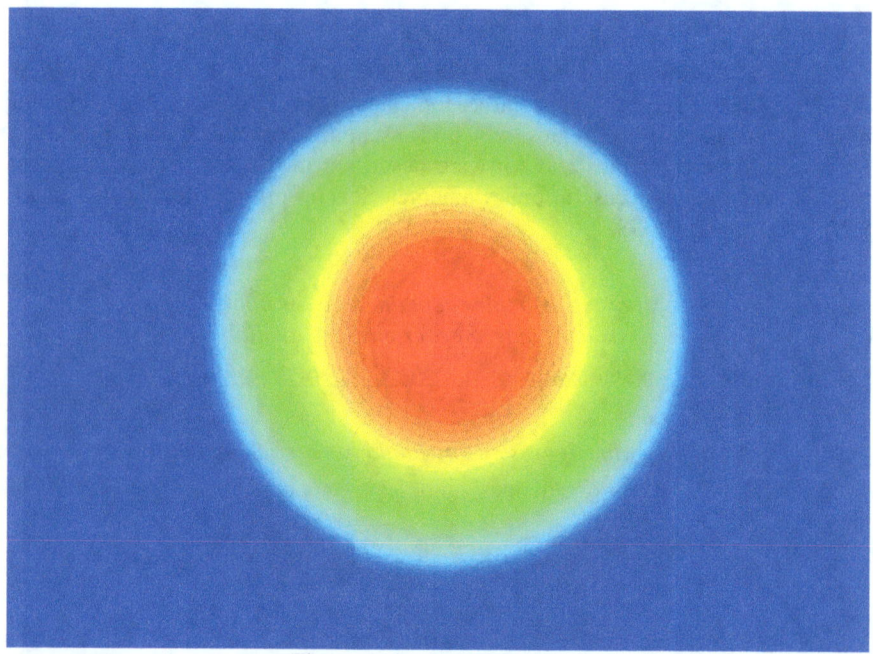

Figure 1. Analytical Solution

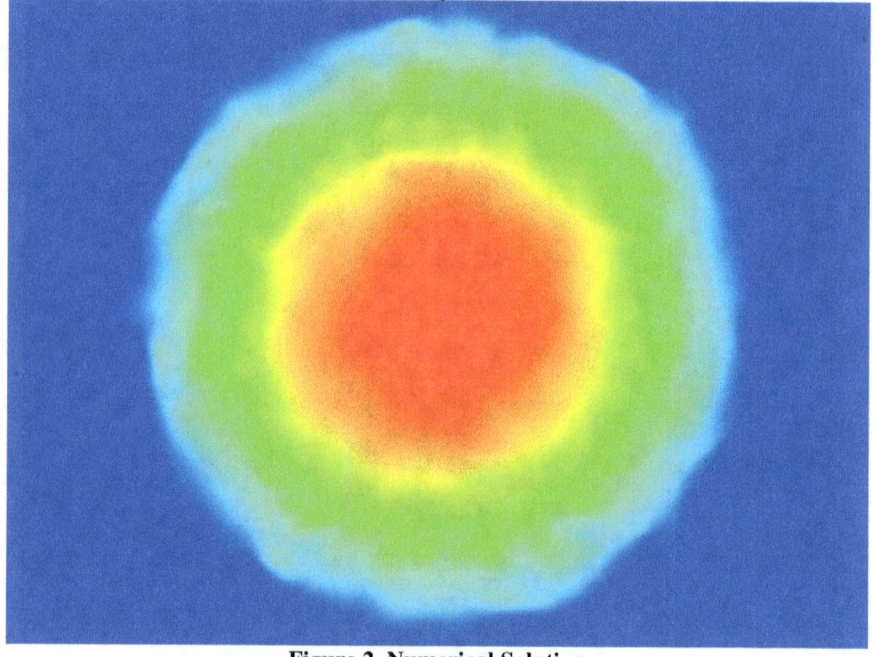

Figure 2. Numerical Solution

Table of Contents

	page
Preface	i
Chapter 1. Introduction	1
Chapter 2. Diffusion in 2D	3
Chapter 3. Dispersion in 2D	13
Chapter 4. Advection in 2D	16
Chapter 5. Contaminant Transport in 3D	27
Chapter 6. MODFLOW Based Models	35
Chapter 7. FRAC3D Based Models	55
Chapter 8. Pump-and-Treat	79
Chapter 9. Reaction and Decay	82
Chapter 10. AT123D Analytical Solution	85
Chapter 11. Two-Dimensional Lagrangian Tracking	90
Chapter 12. Three-Dimensional Lagrangian Tracking	96
Chapter 13. Particle Tracking in Discrete Domains	98
Chapter 14. Hamiltonian Particle Tracking	103
Chapter 15. Diffusion and Dispersion	115
Chapter 16. Flow in Fractures	131
Chapter 17. Contaminant Plumes	135
Chapter 18. Particle Seeds	139
Chapter 19. Animations	146
Chapter 20. Concentration Mappings	149
Chapter 21. Reverse Particle Tracking	151
Chapter 22. Sources and Sinks	153
Chapter 23. Mosquito Tracking	157
Chapter 24. Tracking Particles Inside Pipes	161
Chapter 25. Simple Airborne Contaminant	167
Chapter 26. Point Source Releases	169
Chapter 27. Dispersion	176
Chapter 28. Slot Jets	179
Chapter 29. 3D Thermal Plume	221
Chapter 30. Three-Dimensional Geological Data	230
Chapter 31. Contaminant Plumes in Groundwater	236
Chapter 32. Particle Tracking of Plumes	242
Appendix A. Displaying Data in 3^+D	257
Appendix B. 3D Data files for Tecplot™	261
Appendix C: 3D Data Files for TP2	262
Appendix D. Inverse Distance Interpolation	263
Appendix E. Relaxation Method	267
Appendix F. Kriging	270
Appendix G. BMP to GIF Conversion	273
Appendix H. Potential Fields	274
Appendix I. Boundary Element Method	276
Appendix J. Explicit Runge-Kutta Methods	278
Appendix K. Build3D Model Builder	282
Appendix L. Initial Concentrations	286
Appendix M. Validation of PTRAX	289
Appendix N. PTRAX Coding	294

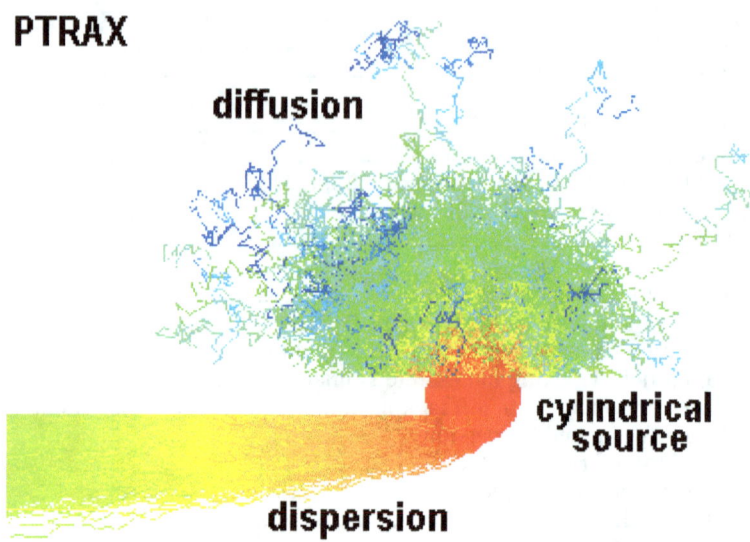

Figure 3. Particle Tracks Showing Diffusion and Dispersion

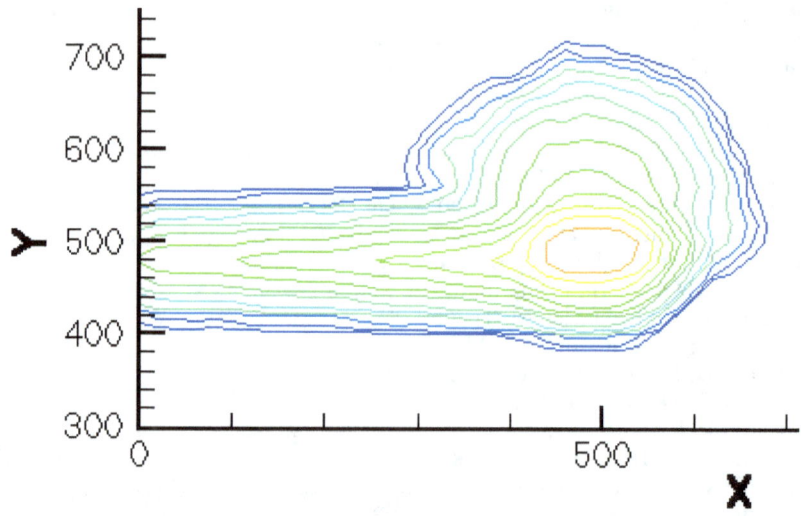

Figure 4. Corresponding Concentration Contours

Chapter 1. Introduction

This text describes the modeling of contaminants in the environment: atmosphere, surface water, and groundwater. Effective design and evaluation of potential designs for remediation requires detailed modeling, include representation of the contaminant at the start of the project and how this may evolve over time with no action and the intended plan. Three aspects of this process are discussed herein: 1) contaminant transport, 2) particle tracking, and 3) plume delineation and representation. We first consider the basics of the approach considered in this text.

Nodes and Elements

As numerical approaches are considered here, we must represent the domains containing contaminants consistent with computational analyses. The domains, therefore, are defined by nodes and elements. Nodes define the corners of the elements, not at the center. Velocities are defined at the center of each element. Nodes and elements can be two- or three-dimensional. Two-dimensional elements include: triangles and quadrangles (i.e., four-sided polygons with no particular constraint, such as square corners or parallel sides). Three-dimensional elements include: tetrahedra and bricks (i.e., six-faced polyhedra with no particular constraint, such as square corners or parallel faces). Outward normal faces are assumed, but are not necessary, as software can easily swap vertices to obtain the preferred orientation. It is assumed that the reader already understands these terms.

Eulerian Point-of-View

How the flow fields are obtained (i.e., analytical solution, finite difference method, finite element method, boundary element method, etc.) is immaterial and will only be discussed in passing (see Appendix I for the boundary element method). Suffice it to say, you will need to generate these somehow, which is the subject of another text. The domains and velocity fields so described are from the Eulerian[1] point-of-view, which basically means that the domain is fixed and forms the frame of reference, while flow passes through the domain, varying with position and perhaps time. An example of this perspective would be standing on a bridge watching the river flow and boats float by.

Lagrangian Point-of-View

The Lagrangian[2] point-of-view is not stationary, but moves through the domain, focused on some particular moving item, such as a particle. An example of this perspective would be a passenger on one of the boats, noting the bridge and first observer as the two move relative to each other. This is the classical approach to particle tracking. Tracking a particle using this methodology

---

[1] Leonhard Euler (1707–1783) Swiss mathematician, physicist, astronomer, geographer, logician, and engineer.
[2] Joseph-Louis Lagrange (1736–1813) Italian mathematician and astronomer.

answers the question: how far will the particle move (and in what direction) over the course of some time step?

Hamiltonian Point-of-View

The Hamiltonian[3] point-of-view is far less common and not so easily illustrated. If you search the Web for descriptions of this approach (e.g., Wikipedia), you will find some rather vague descriptions. Peruse enough of these and a pattern will emerge: the dependent variables discussed are neither space nor time, but quantities like momentum and energy. When it comes to particle tracking, the independent variable is velocity or displacement in the case of diffusion or dispersion. Time is a dependent variable. Tracking a particle using this methodology answers the question: how long will it take the particle to move this far? In either case (Lagrangian or Hamiltonian) we must be concerned with the particles leaving one element and entering another, as this is a discrete and not necessarily continuous domain, although we will begin with simple continuous domains and analytical flow solutions.

---

[3] William Rowan Hamilton (1805–1865) Irish mathematician and physicist.

Chapter 2. Diffusion in 2D

We begin with what is perhaps the simplest meaningful example: diffusion in two dimensions from the Eulerian point-of-view. We have already covered one-dimensional diffusion in **Mass Transfer**. If the diffusion coefficient, D, is constant (spatially and temporally), Fick's second law becomes:

$$\frac{\partial C}{\partial t} = D\left(\frac{\partial^2 C}{\partial x^2} + \frac{\partial^2 C}{\partial y^2}\right) \quad (2.1)$$

The simplest solution to implement is finite differences spatially and Euler's (explicit, single-step) method temporally. Using subscripts i and j to indicate the X and Y directions, respectively, Equation 2.1 becomes:

$$\frac{C_{i,j}^{t+\Delta t} - C_{i,j}^{t}}{\Delta t} = D\left(\frac{C_{i+1,j}^{t} - 2C_{i,j}^{t} + C_{i-1,j}^{t}}{\Delta x^2} + \frac{C_{i,j+1}^{t} - 2C_{i,j}^{t} + C_{i,j-1}^{t}}{\Delta y^2}\right) \quad (2.2)$$

If the problem is not too stiff (rapidly changing spatially or temporally resulting in numerical instability) and we use a sufficiently fine grid and small time steps, we can simply march forth in time to obtain a solution. We must choose a contaminant substance, properties, scale (length and time), and initial concentration. I have provided modeling for the remediation of several sites where Trichloroethylene (TCE) is involved and so we select this for our first example. TCE is a contaminant of particular interest to the USEPA and many articles related to this common solvent can be found on their website.

The solubility of TCE in water is approximately 1000 mg/l[4] and the density of TCE is approximately 1.5 times that of water so an initial concentration of 300 ppm is reasonable. We will begin with an amorphous media before considering porosity and the separate phases (solid and liquid). The diffusion coefficient for TCE in water is approximately 0.00001 cm²/sec. The initially contaminated area extends over 80m and is triangular in shape. The domain of consideration is 640m by 640m. The grid spacing in both directions is 1m.

The ratio $D/\Delta x^2 \approx 0.3$/yr provides an estimate of the appropriate time step. The dimensions (640x640) facilitate graphical representation of the results, which will be two-dimensional, sequential time frames, colored from blue to red based on the log of the concentration with red indicating log(300), blue indicating log(0.3), and gray indicating zero. All of the files can be found in the online archive in folder examples\TCE2D. The initial conditions are shown in the following figure:

---

[4] "Using the Combined SESOIL/AT123D Models to Develop Site-Specific Impact to Ground Water Soil Remediation Standards for Mobile Contaminants," Guidance Document Version 2.1, New Jersey Department of Environmental Protection, Trenton, New Jersey, May, 2014. [As well as providing various properties, this is an important reference for using AT123D, which we discuss in Chapter 9.]

Figure 5. Example 1 Initial Conditions

Using the simple forward Euler method for the time step (Equation 2.2) it should be clear that the contaminant cannot spread any more rapidly than one nodal point per time step. Considering this fact and also the ratio of the diffusion coefficient to the grid size, the time step should be no more than $\Delta t < 0.3$ yr. We initialize the field and then march forward in time. While we could immediately update each concentration while running through a nest of two loops (X and Y), this would produce a sweeping artifact as the cells are updated. Instead, we allocate a working array to contain $\partial C/\partial t$ and then implement $C=C+(\partial C/\partial t)\Delta t$.

It is not necessary to save an image at each time step. One frame per 100 time steps is often enough so that the bitmap image need only be updated every 30 years. As we are coloring the image with the log of the concentration, once the contaminant spreads, it may take longer to update the image than to advance the solution. After 1800 years the contaminant has spread as shown in the following figure. Note that the concentrations are logarithmic.

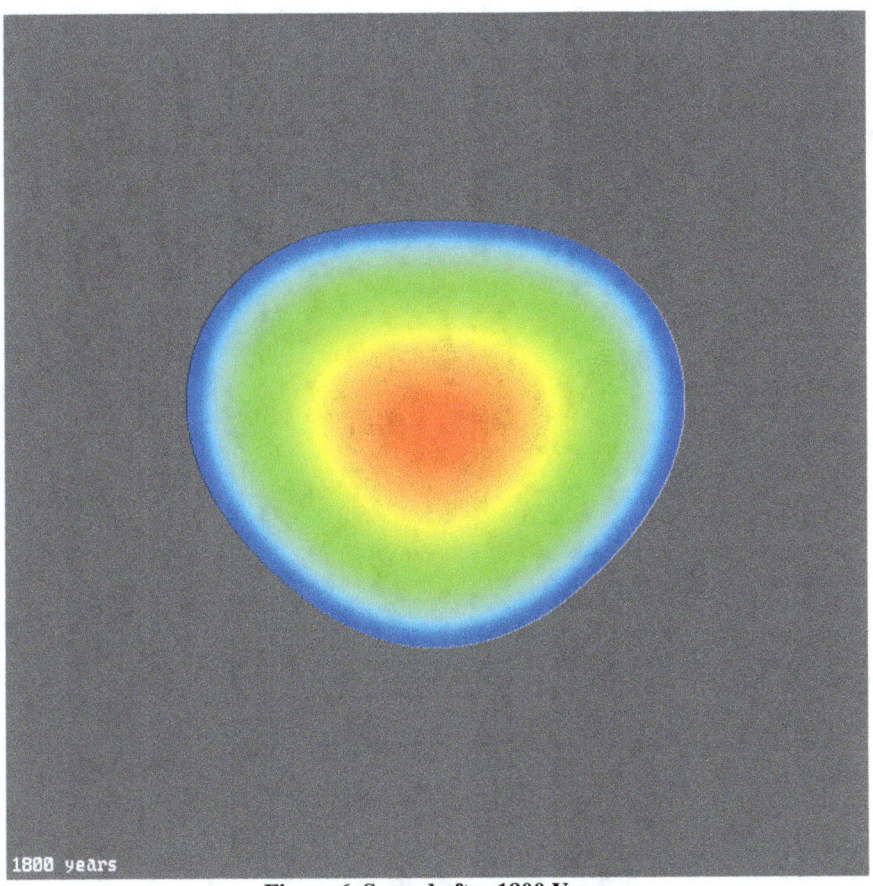

Figure 6. Spread after 1800 Years

You may wonder how small of a time step is small enough and is the forward Euler method adequate. If we increase the time step from 0.3 years (recall $D/\Delta x^2 \approx 0.3$/yr) to 1.0 yr, the solution quickly becomes unstable. If we used a higher order method (for example, 4th Order Runge-Kutta, rather than forward Euler), we could use a somewhat larger Δt, but not much. We would also have the problem of the contaminant not spreading more than one node (in this case 1m) in any direction per time step. This is not to say that higher order methods are without advantage, only that for this simple problem, they do not offer a significant benefit. These will be used in subsequent examples. Pure diffusion is inherently stable. When we introduce flow and turbulence, we will be more concerned with stability and will resort to more clever methods, which will also be more computationally intensive. The unstable result after 300 years with a time step of 1 yr is shown in this next figure:

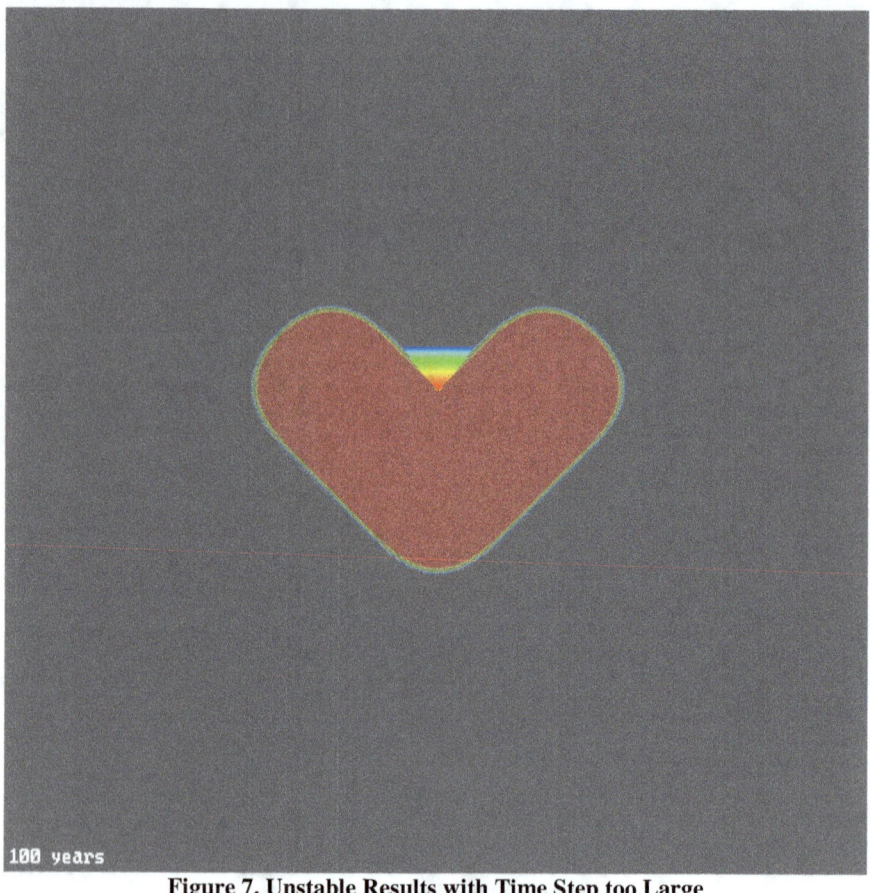

Figure 7. Unstable Results with Time Step too Large

Boundaries

Equation 2.2 can only be applied at the interior points. The code is quite simple. Recall the working array, *W*, used to calculate the change in *C*, as listed below:

```
void AdvanceSolution(double dt)
   {
   int i,x,y;
   for(y=1;y<Ny-1;y++)              /* interior points */
      {
      for(x=1;x<Nx-1;x++)
         {
         i=Nx*y+x;
         dCdt[Nx*y+x]=a*((C[i+ 1]-2.*C[i]+C[i- 1])
                    +(C[i+Nx]-2.*C[i]+C[i-Nx]));
         }
      }
```

```
for(y=1;y<Ny-1;y++)                    /* interior points */
{
    for(x=1;x<Nx-1;x++)
    {
        i=Nx*y+x;
        C[i]+=dt*dCdt[i];
    }
}
```

This still leaves all of the nodes around the edges (top, bottom, left, right, plus four corners). In this case "natural" boundary conditions are applied, which means $\partial C/\partial x=0$ on the left and right sides and $\partial C/\partial y=0$ on the top and bottom sides. The most "natural" way of handling the corners is to set these nodes equal to the average of the three closest interior nodes. Details can be found in the code (tce2d.c).

```
for(y=0,x=1;x<Nx-1;x++)                /* bottom side */
    C[Nx*y+x]=C[Nx*(y+1)+x];
for(y=Ny-1,x=1;x<Nx-1;x++)             /* top side */
    C[Nx*y+x]=C[Nx*(y-1)+x];
for(x=0,y=1;y<Ny-1;y++)                /* left side */
    C[Nx*y+x]=C[Nx*y+x+1];
for(x=Nx-1,y=1;y<Ny-1;y++)             /* right side */
    C[Nx*y+x]=C[Nx*y+x+1];
x=y=0;                                 /* bottom left corner */
C[Nx*y+x]=(C[Nx*y+x+1]+C[Nx*(y+1)+x]+C[Nx*(y+1)+x+1])/3;
x=Nx-1;y=0;                            /* bottom right corner */
C[Nx*y+x]=(C[Nx*y+x-1]+C[Nx*(y+1)+x]
          +C[Nx*(y+1)+x-1])/3.;
x=0;y=Ny-1;                            /* top left corner */
C[Nx*y+x]=(C[Nx*y+x+1]+C[Nx*(y-1)+x]
          +C[Nx*(y-1)+x+1])/3.;
x=Nx-1;y=Ny-1;                         /* top right corner */
C[Nx*y+x]=(C[Nx*y+x-1]+C[Nx*(y-1)+x]+C[Nx*(y-1)+x-
          1])/3.;
```

Impact of Properties

We next consider what happens when we double the diffusion coefficient. For illustration, we will double it in the Y direction but not the X. We must half the time step because we must consider the most restrictive case (X or Y). After the same 1800 years the results are shown in Figure 4, along with the outline of the extent of contamination from Figure 2. Again, recall that the colors represent the log of the concentration. One question that often arises in modeling is, "What if we're not sure about the properties (e.g., diffusion coefficient)?" We may not be sure (especially underground) or precise (properties may vary and "undisturbed" soil samples are hard to obtain), but this doesn't necessarily translate into the same level of uncertainty in the results.

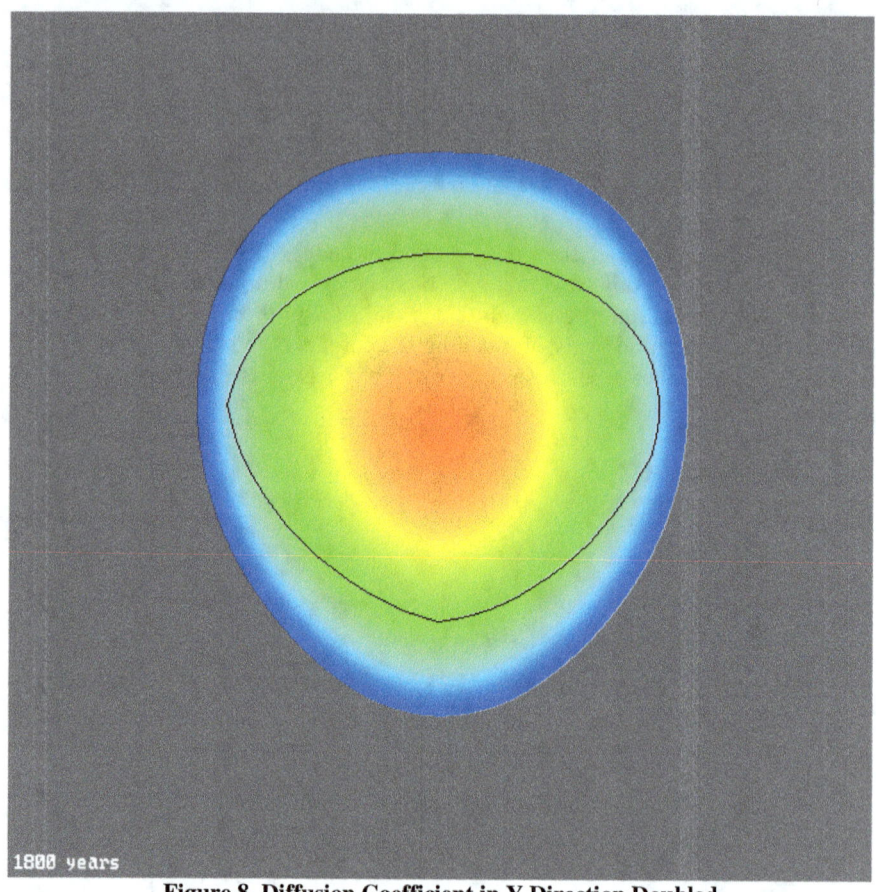

Figure 8. Diffusion Coefficient in Y Direction Doubled

Properties, such as diffusion coefficient, describe the behavior of the media, which is often underground. Soil properties rarely vary linearly. More often these vary with geological formation or alluvial deposition. While these often vary vertically (e.g., layers of different soil types), the planar (X and Y for this example) variation is separated into regions. The model builder, Build3D, which we will discuss in a later chapter, accepts soil types in various geometric shapes and builds the properties so indicated into the corresponding grid elements. In this example, we will just use two polygons and test for above or below and right or left, so as to assign a high, medium, and low value of diffusion coefficient, as shown in the next figure.

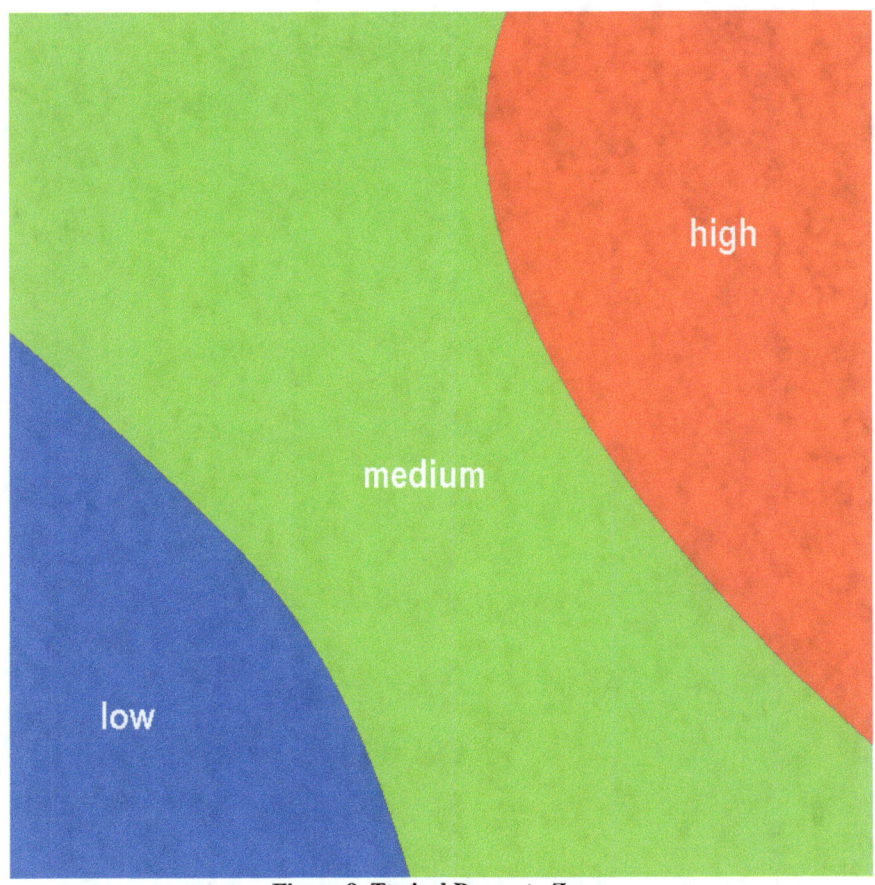

Figure 9. Typical Property Zones

We expect the contaminant will preferentially spread up and to the left into the high area, continue as before in the mid zone and less penetration into the low zone in the lower left. This is exactly what we see with real contaminant plumes. In fact, rate of spread of tracer chemicals is often used to infer properties of the soil. One of the longest studies conducted on this subject is the Macro Dispersion Experiment (MADE) conducted at Columbus Air Force Base. I served as the applied mathematician on the project throughout the 1990s, writing much of the software used. Many reports can be found on this experiment, published by EPRI and also the USGS. I continued working with some members of the same Team for several years after that, using the skills, knowledge, and tools developed during MADE on other projects for the USGS, USEPA, USDoE, USDoD, and USACoE. The resulting contaminated area after 1800 years is shown in this next figure.

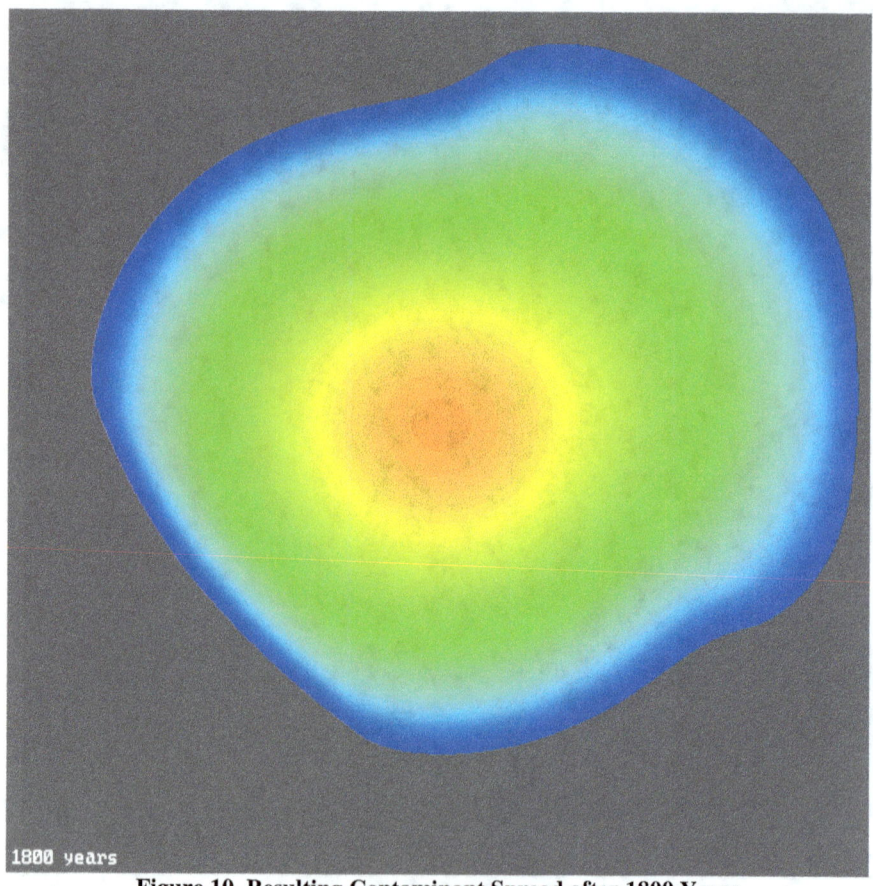

1800 years

Figure 10. Resulting Contaminant Spread after 1800 Years

Time Steps and Stability

For this simulation (using the diffusion field shown in Figure 5 and resulting in the spread after 1800 years shown in Figure 6), we used a time step of Δt of 0.05 (1/20th) year. The highest diffusion coefficient in Figure 5 is 0.0006 cm²/sec. This combined with the grid spacing of 1m yields a characteristic ratio of approximately 1/10th ($\Delta t D/\Delta x^2 = 0.095$). If we double the diffusion coefficient throughout the domain, we should expect to also divide the time step in half to 0.025 (1/40th) year in order to maintain stability. Not surprisingly, the resulting spread is roughly twice that shown in Figure 6.

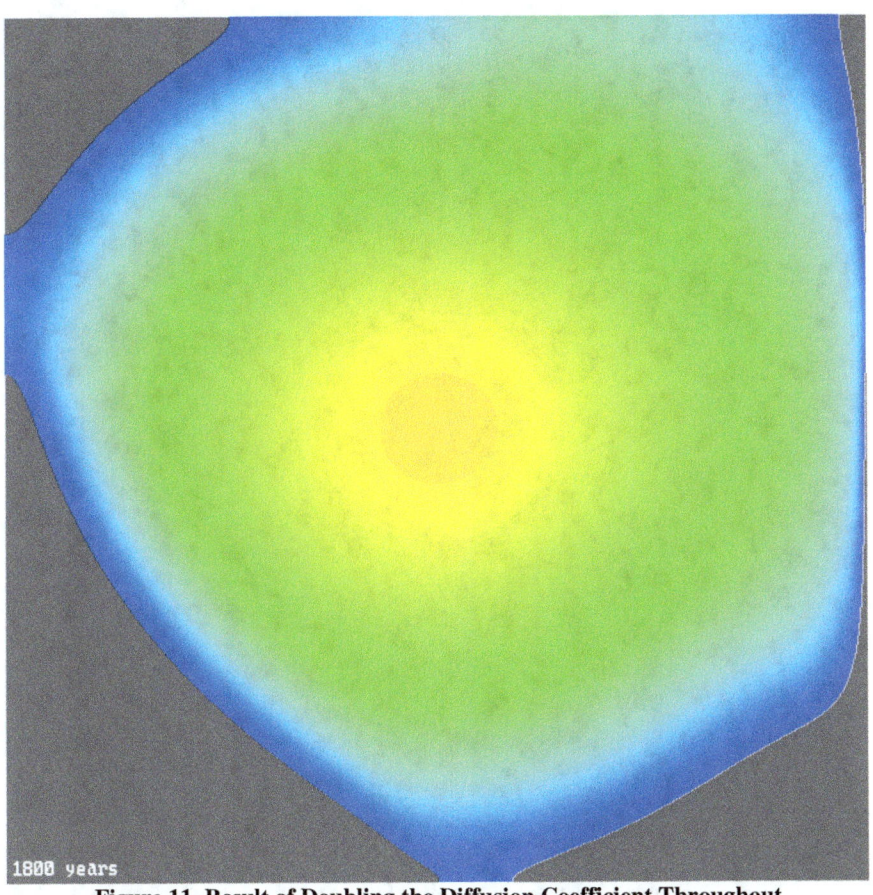

Figure 11. Result of Doubling the Diffusion Coefficient Throughout

If we had not reduced the time step in proportion to the maximum diffusion coefficient, the solution would have eventually become unstable. What happens numerically for this partial differential equation (i.e., Laplace's) is an ever-increasing oscillation (high/low up/down) of values (in this case concentrations) at adjacent nodes, which is a numerical artifact and completely uncharacteristic of the underlying physics described by the governing equation. In other words: garbage. This isn't always easy to detect when running a model. Because this particular model generates images as it steps along through time, we can see this "garbage" show up. A mere factor of 2 (in the wrong way, larger time step or larger diffusion coefficient) will quickly veer of course in 60 years trashing most of the domain, as shown in this next figure:

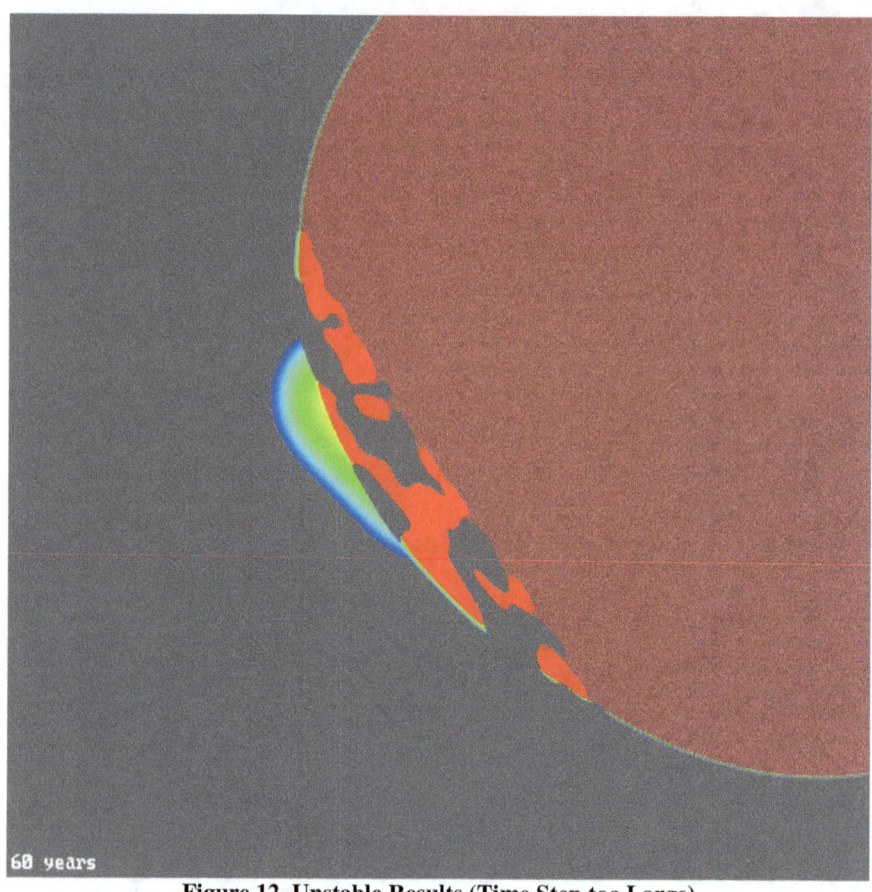

Figure 12. Unstable Results (Time Step too Large)

Chapter 3. Dispersion in 2D

If contaminants only spread by diffusion and the properties used for the examples in Chapter 2 are at all reasonable, contamination in groundwater would be of much lesser concern. In real situations, the contaminants move much faster than this first example would indicate because there are more mechanisms available. While diffusion is mostly omnidirectional (except for markedly different soil types), dispersion exhibits a preferential direction. Increased dispersive spreading (compared to diffusive spreading) perpendicular to the preferential direction is often seen, though noticeably less than that along the preferential direction. The same partial differential equation governs both diffusion and dispersion and the respective coefficients have the same units. A simple variation in dispersion coefficient (which we assign the symbol, A) along the X-axis is shown below (increasing toward the bottom right corner):

Figure 13. Dispersion Coefficient in the X Direction

A different (though still simple) variation in dispersion coefficient in the Y direction (with the same linear coloring scale) is shown in this next figure (decreasing toward the bottom right corner):

Figure 14. Dispersion Coefficient in the Y Direction

These dispersion coefficients are approximately 10x that of the diffusion coefficients (a reasonable ratio), implying that the rate of spreading due to dispersion will be about 10x that of diffusion. The dispersion coefficient is often expressed as the product of the dispersivity (having units of length) times the pore water velocity (having units of length/time) to obtain the dispersion coefficient (having units of length²/time). There will also be more spatial asymmetry arising from the combination of non-uniform properties assigned over the domain. The time step must be adjusted accordingly in order to keep the solution from becoming unstable. We have added two steps in the code (tce2d.c) to account for this automatically: 1) keeping track of the maximum value of A or D, and 2) calculating the value of Δt to make the ratio equal to one-third ($\Delta t D / \Delta x^2 = 1/3$). We also calculate the frequency of saving results so that the frames are written at convenient values of time. After only 180 years under this new scenario, the concentration field becomes:

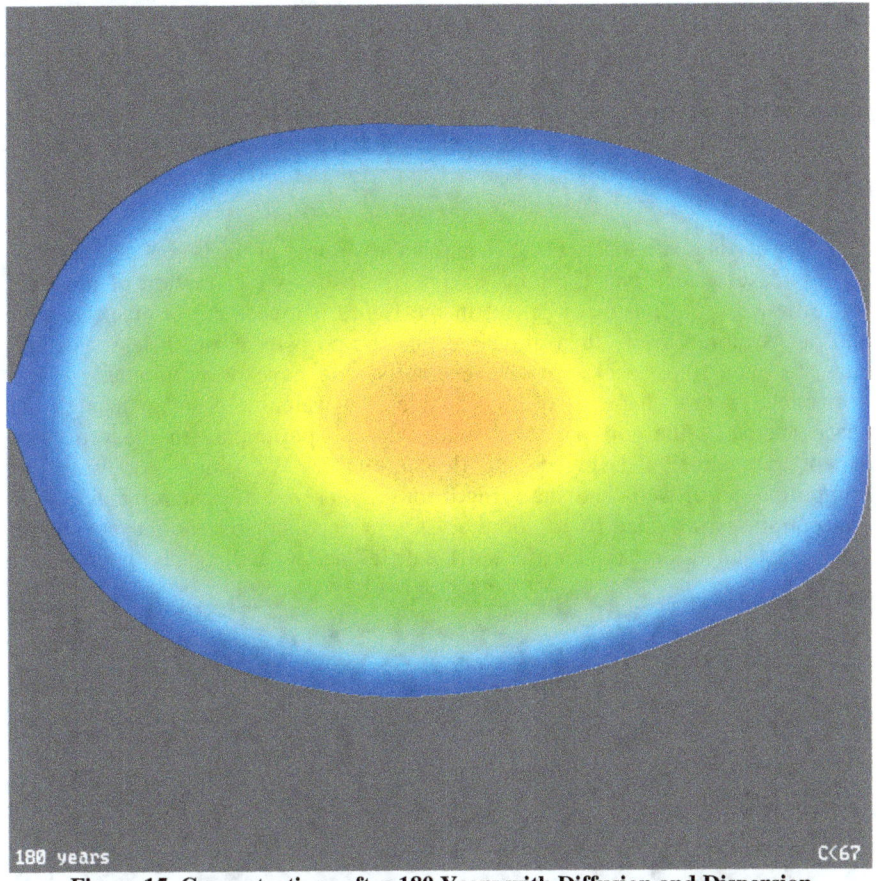

Figure 15. Concentrations after 180 Years with Diffusion and Dispersion

By adding approximately 10x dispersion to diffusion, in one-tenth the time (180 vs. 1800 years), the contaminant has spread so much that the red level has completely disappeared from the figure. The maximum concentration has fallen from 300 ppm to 67 ppm and even that level extends over a smaller area. Keep in mind that mass is conserved (we have not yet considered reactions or decay) so that the same amount of contaminant is still in the environment, only it is much more spread out. If there were any hope of containing or extracting it, this scenario would be a complete failure. Accurate modeling is essential for containment or capture. We must know where the contaminant is going and when it will get there if we are ever to remediate this situation.

Chapter 4. Advection in 2D

We now consider advection; that is, movement of the fluid media containing the contaminant. Before we do this, we must consider porosity, which we ignored in the first two chapters. When considering a contaminant in the ground, there are at least three materials: the contaminant, the soil, and groundwater. We must, therefore, consider spreading (and possibly reaction) of the contaminant in the water and the soil, which may be quite different. We first consider only the contaminant and water, ignoring the soil.

Porosity

Porosity is given the symbol, φ, and is the volume of the secondary material (in this case water) over the total volume. Porosity is the fraction of the total space between the pieces of gravel in the figure below. Sand is much smaller and packs more closely than gravel so that the porosity is much less. Silt and clay are even finer, pack even more closely, and have proportionately lower porosities. In this chapter we will consider the water as passing through the spaces between the soil particles, which do not participate in the processes except to channel and impede the flow of water. We sample the water by pumping it out of wells and analyzing it for the presence of contaminants. These concentrations provided by a laboratory are in the water by weight or volume, not in the total volume (water plus soil), a difference we will discuss later.

Figure 16. Typical Gravel Illustrating Porosity

Porosity can range from 0 to 1, but most often is between 0.2 and 0.3. Typical porosities for some common soil types are listed in Table 1.

Table 1. Typical Soil Porosities[5]

USDA Soil Class	porosity
clay (very fine)	0.20
clay (fine)	0.22
sandy clay	0.24
sandy loam	0.25
silty clay	0.25
sandy clay loam	0.26
silt	0.27
silty clay loam	0.27
loamy sand	0.28
clay loam	0.30
loam	0.30
sand	0.30
sily loam	0.35

Governing Equation Revisited

We now consider the governing partial differential equation (which is the conservation of mass of the contaminant for a differential control volume) in more detail. It is helpful to introduce here the del operator, ∇, to abbreviate the spatial derivatives while implying all three dimensions, which will facilitate our eventual advancement into 3D. This useful operator is defined as:

$$\nabla = \hat{i}\frac{\partial}{\partial x} + \hat{j}\frac{\partial}{\partial y} + \hat{k}\frac{\partial}{\partial z} \qquad (4.1)$$

In Equation 3.1, $\hat{i}$, $\hat{j}$, and $\hat{k}$ are unit vectors in the X, Y, and Z directions, respectively. When applied to a scalar quantity, such as concentration, the del operator yields a vector result. The diffusion and dispersion contributions (to the conservation of mass of the contaminant for a differential control volume) should more accurately be written:

$$\nabla \bullet [(A+D)(\nabla \varphi C)] \qquad (4.2)$$

On the right side of Equation 3.2, the terms A, D, φ, and C are all scalar quantities, but the del operator (in the middle of this group) makes the result a vector. The symbol (•) indicates the dot product. The dot product of two vectors is a scalar so that the entire expression is a scalar. The properties (dispersion coefficient, diffusion coefficient, and porosity) are inside the one or more

---

[5] Bonazountas, M. and Wagner, J. M., "SESOIL: A Seasonal Soil Compartment Model," USEPA Report PB86112406, 1984.

differential operators (one for A and D; two for φ). When we introduced a spatially varying diffusion coefficient in Chapter 2, we should have written the term in the expanded form of Equation 3.3 and also modified the finite difference equations in the code accordingly.

$$\frac{\partial}{\partial x}\left(D\frac{\partial C}{\partial x}\right)+\frac{\partial}{\partial y}\left(D\frac{\partial C}{\partial y}\right) \tag{4.3}$$

Adding the dispersion coefficient and also the porosity, this term expands to:

$$\frac{\partial}{\partial x}\left[(A+D)\frac{\partial(\varphi C)}{\partial x}\right]+\frac{\partial}{\partial y}\left[(A+D)\frac{\partial(\varphi C)}{\partial y}\right] \tag{4.4}$$

The advective contribution to the conservation of mass of the contaminant in vector form using the del operator is given by:

$$\nabla \bullet \left(\varphi C \vec{V}\right) \tag{4.5}$$

Again, the right side of Equation 3.5 is a vector, as porosity, φ, and concentration, C, are scalars and velocity, $\vec{V}$, is a vector. As before, the dot product yields a scalar, which becomes:

$$\frac{\partial}{\partial x}(\varphi C u)+\frac{\partial}{\partial y}(\varphi C v) \tag{4.6}$$

This expands to:

$$u\frac{\partial}{\partial x}(\varphi C)+v\frac{\partial}{\partial y}(\varphi C)+\varphi C\left(\frac{\partial u}{\partial x}+\frac{\partial v}{\partial y}\right) \tag{4.7}$$

The last term in Equation 3.7 is zero by virtue of continuity. Combining these terms (diffusion plus dispersion and advection), we obtain an expression for the conservation of mass for the contaminant:

$$\frac{\partial(\varphi C)}{\partial t}=\frac{\partial}{\partial x}\left[(A+D)\frac{\partial(\varphi C)}{\partial x}\right]+\frac{\partial}{\partial y}\left[(A+D)\frac{\partial(\varphi C)}{\partial y}\right]-\left[u\frac{\partial}{\partial x}(\varphi C)+v\frac{\partial}{\partial y}(\varphi C)\right] \tag{4.8}$$

We will expand and adapt the previous example (tce2d.c) to account for the expanded differentials and also the additional terms. Instead of further complicating this code and obscuring the earlier features, we will slightly modify the name. Tetrachloroethylene (PCE) is often found with or instead of TCE and so this example will be called pce2d.c and retain all the other properties, as these are quite similar for the two solvents. We will continue with the simple finite difference method for the spatial and forward Euler method for temporal dimensions. We will also continue using a one-to-one correspondence between nodal points and pixels when generating the graphics. When we move to 3D, we will adopt a different painting method and use far fewer nodes, which will require some changes to preserve the calculus.

Figure 17. Velocity Field

We will assume a constant porosity of 0.3, though this could also be distributed over the domain in another array. The next thing we will need is a velocity field to represent advection (see figure above). Typical values range from 0.01 to 10 m/day, which roughly correspond to silt and sand, respectively. The longest vector in the figure represents 0.02 m/day. In a real life application, we would use some groundwater flow model to generate this, perhaps FRAC3D[6] or MODFLOW.[7] Here, we will merely fabricate a field, shown as vectors in the preceding figure.

Finite Difference Implementation

We next consider how to implement Equation 3.8 in finite difference form. This expression contains sums $(A+D)$ and products (φC) as well as first and second differentials, making the calculation considerably more complicated. Consider the figure below showing 9 nodes within and surrounding a single differential element or control volume:

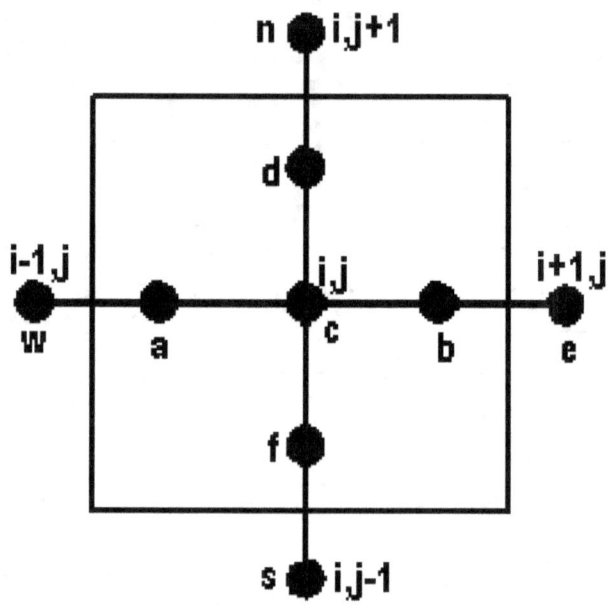

Figure 18. Nodes and Differential Element

The central (i,j), north (i,j+1), south (i,j-1), east (i+1,j), and west (i-1,j) elements are the usual ones for calculating the second derivative and were employed in Chapter 2 (tce2d.c). We have added 4 intermediate nodes labeled a, b, d, and f. The relationships are simple: a=(w+c)/2, b=(e+c)/2, d=(n+c)/2, and f=(s+c)/2. The first term on the right side of Equation 3.8 becomes:

---

[6] Therrien, R., and E.A. Sudicky, "Three Dimensional Analysis of Variably–Saturated Fow and Solute Transport in Discretely-Fractured Porous Media," Journal of Contaminant Hydrology, Vol. 23, No. 2, pp. 1-44, 1996.
[7] McDonald, M. G. and Harbaugh, A. W., "*A Modular Three-Dimensional Finite-Difference Ground-Water Flow Model,*" USGS Report 83-875, 1984.

$$\frac{\partial}{\partial x}\left[(A+D)\frac{\partial(\varphi C)}{\partial x}\right] = \frac{(A+D)_b \frac{\partial(\varphi C)_b}{\partial x} - (A+D)_a \frac{\partial(\varphi C)_a}{\partial x}}{\Delta x} \quad (4.9)$$

The intermediate property values (A_a, A_b, D_a, and D_b) are merely averages:

$$A_a = \frac{A_w + A_c}{2} \quad (4.10)$$

The intermediate partial derivatives are evaluated between two nodes:

$$\frac{\partial(\varphi C)_a}{\partial x} = \frac{\varphi_c C_c - \varphi_w C_w}{\Delta x} \quad (4.11)$$

The Y contributions in the second term on the right side of Equation 3.8 are calculated as for the X contributions with nodes c̲, d̲, f̲, n̲, and s̲. Contributions in the third term on the right side of Equation 3.8 are calculated using nodes a̲, b̲ and d̲, f̲, respectively, as in:

$$u\frac{\partial}{\partial x}(\varphi C) = u_c \frac{\varphi_c C_c - \varphi_w C_w}{\Delta x} \quad (4.12)$$

We calculate each of these quantities and then combine them to arrive at the partial derivative with respect to time of the product of porosity and concentration, $\partial \varphi C/\partial t$. Because we kept the porosity, φ, within the parentheses and differentials, this calculation can accommodate spatially varying porosity, which we will implement in the upcoming examples. While it is plausible that porosity might vary over time, that is rarely considered.

$$\frac{\partial(\varphi C)}{\partial t} = \varphi \frac{\partial C}{\partial t} + C \frac{\partial \varphi}{\partial t} \quad (4.13)$$

<u>Upwind Differences</u>

If we were to simply implement the finite difference expressions for advection as for diffusion and dispersion, we would immediately find that the results are unstable. We might decrease the time step and find that the results are still unstable. In fact, no matter how small we make the time step, the results will be unstable. This observation might cause us to revisit the forward Euler method; perhaps replacing it with an implicit or hybrid method, but the results would still be unstable.

If you peruse the literature on advection, you will find many papers devoted to this issue of instability, which comes down to information. If we use a central difference, we are including information (in this case concentration) from where the flow is coming from and also where it is going to but has not yet gotten there; that is, upwind and downwind, respectively. Information downwind is irrelevant or at best noise. In our code implementation, it is truncation errors and round off, which we don't want to include in our calculation; therefore, we take a spatial finite difference on the side upwind of the differential element (e.g., from the west for positive U and from the south for positive V). This scheme was first suggested by Godunov.[8] The entire code can be found in the online archive in folder examples\pce2d in file pce2d.c an excerpt of which is listed below:

```
for(y=1;y<Ny-1;y++)  /* interior points */
{
  for(x=1;x<Nx-1;x++)
  {
    ic=Nx*y+x;
    ie=ic+1;
    iw=ic-1;
    in=ic+Nx;
    is=ic-Nx;
    Cc=C[ic];
    Ce=C[ie];
    Cn=C[in];
    Cs=C[is];
    Cw=C[iw];
    Fc=F[ic];
    Fe=F[ie];
    Fn=F[in];
    Fs=F[is];
    Fw=F[iw];
    Uc=U[ic];
    Vc=V[ic];
    Aa=(Ah[iw]+Ah[ic])/2.;
    Ab=(Ah[ie]+Ah[ic])/2.;
    Ad=(Av[in]+Av[ic])/2.;
    Af=(Av[is]+Av[ic])/2.;
    Ca=(C[iw]+C[ic])/2.;
    Cb=(C[ie]+C[ic])/2.;
    Cd=(C[in]+C[ic])/2.;
    Cf=(C[is]+C[ic])/2.;
    Da=(D[iw]+D[ic])/2.;
    Db=(D[ie]+D[ic])/2.;
    Dd=(D[in]+D[ic])/2.;
    Df=(D[is]+D[ic])/2.;
    dFCdXa=(Fc*Cc-Fw*Cw)/dX;
    dFCdXb=(Fe*Ce-Fc*Cc)/dX;
    dFCdYd=(Fn*Cn-Fc*Cc)/dY;
    dFCdYf=(Fc*Cc-Fs*Cs)/dY;
    d2ADFCdX2=((Ab+Db)*dFCdXb-(Aa+Da)*dFCdXa)/dX;
    d2ADFCdY2=((Ad+Dd)*dFCdYd-(Af+Df)*dFCdYf)/dY;
    if(Uc>0.)
```

---

[8] Godunov, S. K., "A Difference Scheme for Numerical Solution of Discontinuous Solution of Hydrodynamic Equations",Matematicheskii Sbornik (Transactions of the Moscow Mathematical Society), Vol. 47, pp. 271-306, 1959.

```
      UdFCdX=Uc*(Fc*Cc-Fw*Cw)/dX;
   else
      UdFCdX=Uc*(Fe*Ce-Fc*Cc)/dX;
   if(Vc>0.)
      VdFCdY=Vc*(Fc*Cc-Fs*Cs)/dY;
   else
      VdFCdY=Vc*(Fn*Cn-Fc*Cc)/dY;
   dCdt[ic]=(d2ADFCdX2+d2ADFCdY2-UdFCdX-VdFCdY)/Fc;
```

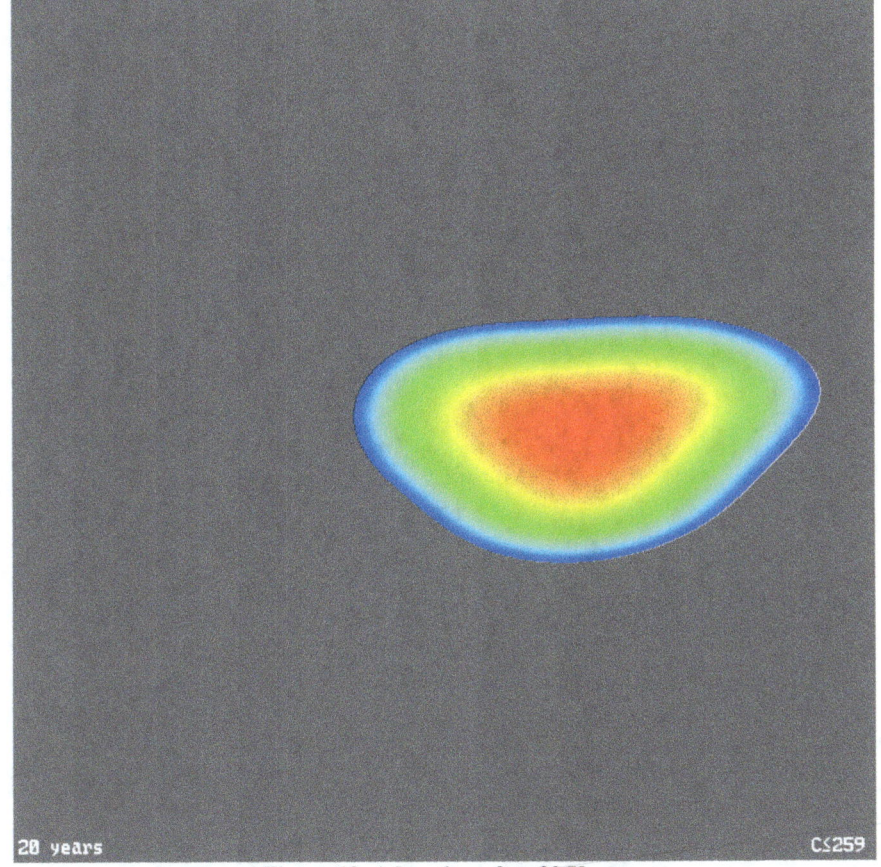

Figure 19. Advection after 20 Years

The solution advances through time as before using the Euler method, only we are now accounting for the spatial variation in properties (diffusion coefficient, dispersion coefficient, and porosity) correctly plus we have added porosity and advection. In just 20 years the center of concentration has moved noticeably to the east as shown in the preceding figure. In 75 years the center of concentration has left the domain:

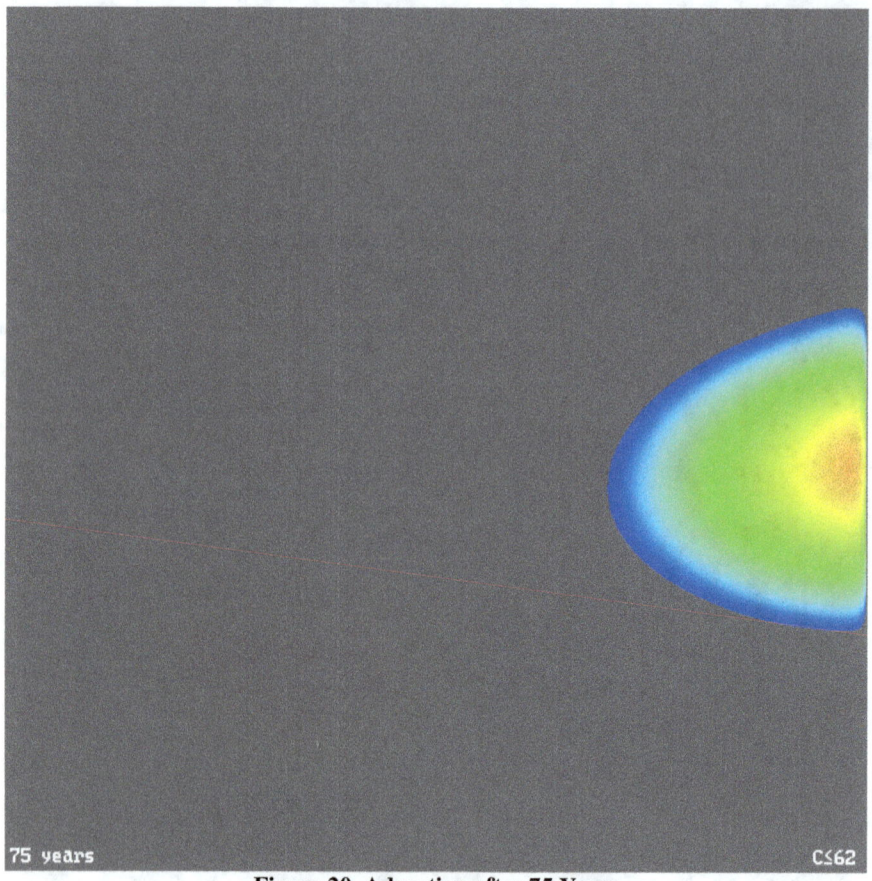

Figure 20. Advection after 75 Years

This domain is only 640m by 640 m, which is rather small compared to the cleanup projects I have worked on, especially the ones created by the DoD. Still, this example does illustrate that, even with a velocity of 0.02 m/day (7.3 m/yr), a contaminant can migrate as well as spread, complicating containment and capture and making remediation difficult.

Some of the largest contaminant plumes I have worked with include those adjacent to Otis Air Force Base and extend over Western Cape Cod, as illustrated in the following figure. These plumes have been remediated. The containment fences of extraction and sampling wells are shown as black circles and plusses, respectively.

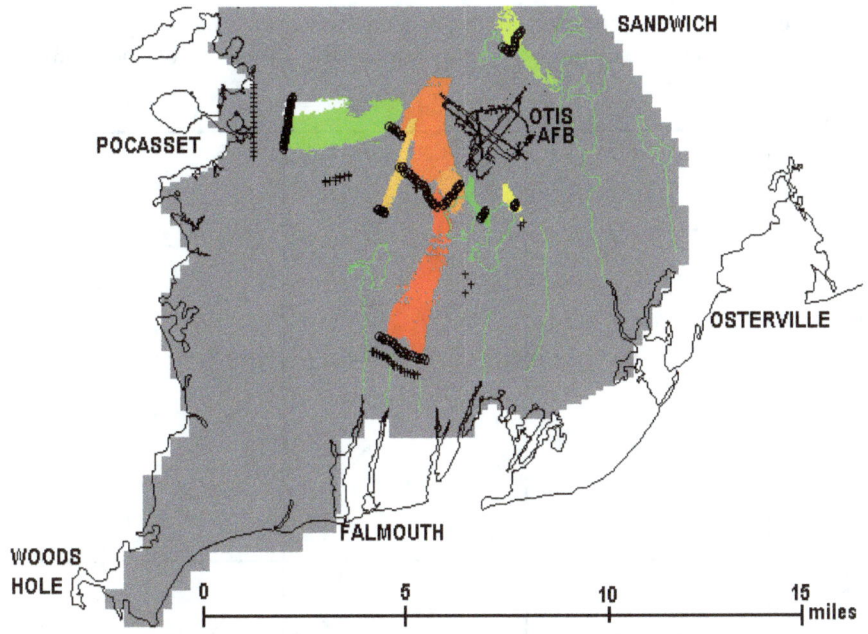

Figure 21. Contaminant Plumes on Western Cape Cod

Spreading and migration of the contaminant can span orders of magnitude, which is why we color the bitmap images in these two example codes (tce2d.c and pce2d.c) based on the log of the concentration.

We can also plot the maximum concentration over time for these various examples, which does somewhat illustrate relative impacts of the available transport mechanisms.

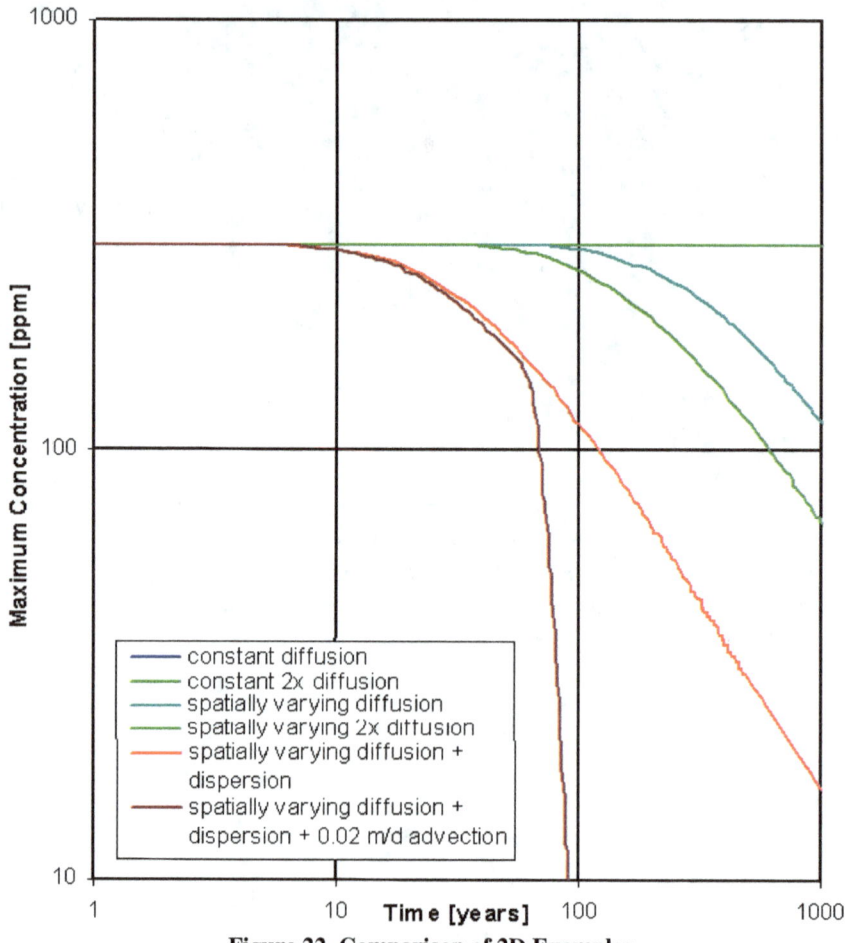

Figure 22. Comparison of 2D Examples

Chapter 5. Contaminant Transport in 3D

The governing partial differential equation that represents the conservation of mass for the contaminant is Equation 3.8 in three dimensions, which is a simple extension from two. While there are many papers and discussions as to which method (e.g., finite difference, finite volume, or finite element) is preferable for modeling fluid flow, those are beyond the scope of this text. The time scale for contaminant transport, whether this be in groundwater, surface water, or atmospheric, is most often much longer than for fluid flow (i.e., water or air). The spatial scale is also much larger. For example, flow through an estuary bordering on the sea will fluctuate somewhat repetitively twice per day because of the tides. There are countless small eddies and separation zones as air flows over a landscape, even a relatively flat one. A contaminant, especially one that will take months—even years—to contain or capture, persists far longer than the finer details of the flowing media in which it resides and with which it migrates. This difference in time scales is why we most often solve such problems in two steps: 1) flow modeling and 2) contaminant transport. We are only concentrating on the second step in this text, as we have covered the first step elsewhere and don't want to detract from the current focus.

Solving the Governing Partial Differential Equation

While the finite volume and finite element methods are indeed useful and may be the most efficient for a particular flow field, these are best suited to obtaining steady state solutions and often involve matrix operations. Peruse the literature on CFD modeling and you will often see references to SOR (successive over-relaxation) and other similar techniques for solving simultaneous linear equations. The version of FRAC3D that I use employs the conjugate gradient method to solve the simultaneous linear equations representing the partial differential equation representing those conservation relationships. In contrast, the finite difference method when applied to contaminant transport begins with an initial condition and steps forward in time. If we had to solve a system of simultaneous linearized equations for each node or integrate over each computational cell applying a Galerkin method at each time step, this would be completely impractical. This is why we use the finite difference method (or particle tracking, as we have discussed elsewhere) to model contaminant transport.

We begin with simple diffusion in 3D, noting that in soil layers, this is often more rapid horizontally than vertically, so we set up our model (tce3d.c in folder examples\tce3d) to handle this. We also consider a different spatial scale (5 km horizontal span 50m vertical). We won't be using a 1m element size, as this would require over two billion nodes. The ratio of discretization in the three directions should be roughly proportional to the rate of change in that direction. If the lateral and horizontal spreading is approximately the same, then Δx and Δy can be the same. If the vertical spreading is approximately 1/10th that in the XY plane, then Δz should be inversely proportional to that. Note that we must also now consider at least XY and Z separately when selecting a time step. In order to get under one million nodes, we would have to have a grid of △x=△y=25m and △z=2.5m or 201x201x21=848,412 nodes. This will take at least twice as long as the 2D models with only 640x640=409,600 nodes.

Estimating Properties

The analytical solution for transient diffusion in one dimension with constant properties was first introduced by Fourier in 1822:

$$C(x,t) = \frac{C_0 \delta}{\sqrt{4\pi D t}} e^{\left(\frac{-x^2}{4Dt}\right)} \quad (5.1)$$

Holding all other parameters constant and differentiating with respect to t yields the time at which (at a given distance, x) the concentration will be a maximum.

$$t_{max} = \frac{x^2}{2D} \quad (5.2)$$

Plugging this expression back into the preceding equation yields the maximum concentration at this time:

$$C_{max} = \frac{C_0 \delta}{x\sqrt{2\pi}} e^{-0.5} \quad (5.3)$$

These simple analytical expressions can provide an estimate of properties and time scales useful in real world applications. For our first 3D example we will combine diffusion and dispersion, keeping the directions separate. We will arbitrarily select a combined diffusion/dispersion coefficient to spread 1 km in the X direction reaching a maximum in 1800 years and 1/100th this rate in the Y direction. The combined coefficient in the Z direction corresponds to 10 m in this same time frame. While this might seem like an unreasonably long time for a remediation project, it would be reasonable for containment in place, confined by a nearly impermeable cap. In this time we would hope toxic chemicals would break down and radioactive ones decay.

The initial concentration is set to 300 ppm in a "wafer" in the center of the domain having a diameter of 800m and a thickness of 12m. All of the code related to bitmap images has been removed, as we are not painting the results, rather writing them out to sequential files. While we could put multiple "snapshots" (i.e., representations of the concentration field at a particular time) in the same file, making it 4D, these would be so large as to make loading them all into memory at the same time problematic and likely beyond the available physical memory. Tecplot™ can accept multiple "zones" which can be turned on and off to produce time series animations.

We have also simplified the allocation of memory within the program plus added functions to find the minimum or maximum value in an array. These facilitate calculation of the time step and also reporting of the maximum concentration. The finite difference equation has also been expanded to incorporate the third (vertical) dimension and the natural boundary conditions have been added for the bottom and top. The log of the initial concentration is shown in the following figure:

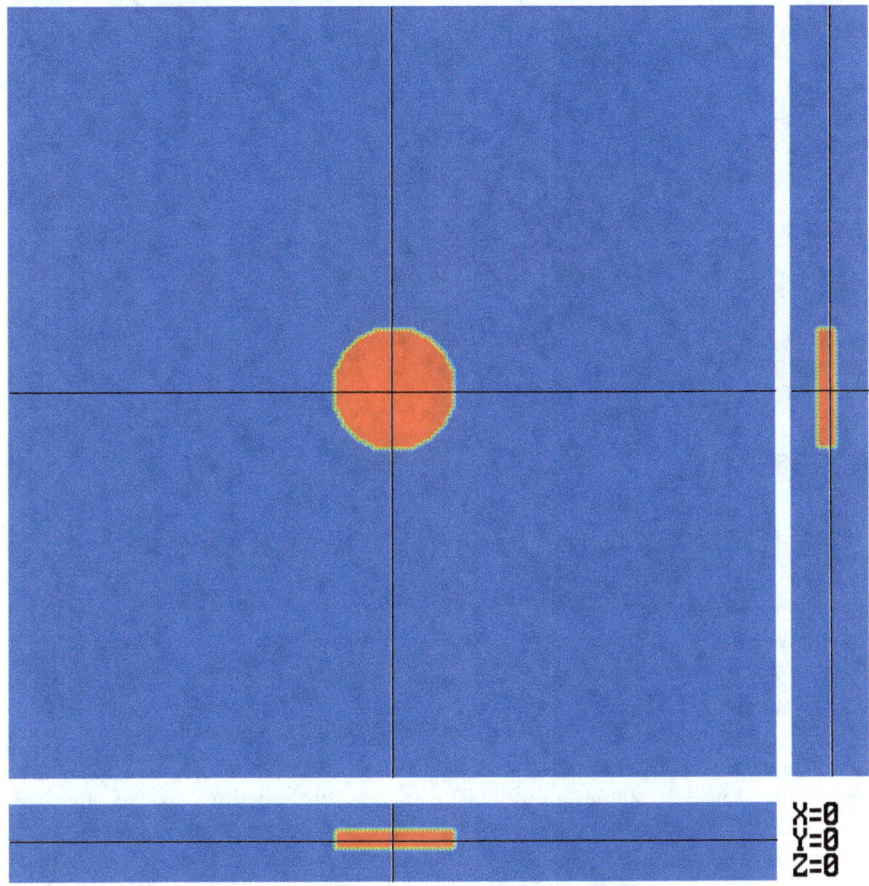

Figure 23. Initial log Concentrations in 3D

We must next consider the time step required to obtain a stable solution. To calculate this, we find the largest value of the combined diffusion/dispersion coefficient in each of the three directions, divide this into the square of the grid spacing in that dimension and take half (e.g.. $\Delta t \leq \Delta x^2/2D_{max}$). This yields 1.1, 113, and 113 years for the X, Y, and Z dimensions, respectively. As we don't want to step forward more than one year for generating output, we round this down to $\Delta t=1$ yr. The results after 100 years are shown in this next figure:

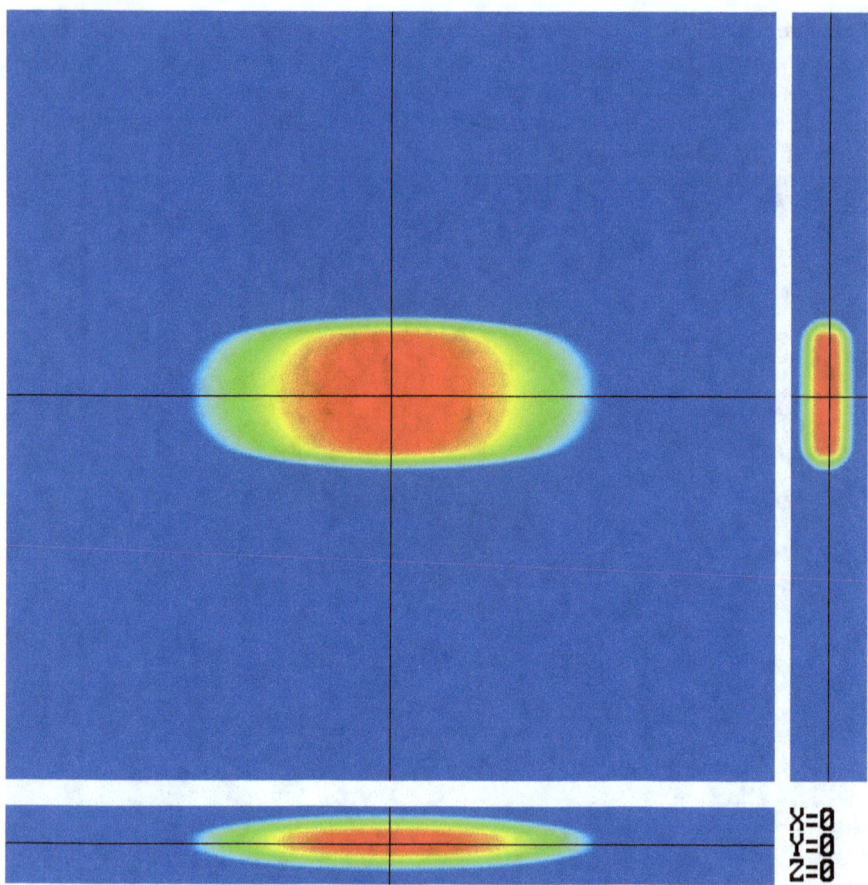
Figure 24. After 100 Years of Diffusion+Dispersion

The time step is small enough and the forward Euler method is adequate to provide a slow but stable solution. We march through time and see that the numerical solution in 3D roughly corresponds to the magnitude of spreading we presumed when selecting the dimensions and properties. This observation is important in that there are many steps in a successful remediation project. Rough estimates are quite useful when evaluating alternatives, such as containment vs. removal (e.g., pump-and-treat). The cost of such options may vary greatly. The time required to implement different options may also vary greatly. These considerations and more should guide the remediation design process if a successful outcome is to be achieved. The concentrations after 1800 years are shown in the following figure:

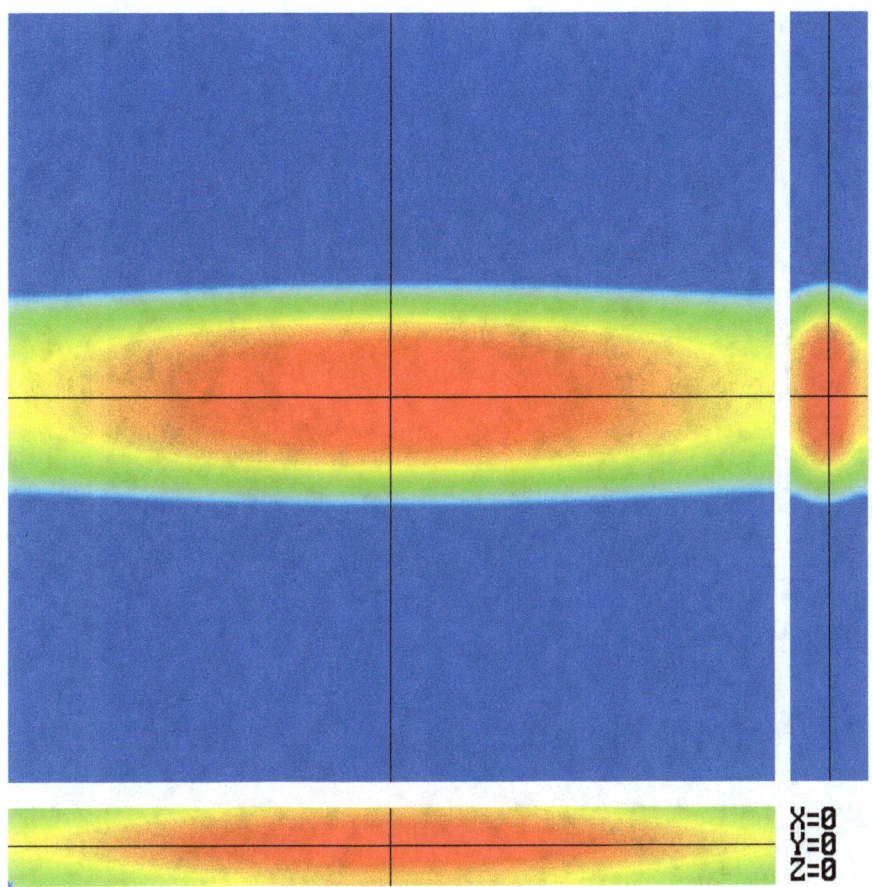

Figure 25. After 1800 Years of Diffusion+Dispersion

Advection

Advection will be present in most sites unless these have been specially prepared to contain the contaminant. We add the same field as before (Figure 17) with a much smaller vertical component and again switch folders and file names to pce3d so as to not overly complicate the original example. As for the 2D example, at this step we also add porosity.

As with 2D advection, we use upwind differences. With the third dimension, we must also add nodes to the arrangement in Figure 18, including an upper and lower plus two more vertical intermediate points, designated g and h. The time step must now consider all transport mechanisms (e.g., $\Delta x^2/2D$ and also $\Delta x/6U$), taking the most restrictive (i.e., smallest) one to avoid instability. After 100 years of combined diffusion/dispersion and advection (approximately 0.02 m/day) the contaminant has spread, migrated to the right and distorted in shape, as shown in this next figure:

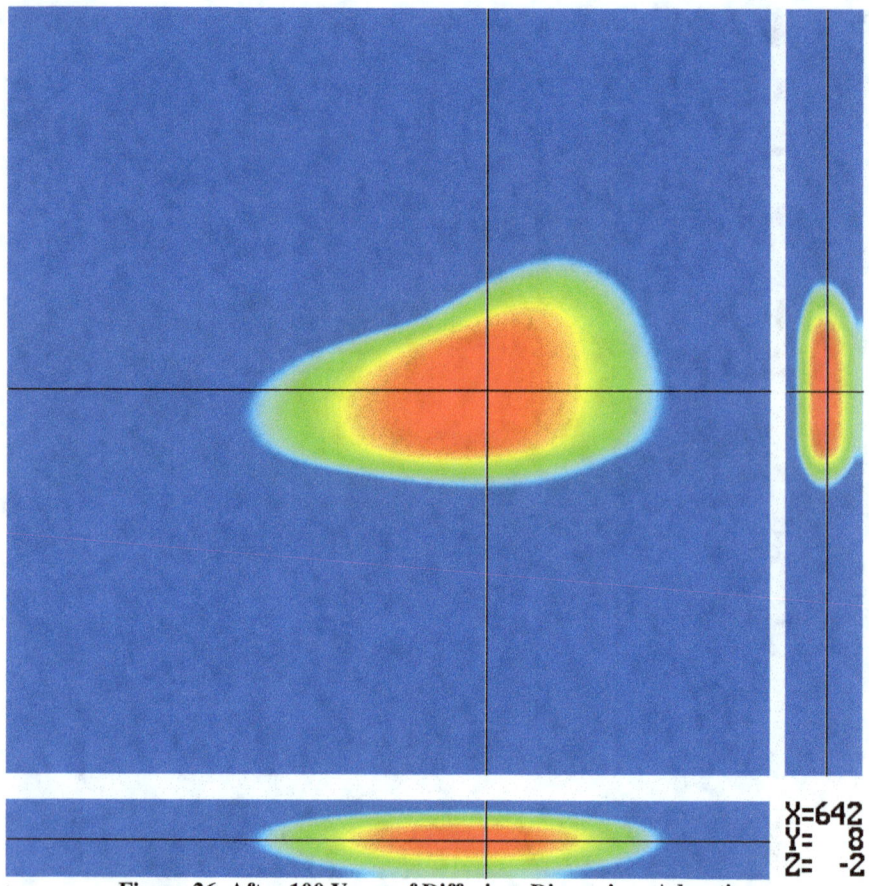

Figure 26. After 100 Years of Diffusion+Dispersion+Advection

The contaminant plume is responding as we would expect, given the properties and flow we have applied to it. It is spreading more laterally in the X direction than horizontally in the Y direction. It is also spreading more downward than upward vertically. We are also starting to see some rotational distortion due to the curling velocity field. After 250 years, the plume has taken on a droplet shape and the center of mass has move to approximately X=1200m, Y=-25m, Z=-5m. We also see more clockwise rotational distortion. The colors (representing the log of the concentration) smoothly change from blue to cyan to green to yellow to orange and finally to red as we approach the center of mass and backward as we move away from the center of mass in any of the three directions. This is evidence that the time stepping (forward Euler method) is still working adequately.

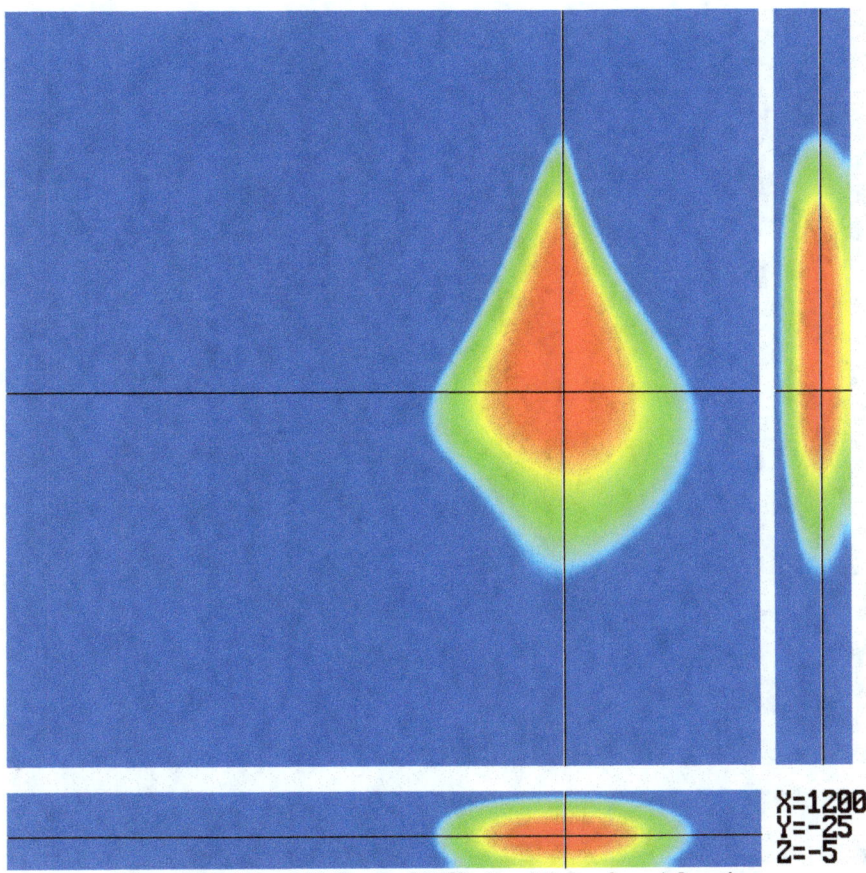

Figure 27. After 250 Years of Diffusion+Dispersion+Advection

This plume is still moving considerably slower than some I have modeled, including those on Western Cape Cod shown in Figure 21, which moved a distance on the order of a kilometer in approximately 25 years. The soil there has a higher hydraulic conductivity and porosity than some other locations within the continental US. In the next example we will adjust the magnitude and also pattern of the flow to see how this impacts the plume.

After 500 years the plume has distorted even more, as seen in the following figure and is even starting to leave the boundary on the right side and also the bottom. We can also see the natural boundary conditions in this figure. The gradients (spatial slope vectors) or directional rate of change of the concentrations, as indicated by the color variation, are smooth at the edges, rather than stacking up colors at the edge.

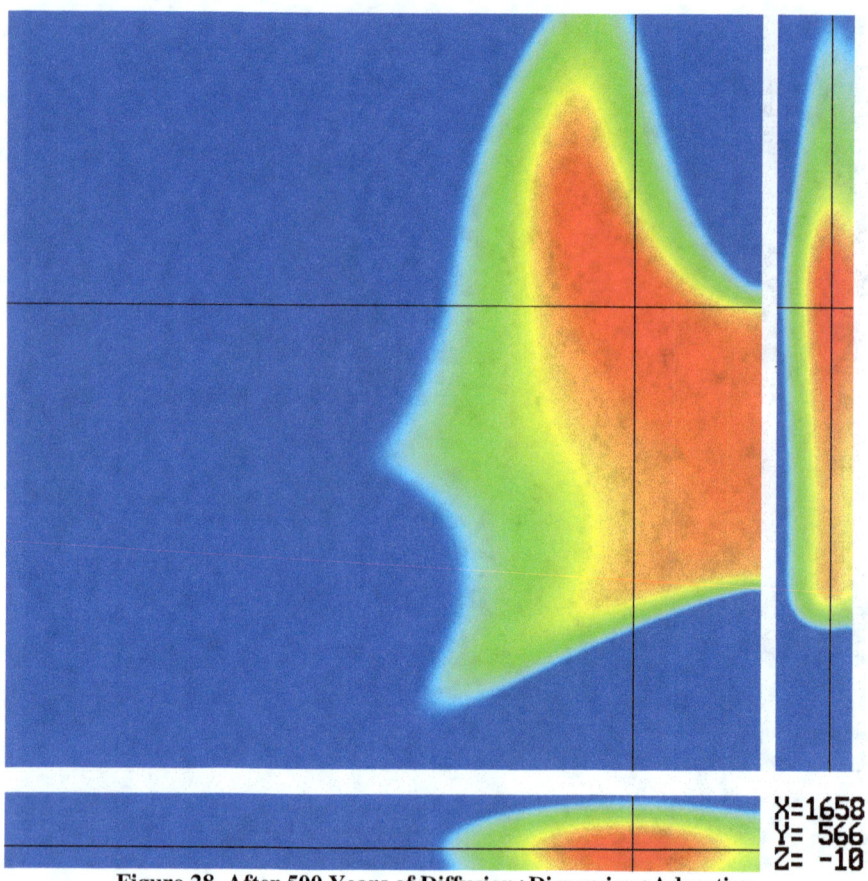
Figure 28. After 500 Years of Diffusion+Dispersion+Advection

Chapter 6. MODFLOW Based Models

The preceding examples have served to illustrate the principles of modeling contaminant transport but are not well suited to actual problems. We need code that will handle a variety of problems conveniently and not require editing and recompilation for each one. That means reading input files, particularly those that have been generated by some other software, such as Build3D (see Appendix K). We will consider two different flow models: MODFLOW in this chapter and FRAC3D in the next. We will read the respective input files (nodes, elements, and properties) and output files (flows) plus a file containing initial concentrations and then march through time calculating the contaminant transport. We will not dwell on the details of MODFLOW in this text. The reader is directed to the many resources available for more specific information.

To accomplish this, we will use an example (ctmod.c), which can be found in the online archive in the examples\modflow folder along with several sets of input files. The output files generated will be compatible with TP2. Should you prefer output compatible with Tecplot™, swap out the function with the code (ctfra.c) discussed in the next chapter.

The prefix for the first example we will consider in this chapter is WINE. Launch the program (ctmod.exe) passing it the parameter WINE. Several files must exist and be in the same folder. The examples are in subfolders. You can either copy the files into the folder with the executable or copy the executable into each of the example folders or go to one of the example folders and type something like the following to launch the executable from the parent folder:

```
..\CTMOD WINE
```

The first four files read for this example will be WINE.BAS (the MODFLOW basic input file), WINE.BCF (the block-centered file), WINE.CBB (the binary flow flow), and WINE.ELP (the property file). This is what you should see:

```
3D Transient Diffusion + Dispersion + Advection
reading MODFLOW input and output files
basic input file: WINE.BAS
  title: PREFIX: WINE
  subtitle: CREATED BY BUILD3D/V1.71
  grid: 45x37x5
  total cells: 8325
  active cells: 5860
  element type: hexahedra (bricks)
  augmented grid: 46x38x6
  resulting nodes: 10488
  active nodes: 7536
block-centered file: WINE.BCF
  0≤X≤7889.85
  0≤Y≤6500.16
  0≤Z≤235.105
binary flow file: WINE.CBB
```

```
-41.0391≤U≤53.6929
-27.8412≤V≤40.8043
-3.3188≤W≤12.3406
element property file: WINE.ELP
  0.3≤F≤0.3
  3≤Dx≤3
  3≤Dy≤3
  0.3≤Dz≤0.3
```

The last file read contains the initial concentrations, WINE.ICC. Note that MODFLOW begins with a full grid (i.e., a block of elements). Some elements may be inactive so as to create an irregular domain. Initial concentrations for this example are shown in the following figure:

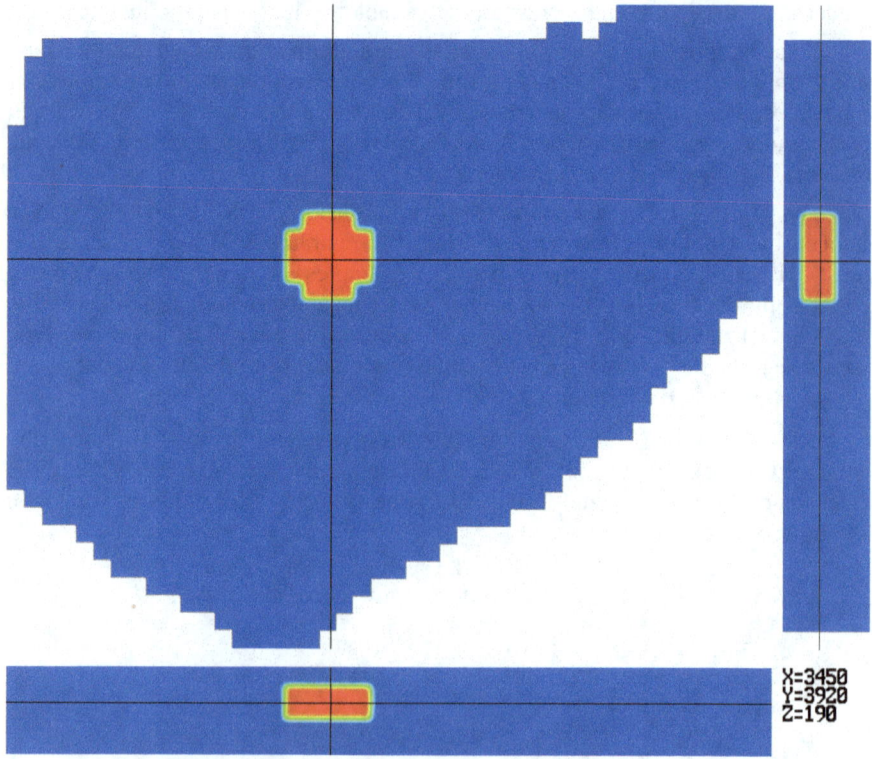

Figure 29. Initial Concentrations for WINE Example

This figure also shows the active and inactive parts of the domain (blue vs. white). As MODFLOW is element-based and ctmod is node-based, the initial concentrations have been relocated to the nearest node (refer to function ReadConcentrations() in source code file ctmod.c).

The vertically-averaged velocity vectors are shown in this next figure:

Figure 30. Vertically-Averaged Velocity Vectors for WINE Example

As before with 3D advection (see pce3d.c), we must consider several things when selecting a time step. A summary of these is listed below:

```
determining time step
    ΔX=175.33,   ΔY=175.68,    ΔZ=14.81
    ΔX²/Dx/2=5123.43,  ΔY²/Dy/2=5143.91,   ΔZ²/Dz/2=365.56
    ΔX/U/6=0.544237,   ΔY/V/6=0.717572,   ΔZ/W/6=0.200017
    Δt=0.121452
```

We advance the solution as before, only working from the elements so that we can only consider those that are active. MODFLOW defines properties within the elements. Note that all of the properties must be consistent. It doesn't matter what properties are used (English or metric), as long as they are consistent. The velocities must in length/time and the diffusion/dispersion coefficients must be in length²/time. The time step is dX^2/D, whatever that works out to be. The same is true for the concentrations. The output will have the same units as the input.

After 25 time units (presumably years), the maximum concentration has dropped from 300 to 285 and spread considerably, as shown in the following figure:

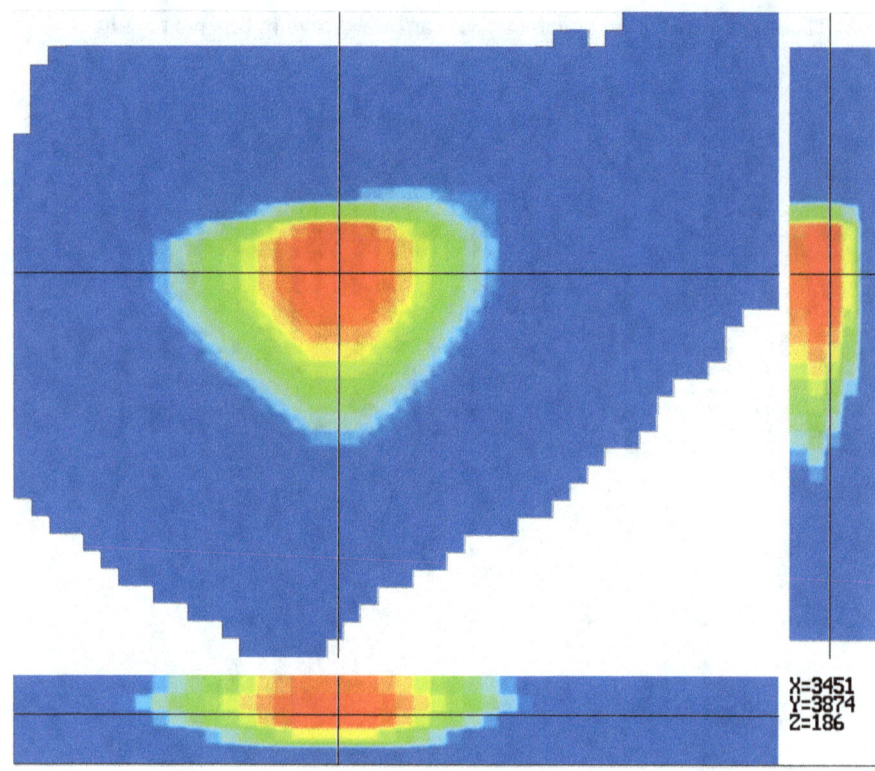

Figure 31. WINE Concentrations (log) after 25 Years

After 100 years the plume has spread even further and noticeably migrated laterally. The impact of the velocity field is also clear after this period.

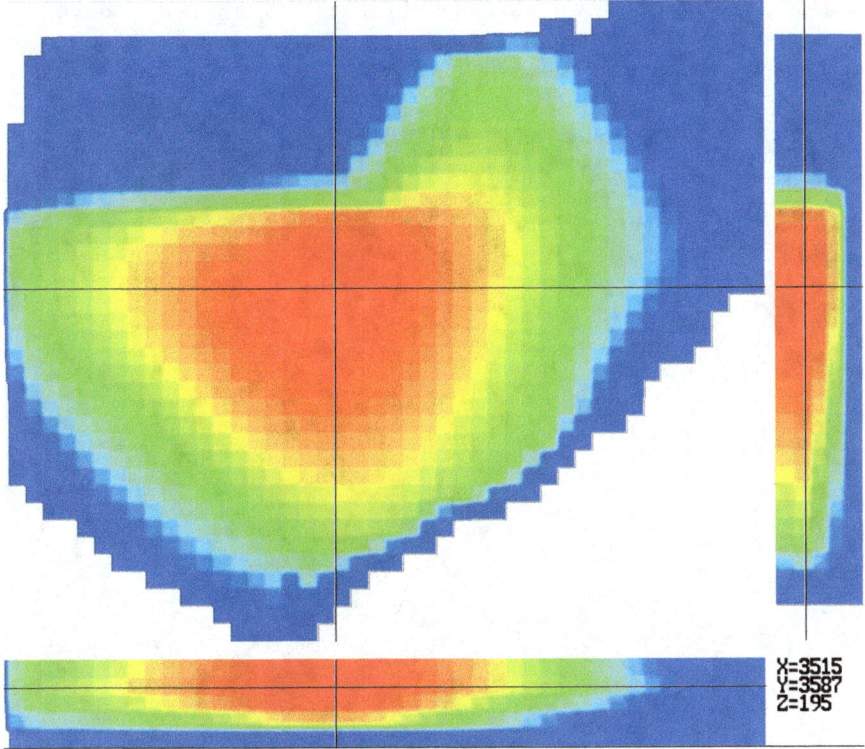

Figure 32. WINE Concentrations (log) after 100 Years

We end the simulation at 650 years, as the maximum concentration has decreased by a factor of 10 (an arbitrary number in this case). This would definitely be consider a complete failure if containment or capture and removal of the contaminant were the objective. This particular scenario is the "no action" alternative, which was not chosen in this case.

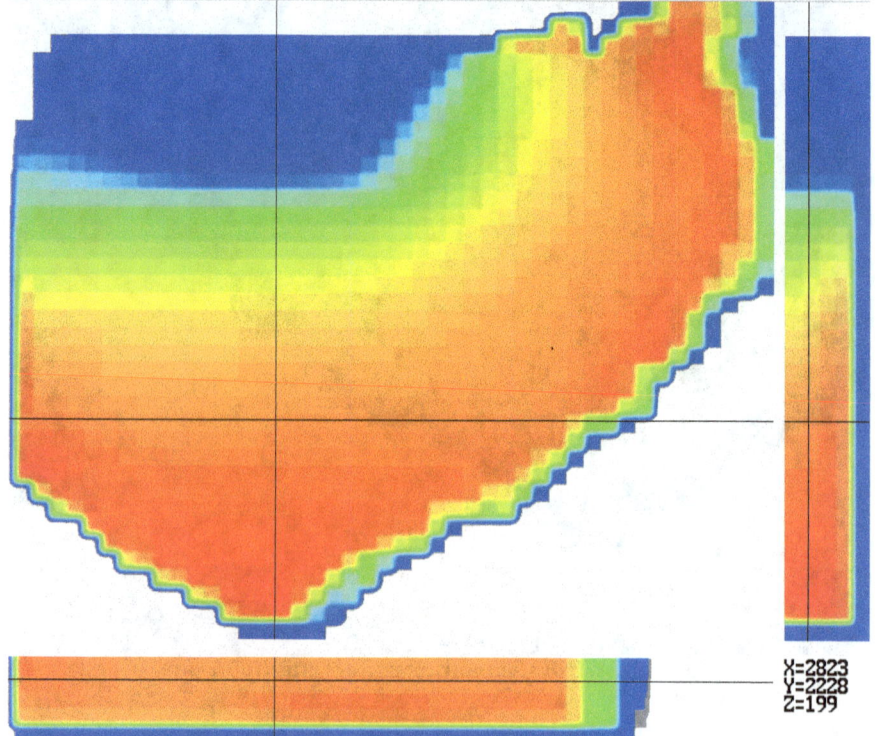

Figure 33. WINE Concentrations (log) after 650 Years

Multiple Plumes and Contaminants

Some remediation projects contain multiple contaminated sites, like where some guy emptied a tanker truck into a vacant lot. [I am not making this up!] Some plumes contain multiple contaminants. This is one reason for using particle tracking instead of the type of transport models we have discussed so far. PTRAX can handle up to 128 different plumes simultaneously with up to 128 different contaminants in each plume. There are 10 different plumes shown in Figure 21. You will find a separate "seed" file for each plume in the folder examples\MODFLOW\OTIS (see files OTIS?.SED). These each contain only a few particle seeds. The full model contained tens of thousands of seeds but those files are quite large and so are not contained in the online archive of examples.

We begin this next MODFLOW-based example (OTIS) with three contaminated spots, as shown in the following figure:

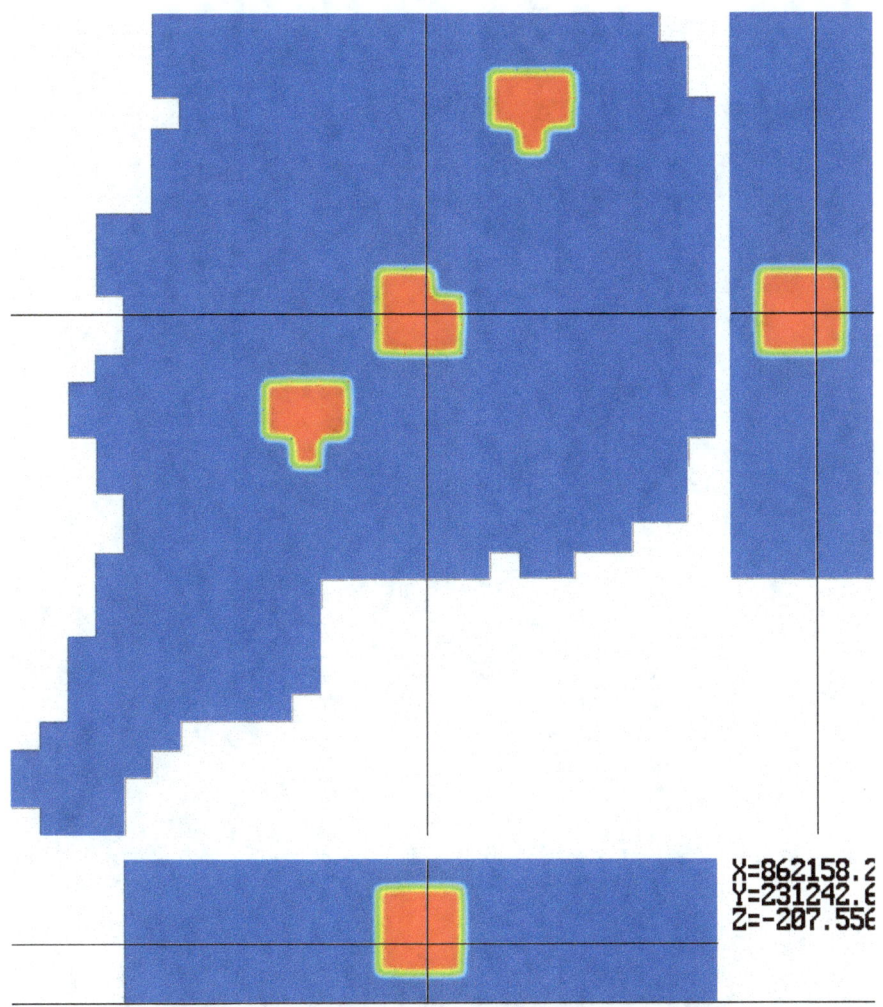

Figure 34. Initial Concentrations OTIS Example OTIS

The vertically-averaged velocity field is shown in this next figure:

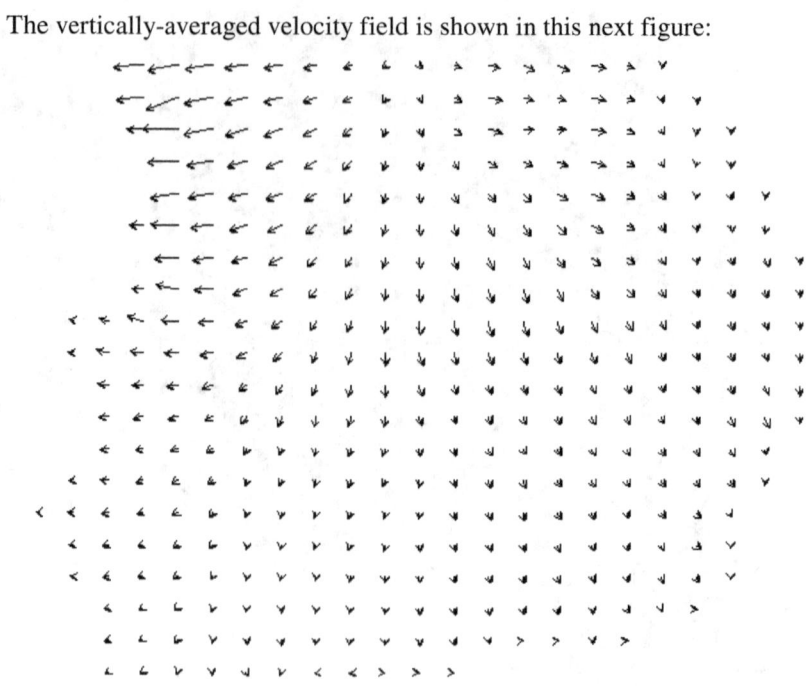

Figure 35. Vertically-Averaged Velocity Vectors for OTIS Example

Launching the program produces the following output:
```
ctmod otis
3D Transient Diffusion + Dispersion + Advection
reading MODFLOW input and output files
prefix: otis
basic input file: OTIS.BAS
  title: PREFIX: OTIS
  subtitle: CREATED BY BUILD3D/V1.73
  Reference Point: X=825900, Y=189450
  grid: 25x29x5
  total cells: 3625
  active cells: 2150
  augmented grid: 26x30x6
  element type: hexahedra (bricks)
  resulting nodes: 4680
  active nodes: 2946
block-centered file: OTIS.BCF
  825900≤X≤887400
  189450≤Y≤255350
  -347.58≤Z≤65.055
element property file: OTIS.ELP
  0.3≤F≤0.3
  3≤Dx≤3
  3≤Dy≤3
  0.3≤Dz≤0.3
binary flow file: OTIS.CBB
  -260.426≤U≤112.026
  -143.778≤V≤31.8607
  -1.65294≤W≤4.03023
vertically-averaged velocities: OTIS.V2D
initial concentrations: OTIS.ICC
determining time step
  ΔX=2460, ΔY=2272.41, ΔZ=21.09
  ΔX²/Dx/2=1.0086E+006, ΔY²/Dy/2=860641,
    ΔZ²/Dz/2=741.313
  ΔX/U/6=1.57435, ΔY/V/6=2.63416, ΔZ/W/6=0.872159
  Δt=0.462376
0 years C≤300 OTIS000.TB3
25 years C≤290 OTIS001.TB3
50 years C≤264 OTIS002.TB3
75 years C≤238 OTIS003.TB3
100 years C≤211 OTIS004.TB3
```

We can see some spreading but no discernable migration after 100 years:

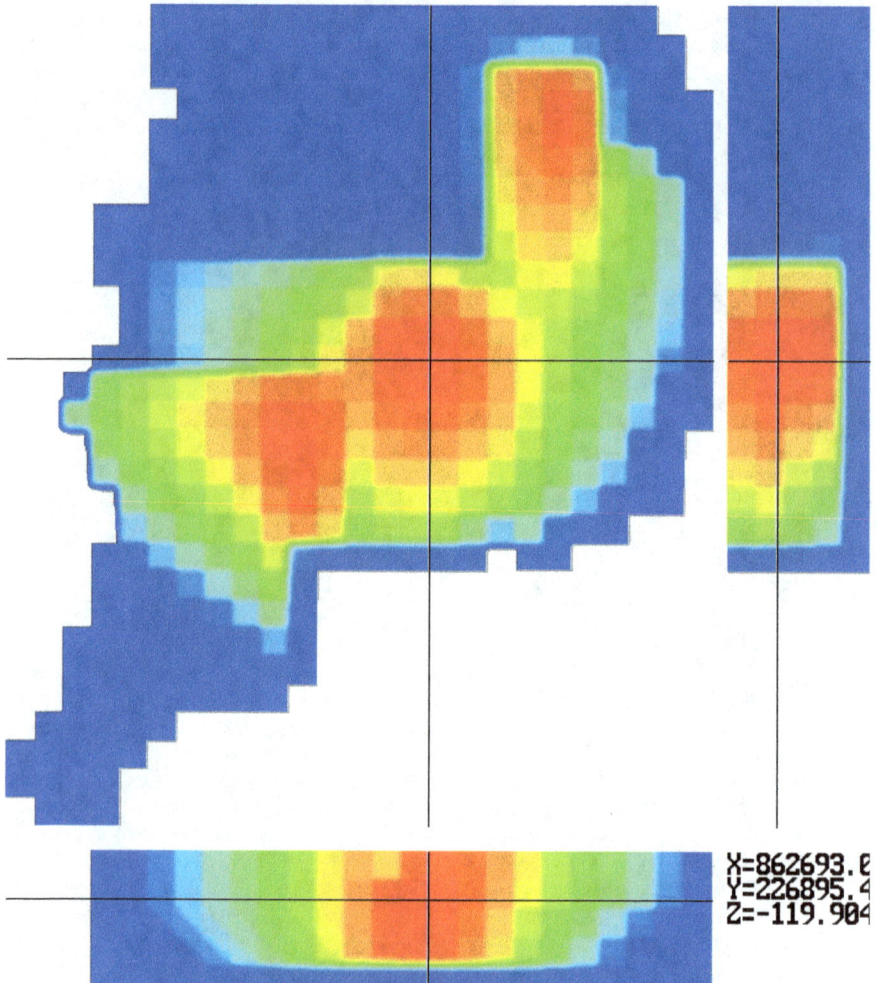

Figure 36. OTIS Concentrations (log) after 100 Years

After 500 years the contaminants have spread considerably and moved appreciably.

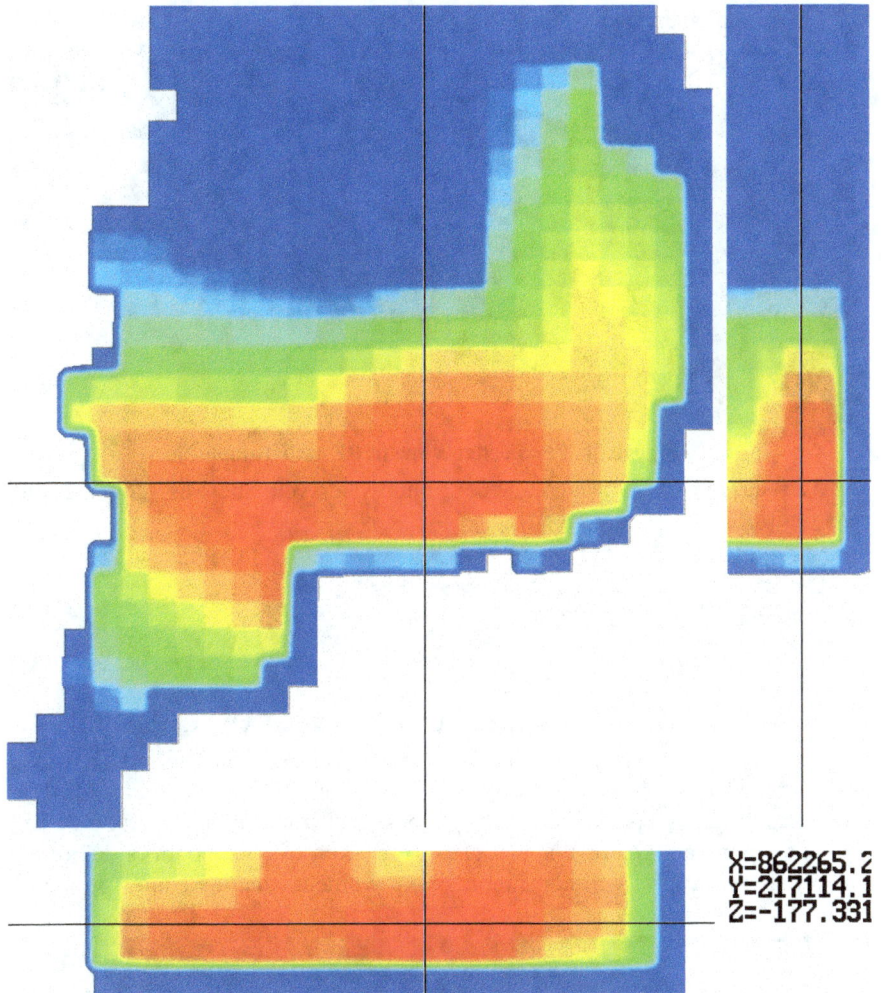

Figure 37. OTIS Concentrations (log) after 500 Years

Impact of Surface Slope

This next example is a valley with considerable surface slope illustrated in this 3D perspective view:

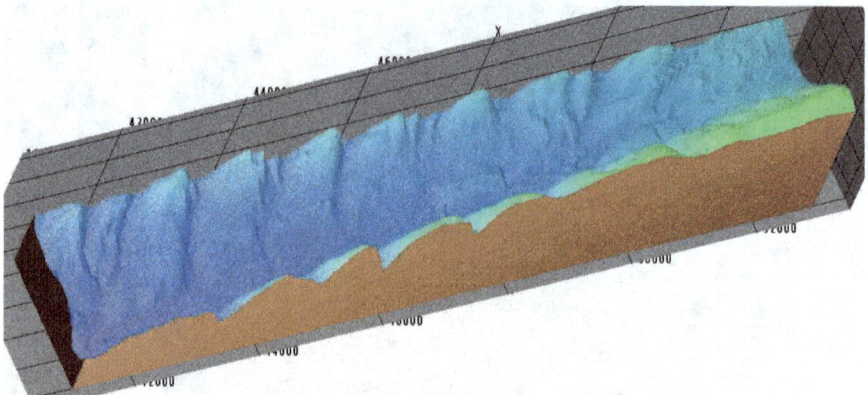

Figure 38. Perspective View of BEAR Example

The surface slope influences the groundwater velocities, as shown in this next figure:

Figure 39. Vertically-Averaged Velocities for BEAR Example

There were numerous contaminants at this remediation site, which was a burial ground for radioactive, toxic, and otherwise hazardous materials, only a few miles from my house. Dozens of plumes were tracked and several different types of capture and removal approaches were taken to clean up this site. For the sake of simplicity, we will consider a single plume emanating from a spot near the top of the valley.

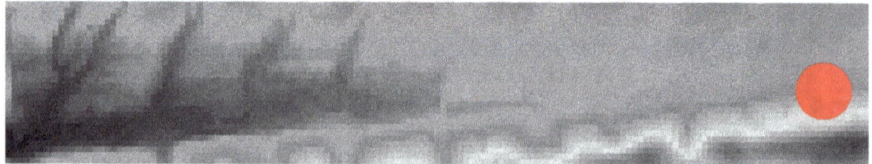

Figure 40. Contaminant Source Location for BEAR Example

Launching the program produces the following output:
```
ctmod bear
3D Transient Diffusion + Dispersion + Advection
reading MODFLOW input and output files
prefix: bear
basic input file: BEAR.BAS
```

```
title: PREFIX: BEAR
subtitle: CREATED BY BUILD3D/V1.71
grid: 60x15x5
total cells: 4500
active cells: 4500
augmented grid: 61x16x6
element type: hexahedra (bricks)
resulting nodes: 5856
active nodes: 5856
block-centered file: BEAR.BCF
   0≤X≤11925
   0≤Y≤2224.95
   0≤Z≤1189.1
element property file: BEAR.ELP
   0.3≤F≤0.3
   3≤Dx≤3
   3≤Dy≤3
   0.3≤Dz≤0.3
binary flow file: BEAR.CBB
   -5975.29≤U≤591.701
   -3032.46≤V≤1861.64
   -1788.07≤W≤1160.5
vertically-averaged velocities: BEAR.V2D
initial concentrations: BEAR.ICC
determining time step
   ΔX=198.75, ΔY=148.33, ΔZ=160.45
   ΔX²/Dx/2=6583.59, ΔY²/Dy/2=3666.96, ΔZ²/Dz/2=42907
   ΔX/U/6=0.00554366, ΔY/V/6=0.00815234, ΔZ/W/6=0.0149556
   Δt=0.00270332
0   years C≤300 BEAR000.TB3
25  years C≤300 BEAR001.TB3
50  years C≤299 BEAR002.TB3
75  years C≤298 BEAR003.TB3
100 years C≤298 BEAR004.TB3
125 years C≤297 BEAR005.TB3
150 years C≤297 BEAR006.TB3
175 years C≤296 BEAR007.TB3
200 years C≤296 BEAR008.TB3
225 years C≤295 BEAR009.TB3
250 years C≤294 BEAR010.TB3
275 years C≤294 BEAR011.TB3
300 years C≤293 BEAR012.TB3
325 years C≤292 BEAR013.TB3
350 years C≤292 BEAR014.TB3
375 years C≤291 BEAR015.TB3
400 years C≤291 BEAR016.TB3
425 years C≤290 BEAR017.TB3
450 years C≤289 BEAR018.TB3
475 years C≤289 BEAR019.TB3
500 years C≤288 BEAR020.TB3
```

After 100 years we see that the contaminant has spread down hill for almost the entire length of the valley, yet there is still a significant concentration at the source. This is characteristic of some sites that have significantly different properties and/or velocities in the three different spatial directions (horizontally, laterally, and vertically).

Figure 41. BEAR Concentrations (log) after 100 Years

We see very little change between 100 and 500 years:

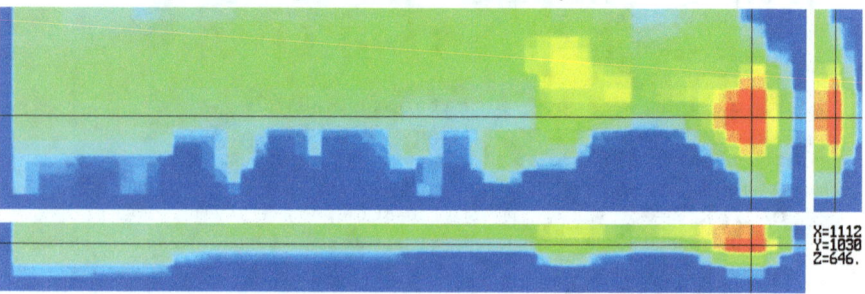

Figure 42. BEAR Concentrations (log) after 500 Years

The MADE Site

The last MODFLOW-based example we will consider comes from the MADE site, for which I provided modeling support and software development for the first decade. We learned much from this site, not only about the soil properties and interactions with the contaminants, but also about the software required to manage the data collected and build models. The 3D hydraulic conductivity field was one of the most complex tasks I tackled early on. This next figure shows the original hydraulic conductivity field for the entire site.

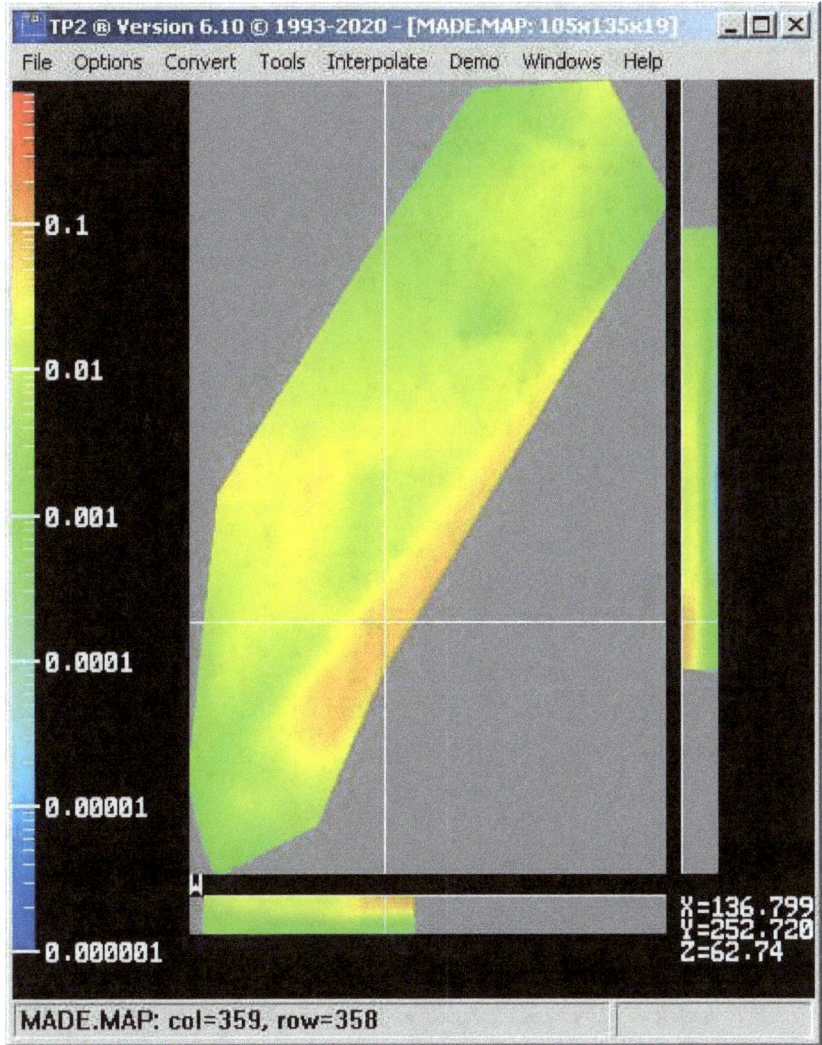

Figure 43. Original Hydraulic Conductivity Field for the MADE Site

I developed a program (Field3D) to handle this data, as nothing was available commercially at that time capable of dealing with it. This functionality was later built into TP2. More recent versions (7.04 and later) of Tecplot™ can also handle this type of data. Many studies were conducted at the MADE site and sections of the entire area were the subject of specific focus, including the cutout region in the MADE example that can be found in the online archive in the examples\MODFLOW\MADE folder. The vertically-averaged velocity vectors for this cutout region are shown in the following figure:

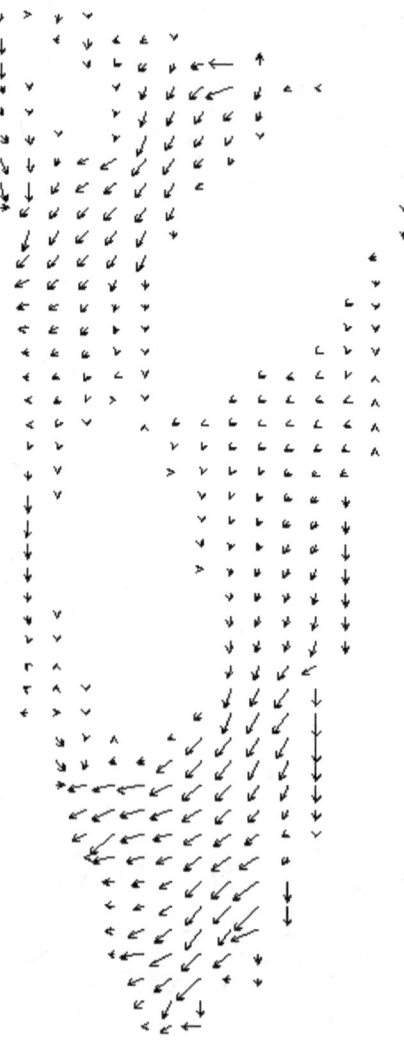

Figure 44. Vertically-Averaged Velocity Vectors for MADE Example

While many plumes were created, we will only consider a few in this example. The initial concentrations are shown in this next figure:

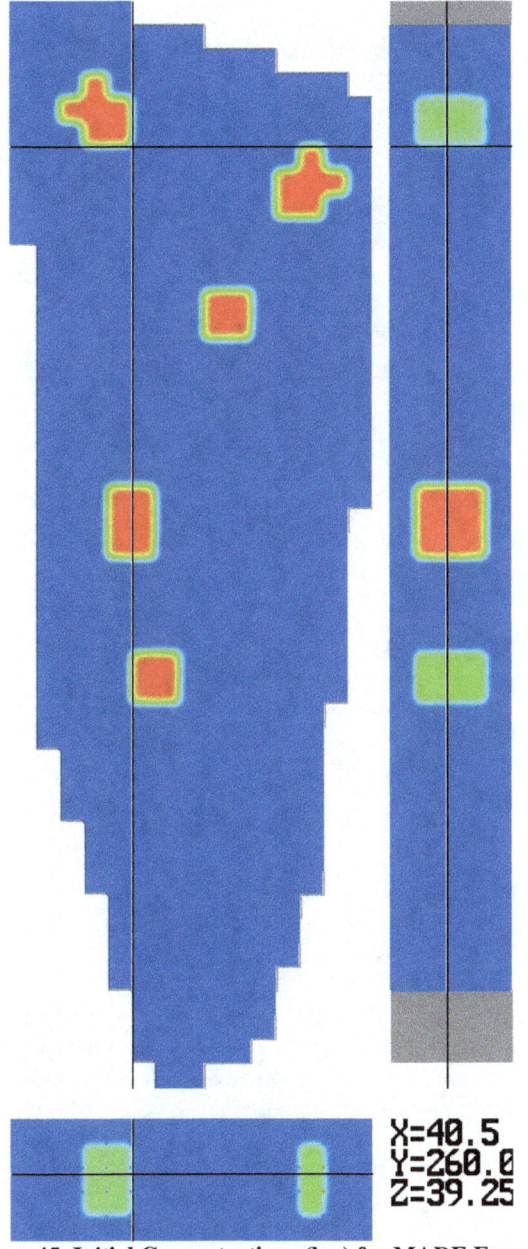

Figure 45. Initial Concentrations (log) for MADE Example

Launching the program produces the following output:

```
ctmod made
3D Transient Diffusion + Dispersion + Advection
reading MODFLOW input and output files
prefix: made
basic input file: MADE.BAS
   title: PREFIX: MADE
   subtitle: CREATED BY BUILD3D/V1.71
   grid: 15x45x5
   total cells: 3375
   active cells: 2350
   augmented grid: 16x46x6
   element type: hexahedra (bricks)
   resulting nodes: 4416
   active nodes: 3180
block-centered file: MADE.BCF
   0≤X≤120
   0≤Y≤300.001
   0≤Z≤118.205
element property file: MADE.ELP
   0.3≤F≤0.3
   3≤Dx≤3
   3≤Dy≤3
   0.3≤Dz≤0.3
binary flow file: MADE.CBB
   -28.558≤U≤13.8079
   -30.3216≤V≤10.8248
   -9.41612≤W≤11.996
vertically-averaged velocities: MADE.V2D
initial concentrations: MADE.ICC
determining time step
   ΔX=8, ΔY=6.6667, ΔZ=14.023
   ΔX²/Dx/2=10.6667, ΔY²/Dy/2=7.40748, ΔZ²/Dz/2=327.741
   ΔX/U/6=0.0466885, ΔY/V/6=0.0366444, ΔZ/W/6=0.194829
   Δt=0.0184937
0 years   C≤300  MADE000.TB3
25 years  C≤91   MADE001.TB3
50 years  C≤50   MADE002.TB3
75 years  C≤36   MADE003.TB3
100 years C≤29   MADE004.TB3
```

The **MADE** site was quite small compared to the other cleanup projects the Team worked on. It was chosen because it was inside the guarded perimeter of the Air Force Base and wouldn't be disturbed. It also provided a much faster reaction time, which was more conducive to study than the much longer remediation projects.

The concentrations after 25 years are shown in the figure below. This was a small site compared to

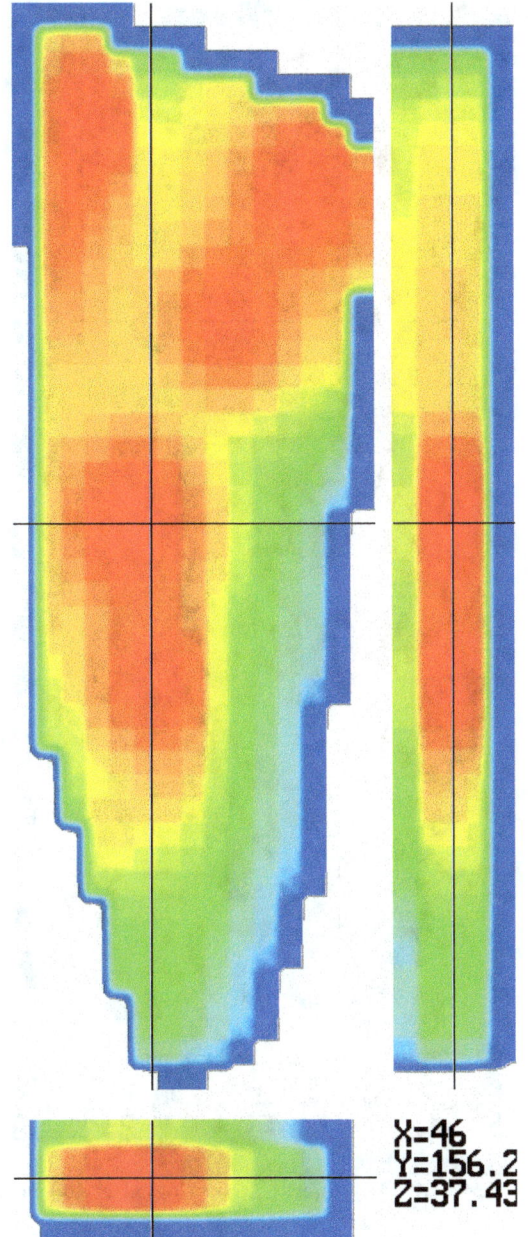

Figure 46. MADE Concentrations (log) after 25 Years

The concentrations after 100 years are shown in the following figure:

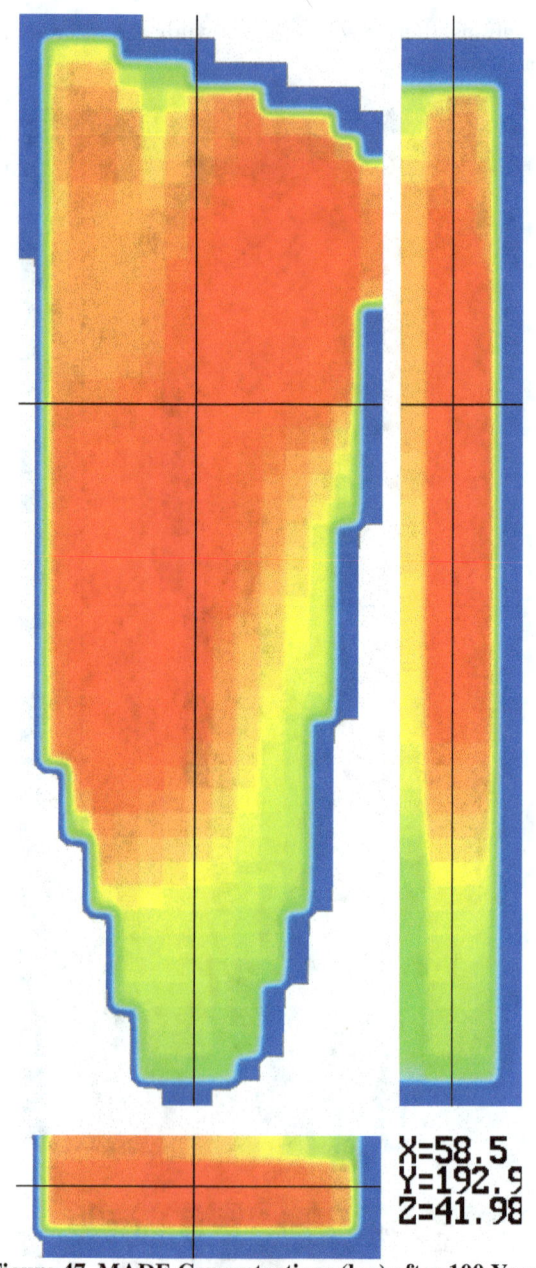

Figure 47. MADE Concentrations (log) after 100 Years

Chapter 7. FRAC3D Based Models

We will now modify the code from the previous chapter to read FRAC3D input and output files, which are very different from MODFLOW files. The code (ctfrac.c), along with several examples in this second format, can be found in the online archive in the examples\frac3d folder. The output files generated will be compatible with Tecplot™. Should you prefer output compatible with TP2, swap out the function with the code (ctmod.c) discussed in the previous chapter or convert the files, as described in the corresponding appendices.

The MODFLOW-based examples in the previous chapter began with a full grid with some cells deactivated in order to create irregular domains. FRAC3D works differently by reading nodes and elements, which are connected by virtue of common nodes. While the MODFLOW models all had regularly-spaced grids, this is not necessarily the case with FRAC3D models. Lacking this regular pattern of nodes and elements, we must figure out which nodes are connected and how. As shown in the following figure, with stacked hexahedral elements, a node may have up to 26 neighbors.

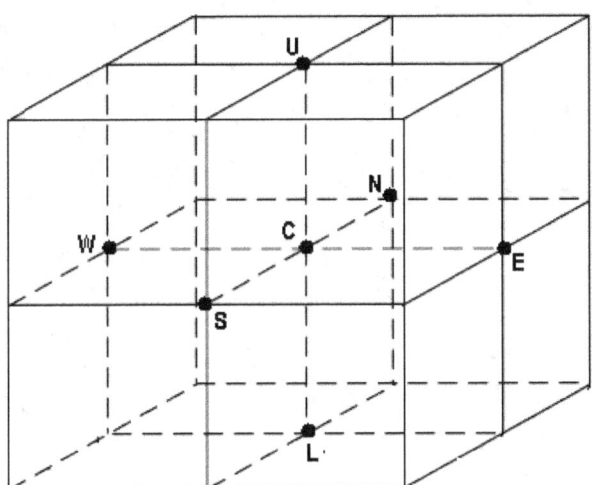

Figure 48. Stacked Hexahedra (Bricks) Showing Adjacent Nodes

We first build a list of adjacent nodes. [It is assumed here that the reader understands how to loop through elements associating each pair of nodes. If not, refer to ctfra.c function NodeNeighbors().] If a node doesn't have 26 neighbors, then it is on a boundary. The natural boundary conditions we applied previously were the partial derivatives of the concentration along a direction perpendicular to any boundary was assumed to be zero. This same natural boundary condition will result with the current implementation, as the finite difference contribution in that direction (toward the boundary) will be zero. We must also move the properties from the elements to the nodes, which is a simple averaging process.

The properties (Dx, Dy, Dz, and φ) for each node will be the average of all the elements containing this node.

The elements for the first FRAC3D-based example (LAFB) are shown in the following figure:

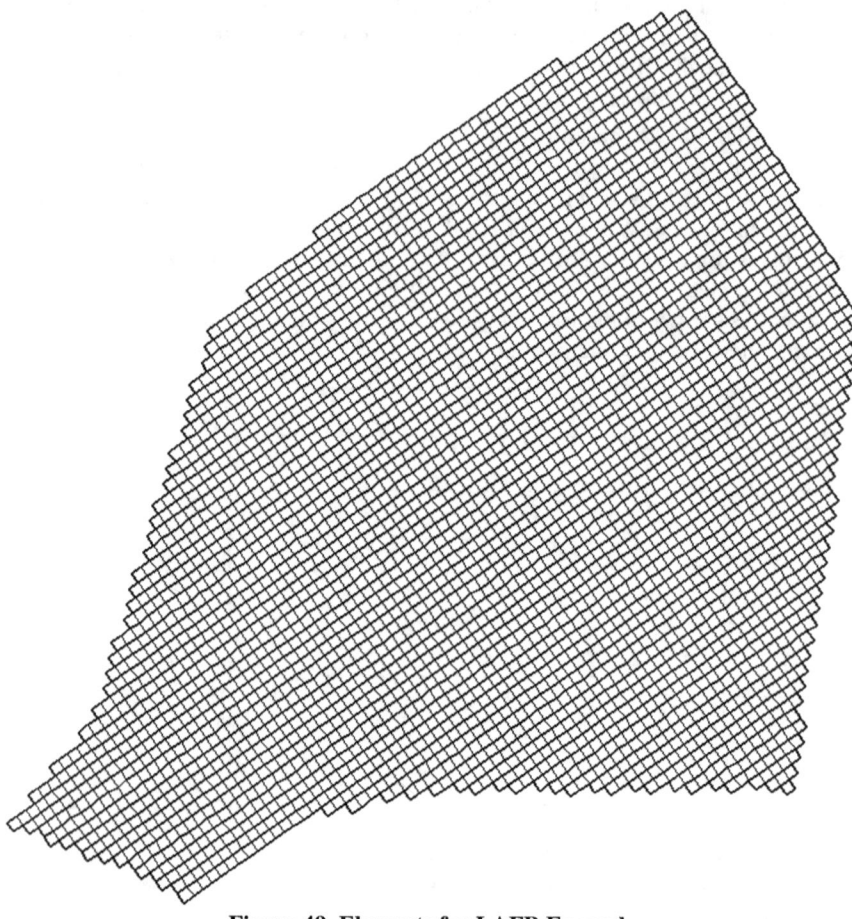

Figure 49. Elements for LAFB Example

The element outlines can be found in file LAFB.P2D and can be drawn with TP2. The nodes are in LAFB.NDE, the elements are in LAFB.ELM, the properties are in LAFB.PRP, the (porous media) velocities are in LAFB.VEP, and the initial concentrations in LAFB.ICN.

The velocity vectors are written to LAFB.VEC formatted for Tecplot™ and are shown in this next figure:

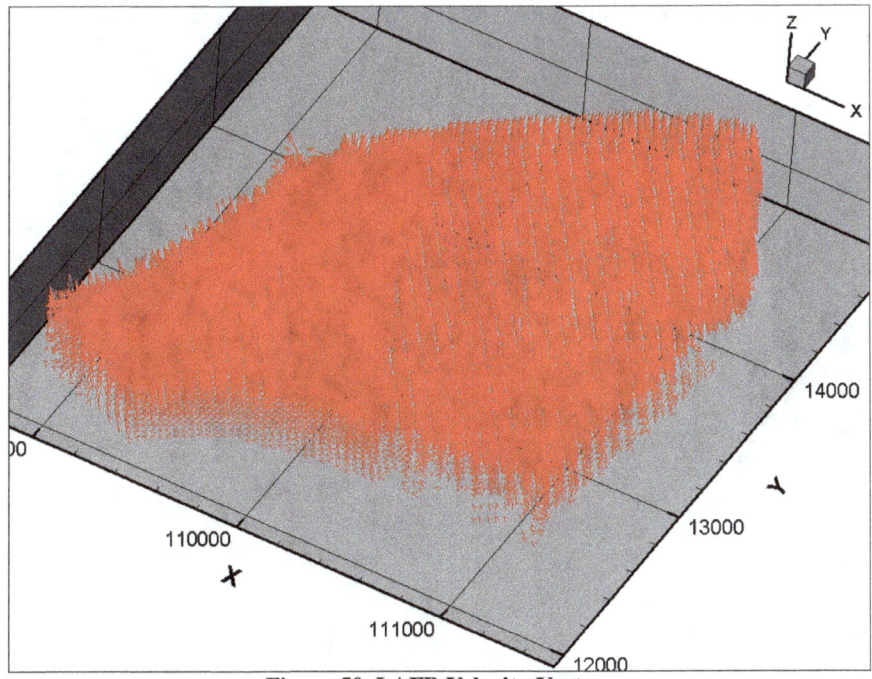

Figure 50. LAFB Velocity Vectors

These velocities may be found in file LAFB.VEC, which looks like:
```
VARIABLES="X", "Y", "Z", "U", "V", "W"
109373 12074.8 385.425 -0.22144 -0.078174 0.0270086
109401 12094.3 386.137 -0.59286 -0.35286 0.022306
109354 12102.5 385.875 -0.4435 0.055038 0.0228016
etc.
```

A corresponding layout file is provided, LAFB.LAY, to facilitate displaying this information with Tecplot™. The data file is listed at the top along with the variable names, as shown below:
```
$!VarSet |LFDSFN1| = '"LAFB.VEC"'
$!VarSet |LFDSVL1| = '"X" "Y" "Z" "U" "V" "W"'
$!SETSTYLEBASE FACTORY
$!GLOBALTHREEDVECTOR
  UVAR = 4
  VVAR = 5
  WVAR = 6
  RELATIVELENGTH = 2E-005
  HEADSIZEINFRAMEUNITS = 1
  SIZEHEADBYFRACTION = NO
$!FIELDLAYERS
  SHOWMESH = NO
```

57

```
SHOWVECTOR = YES
SHOWBOUNDARY = NO
```
The initial concentrations (two plumes) are shown in this next figure:

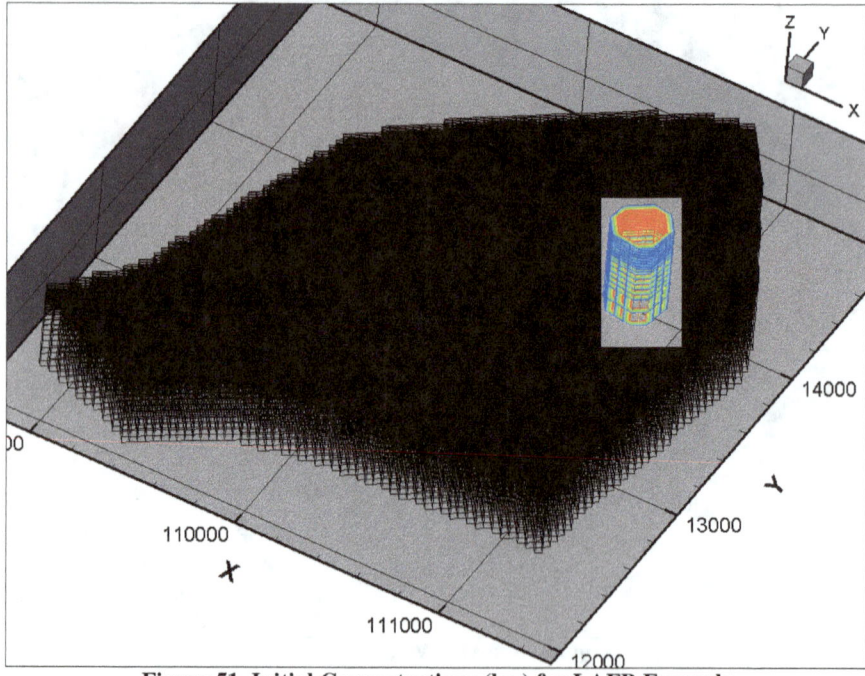

Figure 51. Initial Concentrations (log) for LAFB Example

Concentrations are written to files are named LAFB???.DAT, which can be imported by Tecplot™. The prefix is used to create the sequential output files. The same layout file (LAFB.LAY) can be used to display the concentrations. Refer to the Tecplot™ documentation for how to turn the elements on and off and control the display of concentrations, including colors. You will need to manually load the concentration file for each time step (File => Load DataFile => Replace Dataset and retain plot style).

Before we move ahead numerical solutions, we first compare this approach to the analytical solution, beginning with diffusion alone. The analytical results after 100 years is shown in this next figure:

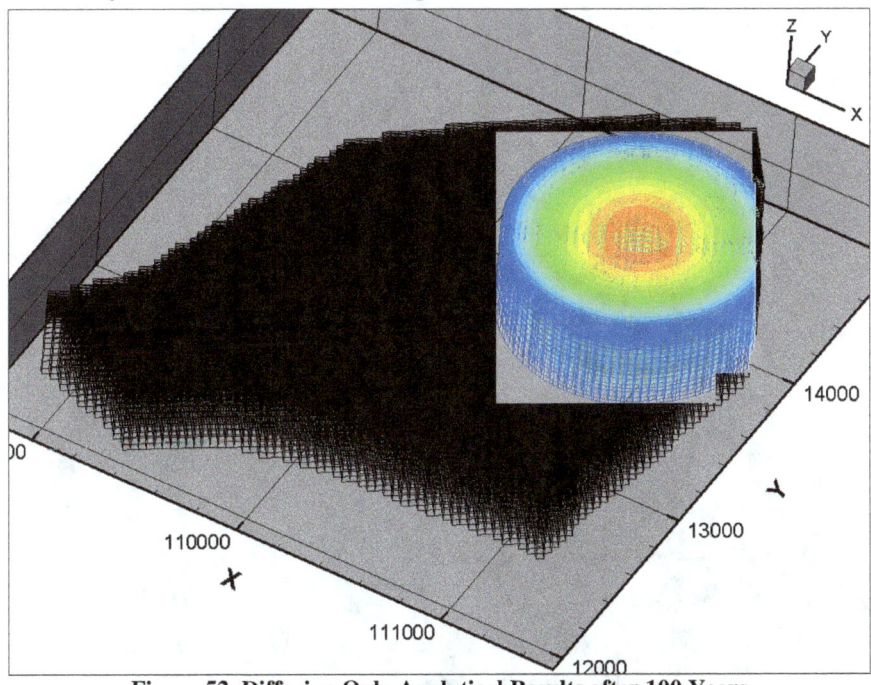

Figure 52. Diffusion Only Analytical Results after 100 Years

We use linear-least squares regression to approximate the concentration over each internal node (those having 26 neighbors) and then take the partial derivatives of the approximation:

```
for(n=0;n<Nn;n++)
{
if(Ln[n]<26)
  continue;
for(i=0;i<Ln[n];i++)
  {
  l=La[26*n+i];
  C[i]=Node[l].C;
  X[i]=Node[l].X-Node[n].X;
  Y[i]=Node[l].Y-Node[n].Y;
  Z[i]=Node[l].Z-Node[n].Z;
  }
X[i]=Y[i]=Z[i]=0.;
C[i++]=Node[n].C;
SecondOrderRegression(X,Y,Z,C,i,A);
dCdX=A[1];
dCdY=A[2];
```

```
            dCdZ=A[3];
            d2CdX2=2.*A[4];
            d2CdY2=2.*A[7];
            d2CdZ2=2.*A[9];
            dCdt[n]=Node[n].Dx*d2CdX2
                   +Node[n].Dy*d2CdY2
                   +Node[n].Dz*d2CdZ2
                   -Node[n].U*dCdX
                   -Node[n].V*dCdY
                   -Node[n].W*dCdZ;
    }
```

The numerical results are shown in this next figure:

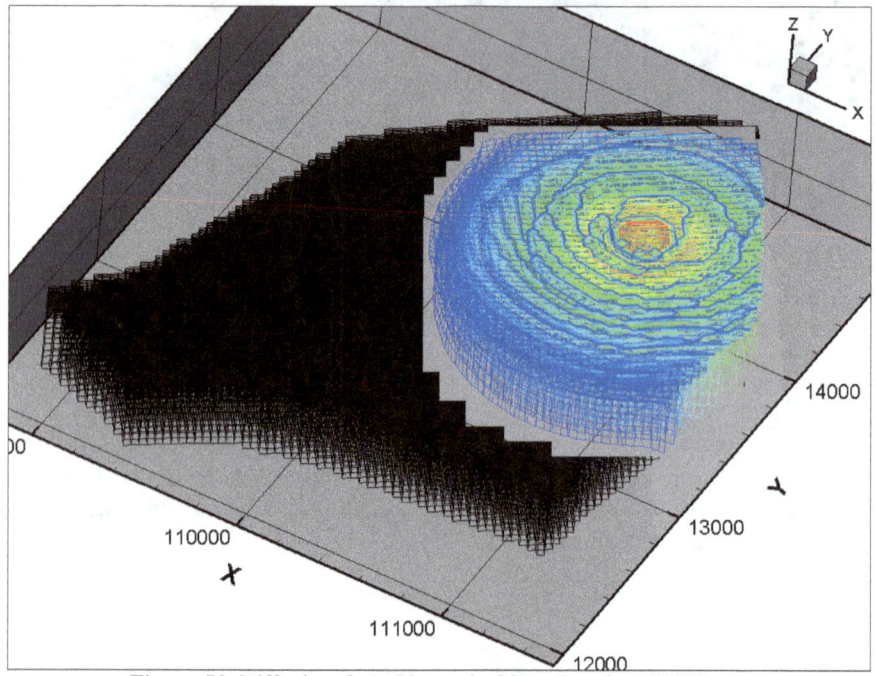

Figure 53. Diffusion Only Numerical Results after 100 Years

The colors are the same in Figures 53 and 54 (see contour levels in Tecplot™). Note that the advance of the outer blue ring corresponding to $10^{-7.3}$ ppm is the same for the analytical and numerical solutions even though there is some blotchiness toward the center of the plume.

Check and Restart

If you read through the main function in each of these example codes, you will find a check: if(Cx>Co). Unless we are adding mass or increasing toxicity over time, the maximum concentration in the domain shouldn't increase. This is an indication of a numerically unstable condition. In the previous example codes we simply wrote an error message and exited the program. In ctfra.c we restart and try again several times. Before restarting we do two things: 1) divide the time step by 2; and 2) increase the relaxation parameter.

You may be familiar with successive over relaxation (SOR) and possibly under relaxation. Over relaxation only works when solving systems of linear equations. When solving nonlinear equations, especially stiff ones, under relaxation is often necessary. The differencing scheme used here (multivariate linear regression of second order) there is a tendency to over estimate the partial derivatives. When we are simply curve-fitting and considering the target value this may not be so noticeable but when we're estimating the derivatives, it is. To quash this over estimation, we implement a simple relaxation (i.e., averaging the surrounding nodes):

```
for(r=0;r<relax;r++)
{
    for(n=0;n<Nn;n++)
    {
        for(i=0;i<Ln[n];i++)
        {
            l=La[26*n+i];
            dCdt[n]+=dCdt[l];
        }
        dCdt[n]/=Ln[n]+1;
    }
}
```

Launching the program (ctfra.c) for this example produces:

```
ctfra lafb
3D Transient Diffusion + Dispersion + Advection
reading FRAC3D input and output files
prefix: lafb
node file: LAFB.NDE
  nodes: 99372
  108830≤X≤111418
  12051≤Y≤14764
  359.5≤Z≤701.8
element file: LAFB.ELM
  elements: 91575
  rotation: 36°
property file: LAFB.PRP
  73.05≤Dx≤73.05
  73.05≤Dy≤73.05
  7.305≤Dz≤7.305
  0.05≤F≤0.05
velocity file: LAFB.VEP
  -186.66≤U≤262.608
  -143.844≤V≤29.583
  -25.2326≤W≤30.5568
velocities: LAFB.VEC
determining time step
```

```
  ΔX=47,   ΔY=47,   ΔZ=5.4
  ΔX²/Dx/2=15.1198,   ΔY²/Dy/2=15.1198,   ΔZ²/Dz/2=1.99589
  ΔX/U/6=0.029829,   ΔY/V/6=0.0544571,   ΔZ/W/6=0.0294533
  Δt=0.0115643
neighbors: 2442220
  internal nodes: 84144
  boundary nodes: 15228
concentration file: LAFB.ICN
  C≤300
0 years  C≤300  LAFB000.DAT
1 years  C≤300  LAFB001.DAT
```

The concentrations after 100 years with diffusion/dispersion:

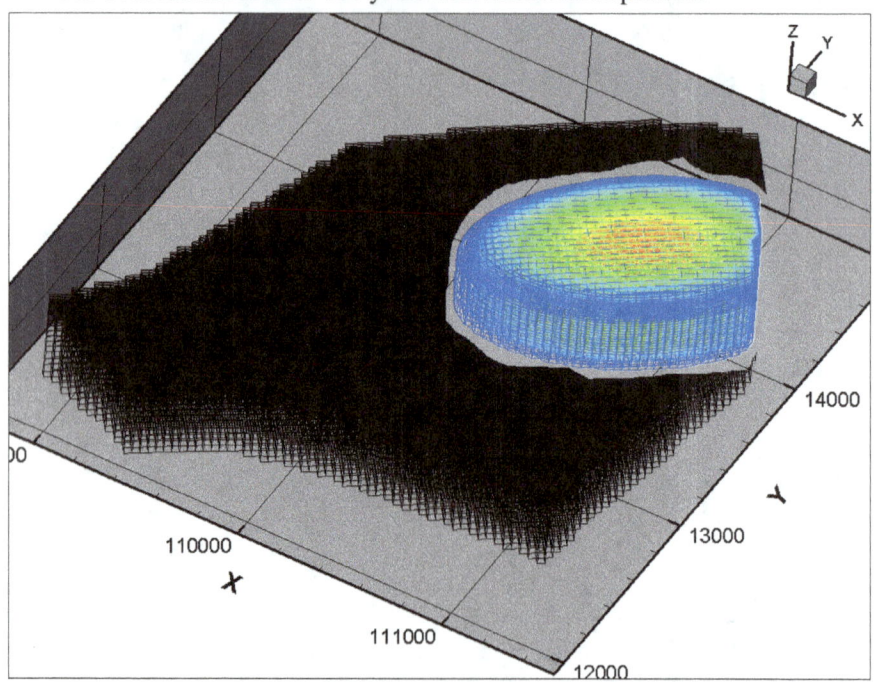

Figure 54. Full Numerical Results after 100 Years

PORTS Example

This next FRAC3D based example can be found in the online archive in folder examples\FRAC3D\PORTS. The elements are shown in this first figure:

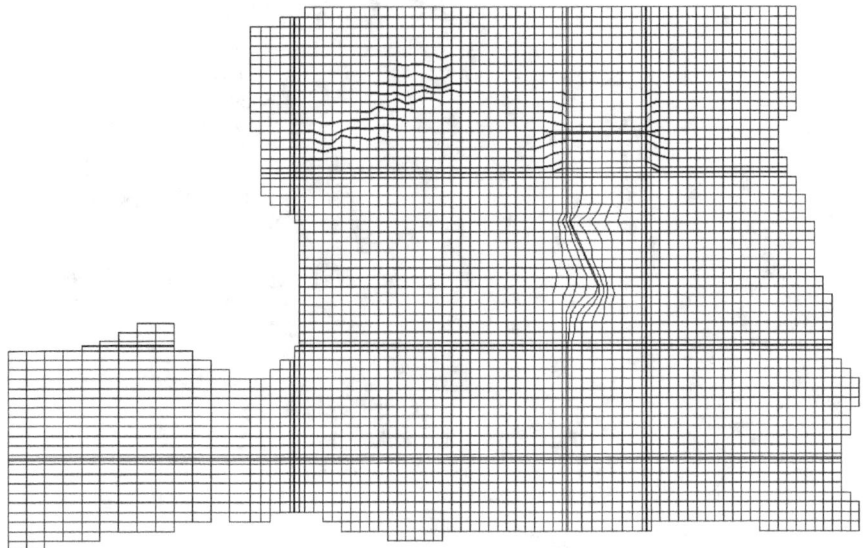

Figure 55. Elements for PORTS Example

The elements are of non-uniform size and a few are rotated but this will not greatly impact our calculations here. These adjustments were made to the original model to investigate the impact of fractures using particle tracking, as karst formations are common at the site. Further drilling and investigations were conducted to better characterize the ground properties. Many model revisions were created before settling on one that was used to guide the remediation effort.

The velocity vectors are shown in this next figure:

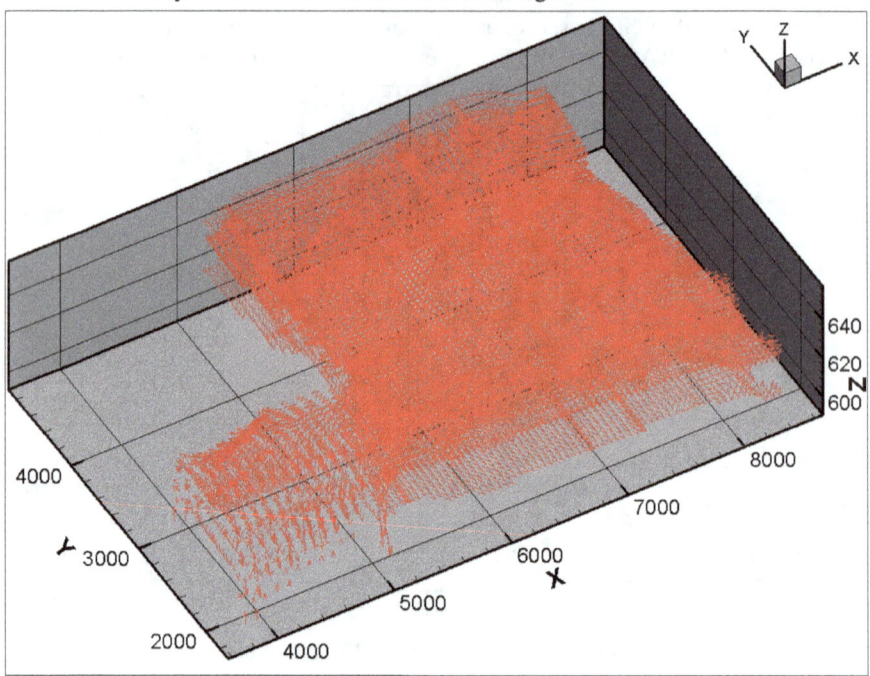

Figure 56. PORTS Example Velocity Vectors

Velocities varied throughout the domain for this particular site. Notice the changes at the edge on the southwest side in particular. This odd shape arose from construction in the area, including buildings and roads. All of these factors were considered in building the model and designing the remediation strategy.

While there were several plumes at this site and several contaminants involved, we will consider a single plume for our example here. We could add a term for each contaminant and calculate them simultaneously or run the model several times, once for each contaminant species. The initial concentrations are shown in this next figure:

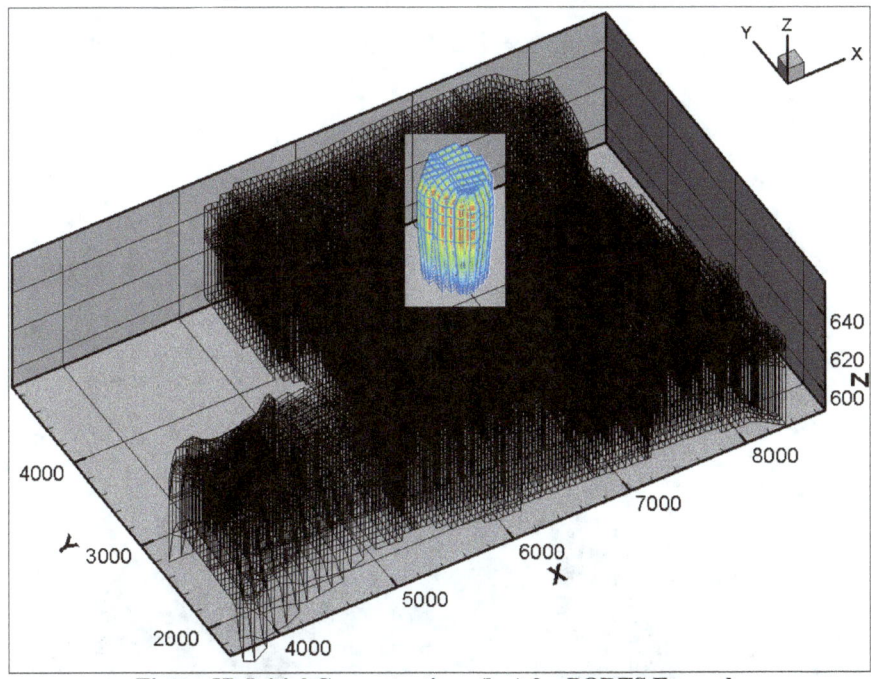

Figure 57. Initial Concentrations (log) for PORTS Example

Launching the program produces the following output:

```
ctfra ports
3D Transient Diffusion + Dispersion + Advection
reading FRAC3D input and output files
prefix: ports
node file: PORTS.NDE
  nodes: 26844
  3750≤X≤8460
  1795≤Y≤4792
  575.4≤Z≤665
element file: PORTS.ELM
  elements: 21500
  rotation: 0°
property file: PORTS.PRP
  365.25≤Dx≤365.25
  365.25≤Dy≤365.25
  36.525≤Dz≤36.525
  0.1≤F≤0.1
```

```
velocity file: PORTS.VEP
  -305.66≤U≤351.18
  -768.92≤V≤185.26
  -23.56≤W≤7.84
velocities: PORTS.VEC
determining time step
  ΔX=50,  ΔY=53,  ΔZ=4.6
  ΔX²/Dx/2=3.42231,  ΔY²/Dy/2=3.84531,  ΔZ²/Dz/2=0.289665
  ΔX/U/6=0.0237295,  ΔY/V/6=0.011488,  ΔZ/W/6=0.032541
  Δt=0.00610038
neighbors: 600676
  internal nodes: 16512
  boundary nodes: 10332
concentration file: PORTS.ICN
  C≤300
0 years C≤300 PORTS000.DAT
1.000 years C≤284 PORTS001.DAT
36.017 years C≤3 PORTS036.DAT
```

The concentrations after 10 years are shown in this next figure:

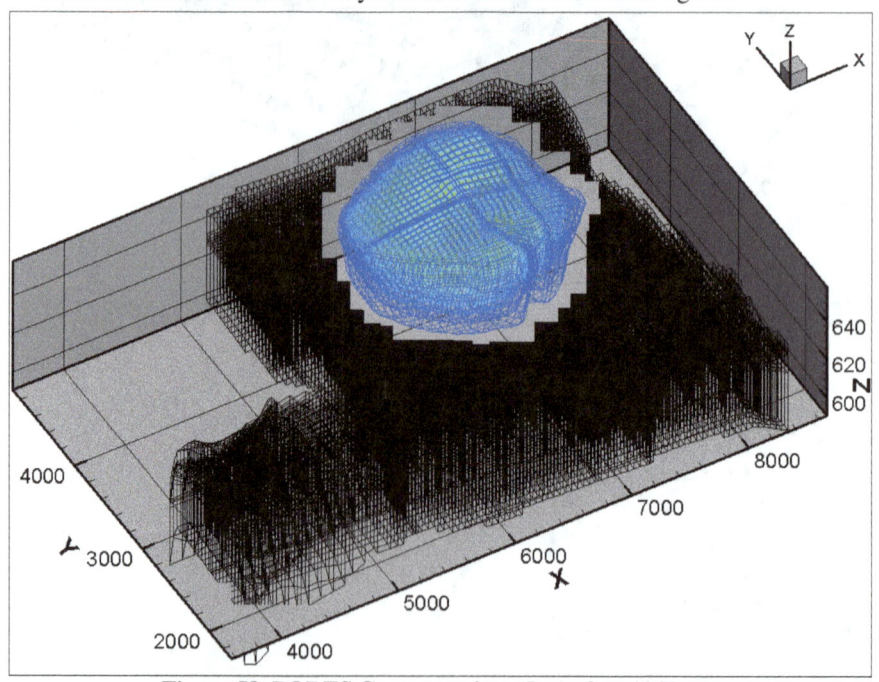

Figure 58. PORTS Concentrations (log) after 10 Years

The interior of the plume is obscured by the top. Tecplot™ has a feature (menu => style => value blanking) that can turn off drawing the top of the plume, thus exposing the interior. The selection and result are shown in this next figure:

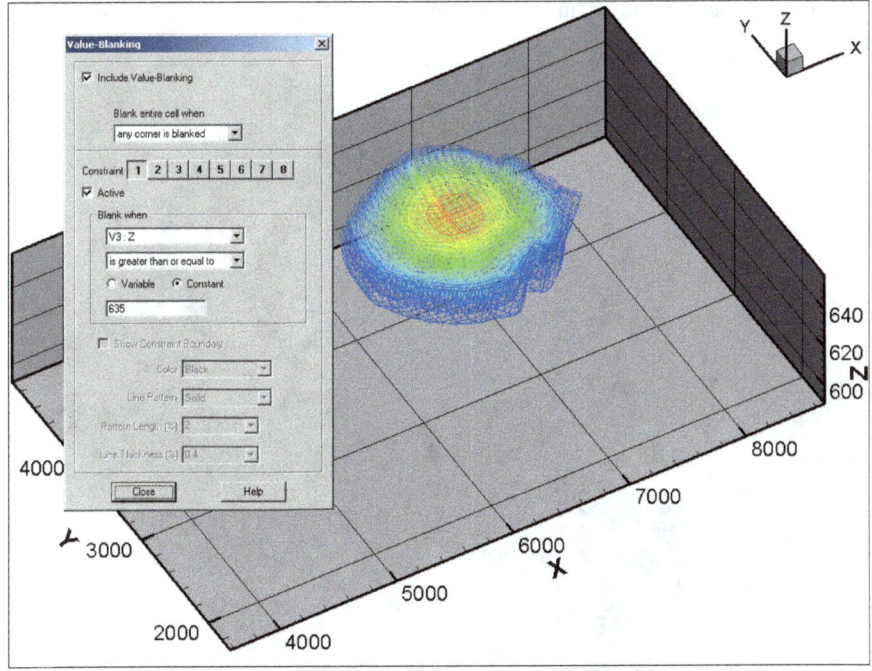

Figure 59. Concentrations with Value Blanking

Value blanking can also be used with a macro to slice through the plume in any of the three directions, exporting the graphic each time to sequential files. The following is an example of such a Tecplot™ macro:

```
#!MC 800
$!REDRAWALL
$!EXPORTSETUP EXPORTFORMAT = RASTERMETAFILE
$!EXPORTSETUP EXPORTFNAME = 'plume.rm'
$!FIELD [1]   IJKMODE{PLANES = K}
$!FIELD [1]   IJKMODE{KRANGE{MIN = 1}}
$!FIELD [1]   IJKMODE{KRANGE{MAX = 1}}
$!REDRAWALL
$!EXPORT
  APPEND = NO
$!FIELD [1]   IJKMODE{KRANGE{MIN = 2}}
$!FIELD [1]   IJKMODE{KRANGE{MAX = 2}}
$!REDRAWALL
$!EXPORT
  APPEND = YES
$!FIELD [1]   IJKMODE{KRANGE{MIN = 3}}
$!FIELD [1]   IJKMODE{KRANGE{MAX = 3}}
$!REDRAW
```

```
$!EXPORT
  APPEND = YES
```

The plume has spread and dissipated considerably so that after 36 years, this is what's left:

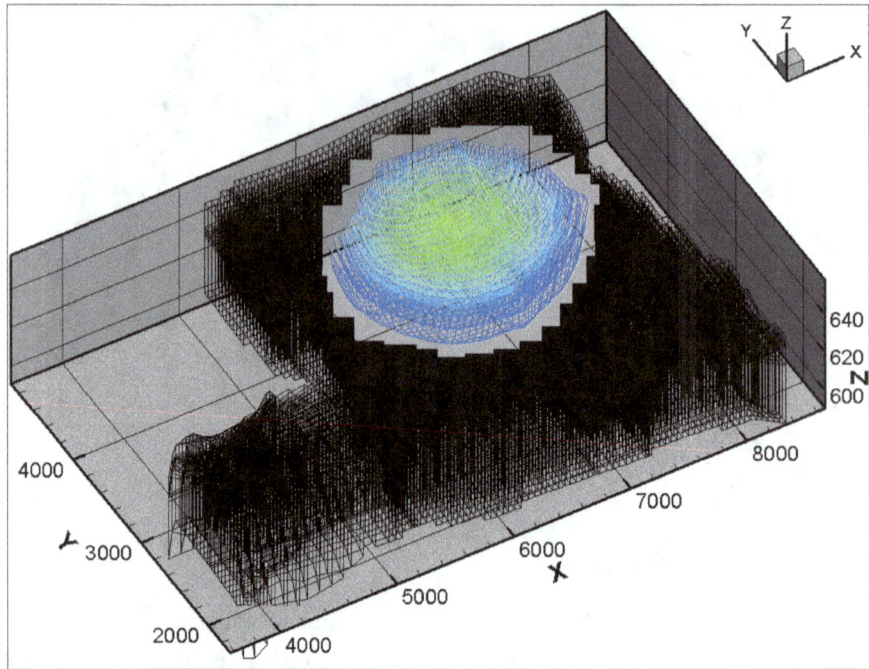

Figure 60. PORTS Concentrations (log) after 36 Years

ORNL Example

The elements for this next example are shown in the following figure:

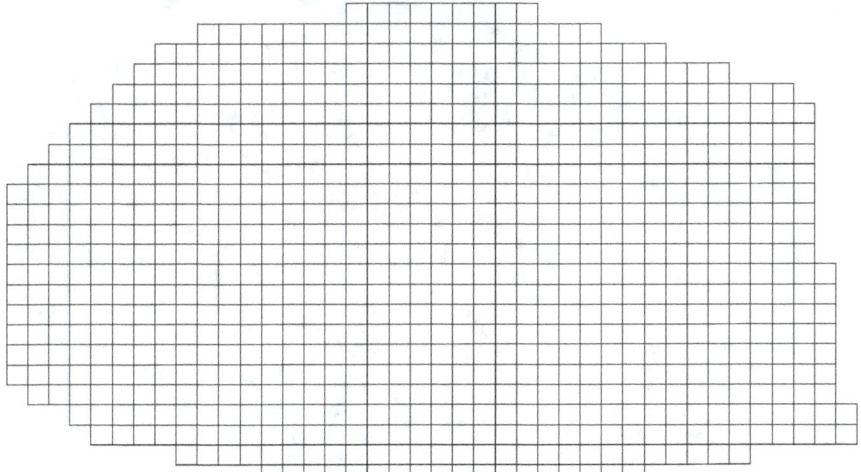

Figure 61. ORNL Example Elements

The velocity vectors are shown below:

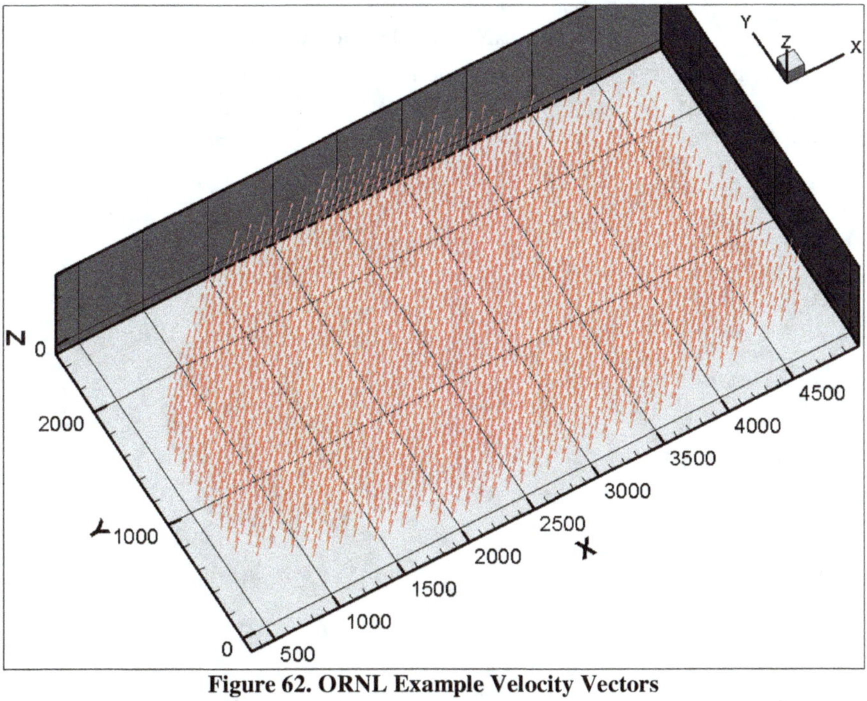

Figure 62. ORNL Example Velocity Vectors

There were also several plumes at this site but we consider only one here:

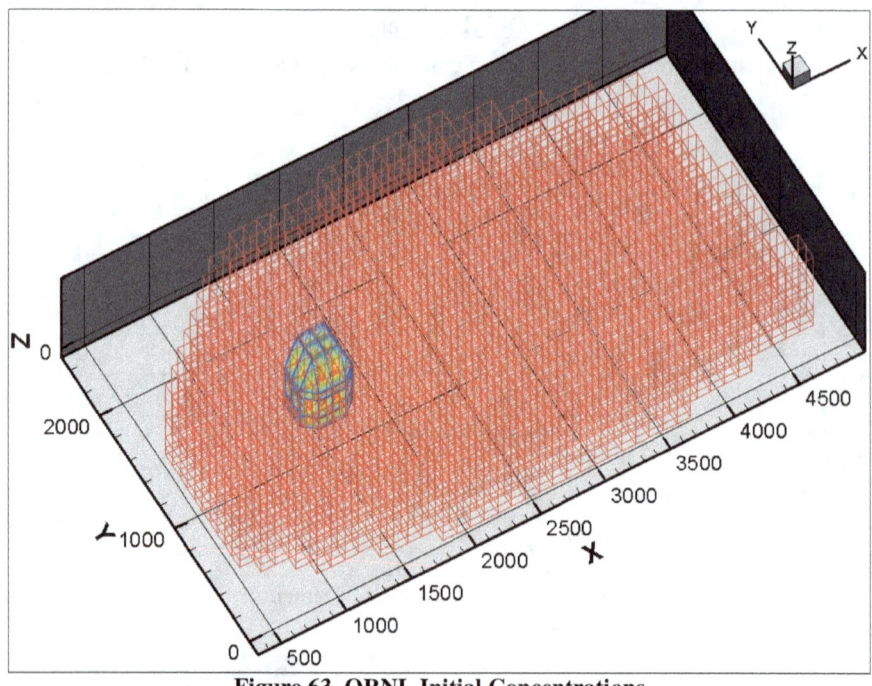

Figure 63. ORNL Initial Concentrations

Launching the program produces the following output:
```
ctfra ornl
3D Transient Diffusion + Dispersion + Advection
reading FRAC3D input and output files
prefix: ornl
node file: ORNL.NDE
  nodes: 5166
  532≤X≤4787
  0≤Y≤2400
  -23.7≤Z≤294.3
element file: ORNL.ELM
  elements: 3980
  rotation: 0°
property file: ORNL.PRP
  365.25≤Dx≤365.25
  365.25≤Dy≤365.25
  36.525≤Dz≤36.525
  0.1≤F≤0.1
velocity file: ORNL.VEP
  500≤U≤500
  100≤V≤100
  10≤W≤10
velocities: ORNL.VEC
determining time step
```

```
ΔX=106, ΔY=100, ΔZ=125.6
ΔX²/Dx/2=15.3812, ΔY²/Dy/2=13.6893, ΔZ²/Dz/2=215.953
ΔX/U/6=0.0353333, ΔY/V/6=0.166667, ΔZ/W/6=2.09333
Δt=0.028635
neighbors: 112546
  internal nodes: 2932
  boundary nodes: 2234
concentration file: ORNL.ICN
  C≤300
0 years C≤300 ORNL000.DAT
1.002 years C≤164 ORNL001.DAT
```

This is a small, rapidly spreading plume. After 1 year the results are:

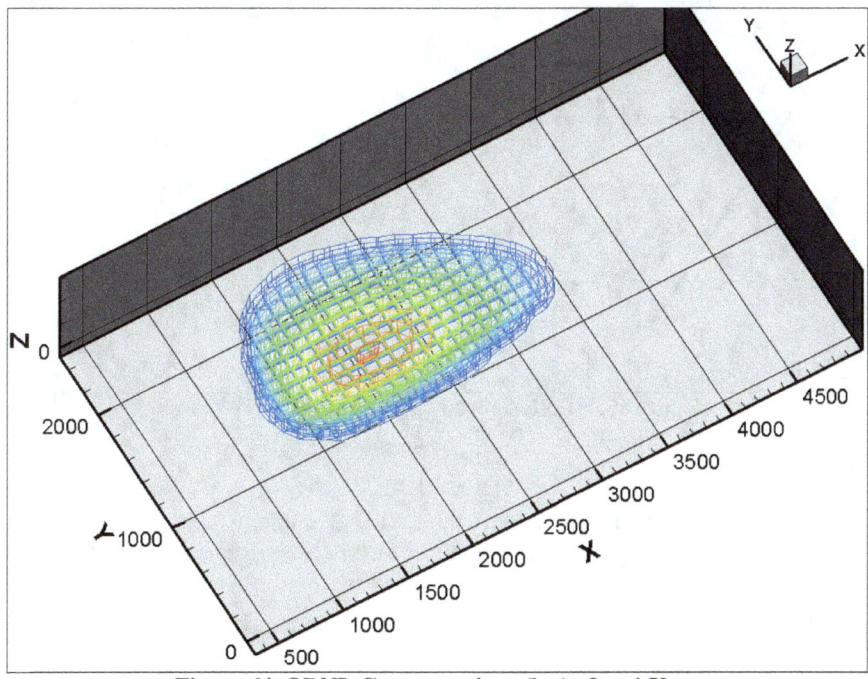

Figure 64. ORNL Concentrations (log) after 1 Year

The plume quickly dissipates and exits the site (we lost this one). After 6 years it looks like this:

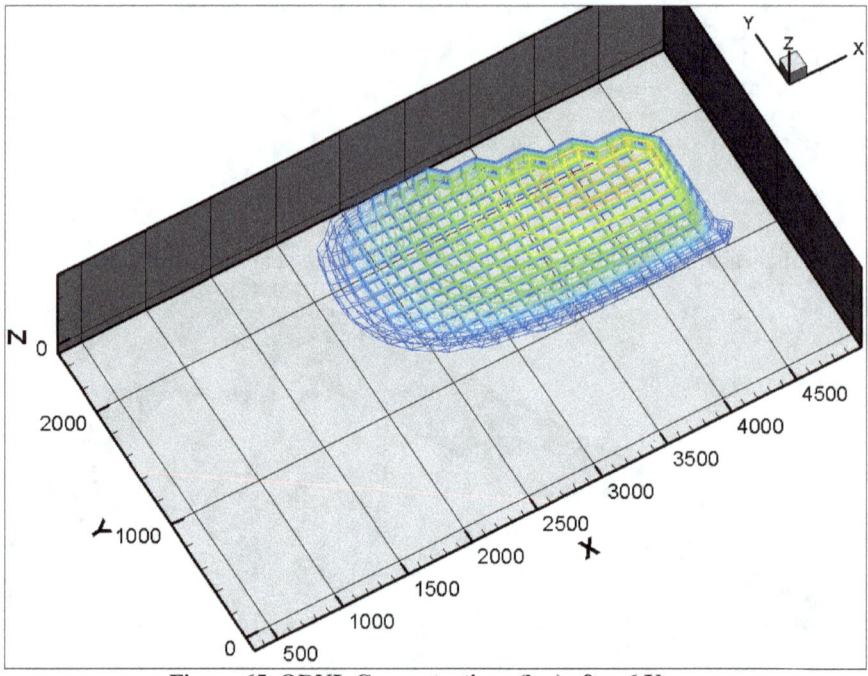

Figure 65. ORNL Concentrations (log) after 6 Years

Rotated Elements

We introduced grid rotation with the LAFB example, which in that case was 36°. Each element has a different rotation in this next example. As before, we calculate the finite differences with respect to the two planar element coordinates $\bar{P}$ and $\bar{Q}$, then multiply by the rotation matrix:

$$\begin{bmatrix} \cos\theta & -\sin\theta \\ \sin\theta & \cos\theta \end{bmatrix} \quad (7.1)$$

We use the rotation angles with the nodal equations and so we transfer the angle from each element to the first node. The rotated difference equations become:

```
for(n=0;n<Nn;n++)
  {
  if(Ln[n]<26)  /* only consider internal nodes */
    continue;
  dP=hypot(Node[Ie[n]].Y-Node[Iw[n]].Y,Node[Ie[n]].X-
    Node[Iw[n]].X)/2.;
```

```
    dQ=hypot(Node[In[n]].Y-Node[Is[n]].Y,Node[In[n]].X-
Node[Is[n]].X)/2.;
    dZ=(Node[Iu[n]].Z-Node[Il[n]].Z)/2.;
    Cc=Node[n].C;
    Ce=Node[Ie[n]].C;
    Cn=Node[In[n]].C;
    Cs=Node[Is[n]].C;
    Cw=Node[Iw[n]].C;
    Cu=Node[Iu[n]].C;
    Cl=Node[Il[n]].C;
    Ca=(Cw+Cc)/2.;
    Cb=(Ce+Cc)/2.;
    Cd=(Cn+Cc)/2.;
    Cf=(Cs+Cc)/2.;
    Cg=(Cu+Cc)/2.;
    Ch=(Cl+Cc)/2.;
    dCdPa=(Cc-Cw)/dP;
    dCdPb=(Ce-Cc)/dP;
    dCdQd=(Cn-Cc)/dQ;
    dCdQf=(Cc-Cs)/dQ;
    dCdZg=(Cu-Cc)/dZ;
    dCdZh=(Cc-Cl)/dZ;
    d2CdP2=(dCdPb-dCdPa)/dP;
    d2CdQ2=(dCdQd-dCdQf)/dQ;
    d2CdZ2=(dCdZg-dCdZh)/dZ;
    Dx=Node[n].Dx;
    Dy=Node[n].Dy;
    Dz=Node[n].Dz;
    U=Node[n].U;
    V=Node[n].V;
    W=Node[n].W;
    if(U>0.)
       UdCdX=U*(cos*(Cc-Cw)/dP-sin*(Cc-Cs)/dQ);
    else
       UdCdX=U*(cos*(Ce-Cc)/dP-sin*(Cn-Cc)/dQ);
    if(V>0.)
       VdCdY=V*(sin*(Cc-Cw)/dP+cos*(Cc-Cs)/dQ);
    else
       VdCdY=V*(sin*(Ce-Cc)/dP+cos*(Cn-Cc)/dQ);
    if(W>0.)
       WdCdZ=W*(Cc-Cl)/dZ;
    else
       WdCdZ=W*(Cu-Cc)/dZ;
    dCdt[n]=Dx*(co*d2CdP2-si*d2CdQ2)
           +Dy*(si*d2CdP2+co*d2CdQ2)
           +Dz*d2CdZ2
           -UdCdX-VdCdY-WdCdZ;
```

The Z direction is not rotated so no modification is required. This rotation and also the relaxation and restart procedure are two examples of how you might need to modify a model to handle numerical issues as they arise. The elements for this last FRAC3D-based example are all slightly rotated:

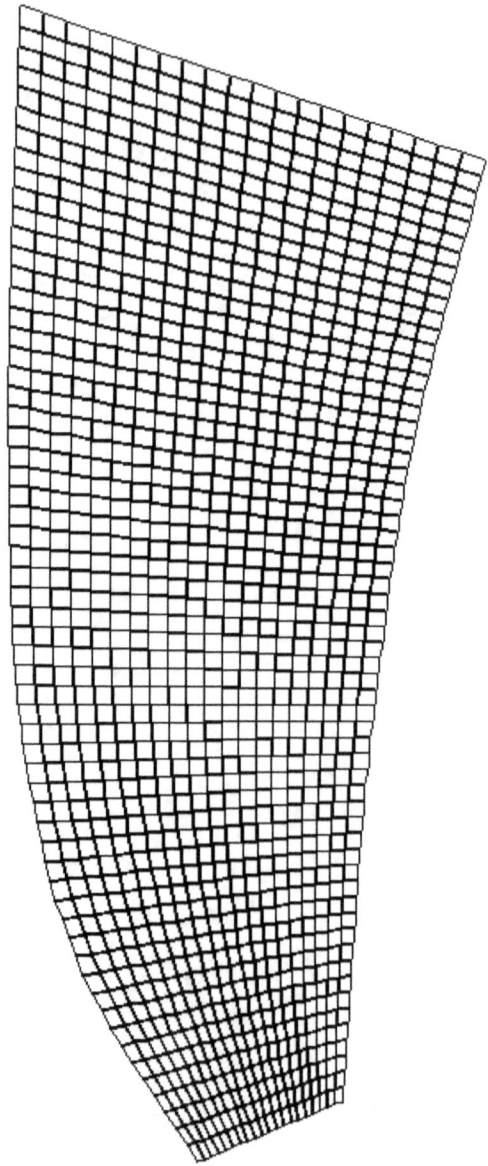

Figure 66. CAFB Example Elements

The velocity vectors for this example are shown in the following figure:

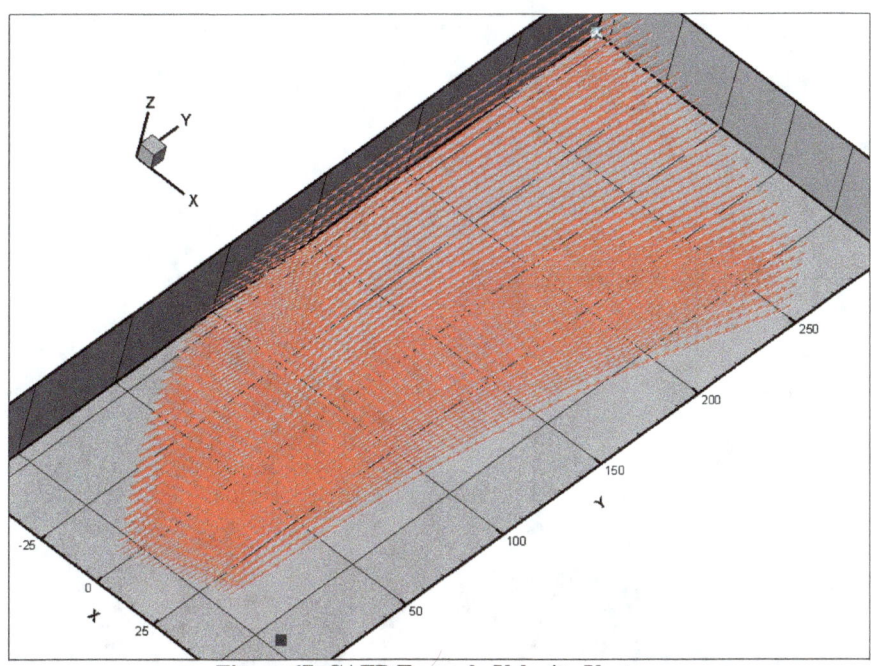

Figure 67. CAFB Example Velocity Vectors

Launching the program produces the following output:

```
ctfra cafb
3D Transient Diffusion + Dispersion + Advection
reading FRAC3D input and output files
prefix: cafb
node file: CAFB.NDE
  nodes: 10248
  -52.2≤X≤71.5
  -11.6≤Y≤286.9
  0≤Z≤35
element file: CAFB.ELM
  elements: 8400
  rotation: 124°
property file: CAFB.PRP
  73.05≤Dx≤73.05
  73.05≤Dy≤73.05
  3.6525≤Dz≤3.6525
  0.1≤F≤0.1
velocity file: CAFB.VEP
  20≤U≤20
  200≤V≤200
  2≤W≤2
determining time step
  ΔX=7.1, ΔY=5.9, ΔZ=5
  ΔX²/Dx/2=0.345038, ΔY²/Dy/2=0.238261, ΔZ²/Dz/2=3.42231
```

```
ΔX/U/6=0.0591667, ΔY/V/6=0.00491667, ΔZ/W/6=0.416667
Δt=0.00434634
neighbors: 232654
  internal nodes: 6726
  boundary nodes: 3522
concentration file: CAFB.ICN
  C≤300
0 years C≤300 CAFB000.DAT
1 years C≤171 CAFB001.DAT
```

The initial concentrations are shown below:

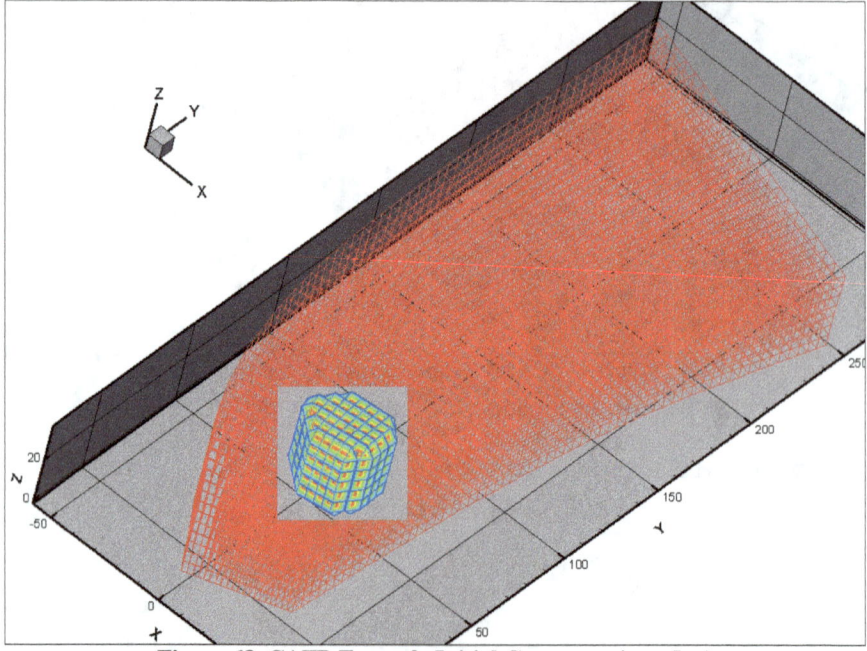

Figure 68. CAFB Example Initial Concentrations (log)

This is a small domain with a rapidly dissipating and sweeping plume. After just 5 years it looks like this:

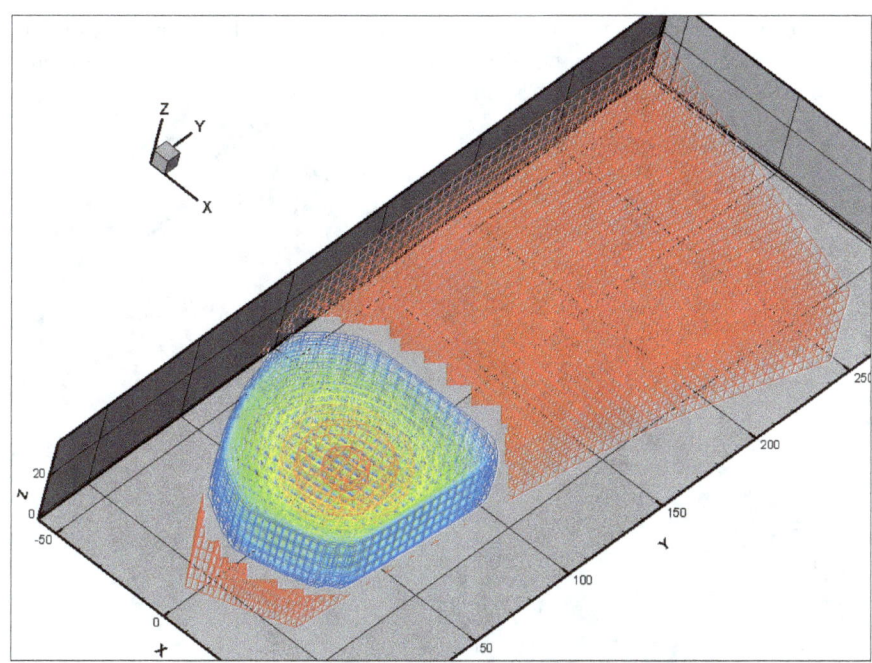

Figure 69. CAFB Concentrations (log) after 5 Years

The plume continues to sweep toward the northern boundary. Extremely small time steps were required to achieve stability, which is why the contours are so smooth. After 25 years the plume looks like this:

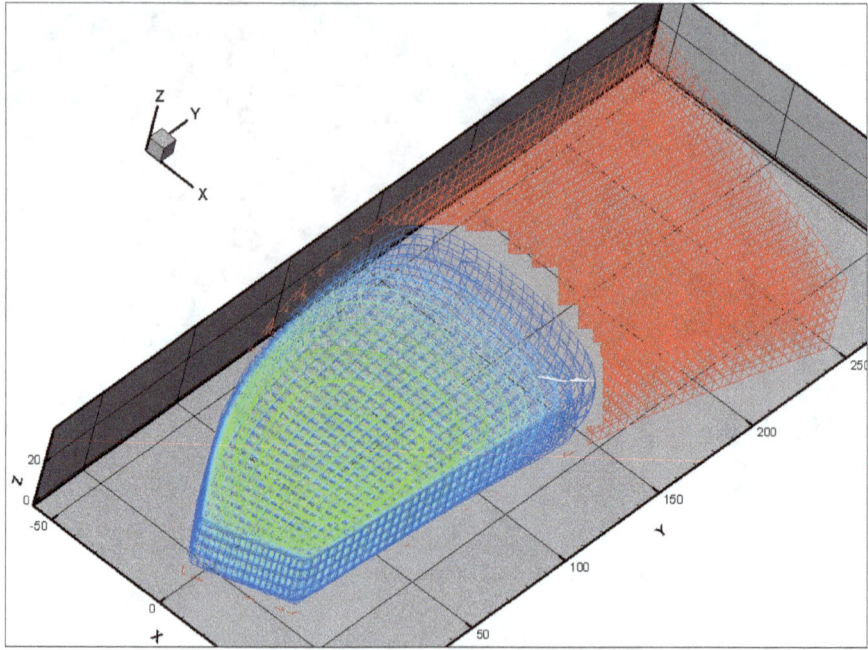

Figure 70. CAFB Concentrations (log) after 25 Years

Chapter 8. Pump-and-Treat

One of the most effective remediation strategies for removing contaminants from the ground is pump-and-treat. Groundwater is pumped from one or more wells and fed through a series of drums containing "magic dust" (i.e., some reacting and/or absorbing media that will remove the contaminant and can be disposed of properly or incinerated). In order for this approach to be effective, the pattern of movement of the contaminant through the ground must be accurately understood, which requires not only calculating but also verifying the flow. Once this has been sufficiently established, you don't want to change anything, so simply pumping water out of the ground might alter the flow pattern and decrease effectiveness. That is why the contaminated water is most often treated and then injected back into the ground nearby, as shown in this figure:

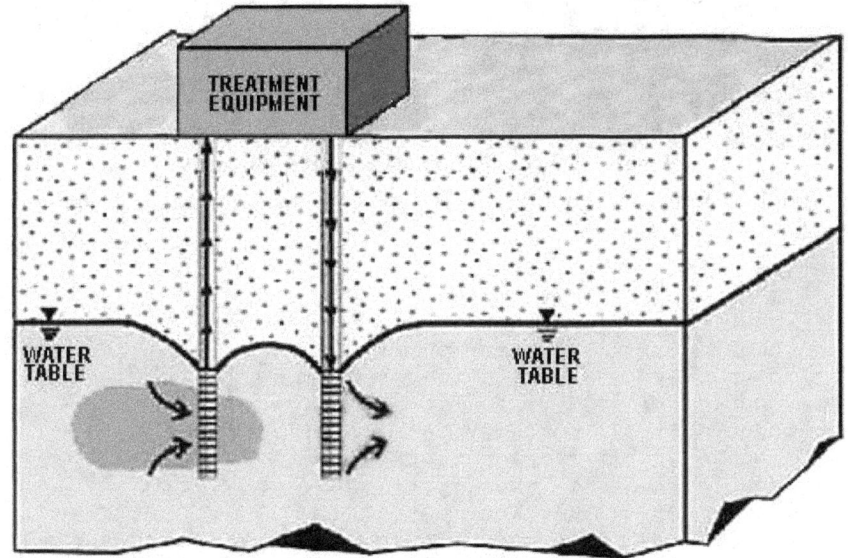

Figure 71. Pump-and-Treat Concept

The equipment to accomplish this varies but most often includes one or more pumps and one or more barrels of magic dust. One project I worked on required three sequential passes through the extraction material to adequately reduce the concentration. The first barrel became fouled most quickly so that each month the three barrels were rotated, discarding the first one and adding a fresh one to the end of the process. Four sets of barrels were used so that twelve were in use at all times with four waiting deployment. The pumps were run by dozens of automotive batteries, which were recharged weekly. All of the equipment was sunk in a concrete vault in the ground with a locked steel door to prevent tampering and theft, which occurred on a previous job. The equipment looked something like this:

Figure 72. Typical Pump-and-Treat Equipment

Implementing this in the model is simple. A switch is added to the node structure:

```
typedef struct{
  int well;    /* extraction well */
  double C;    /* concentration */
  double Dx;   /* X diffusion coefficient */
  double Dy;   /* Y diffusion coefficient */
  double Dz;   /* Z diffusion coefficient */
  double F;    /* porosity */
  double U;    /* X velocity */
  double V;    /* Y velocity */
  double W;    /* Z velocity */
  double X;    /* location */
  double Y;    /* location */
  double Z;    /* location */
  double co;   /* cos(rotation) */
  double si;   /* sin(rotation) */
}NODE;
NODE*Node;
```

Nodes are flagged by their position:

```
void AddWell(double X,double Y,double R)
  {
  int i,n;
  Nw++;
  for(i=n=0;n<Nn;n++)
    if(hypot(Node[n].X-X,Node[n].Y-Y)<=R)
```

```
{
    Node[n].well=Nw;
    i++;
}
printf("%i nodes in well %i\n",i,Nw);
}
```

Wells are added as necessary:

```
AddWell(0.,100.,5.);
```

As the solution progresses, the concentration at all flagged nodes is set to zero:

```
for(n=0;n<Nn;n++)
    if(Node[n].well)
        Node[n].C=0.;
    else if(Ln[n]>=26)
        Node[n].C=dCdt[n]/(Ln[n]+1);
```

This is what a well looks like in the CAFB example:

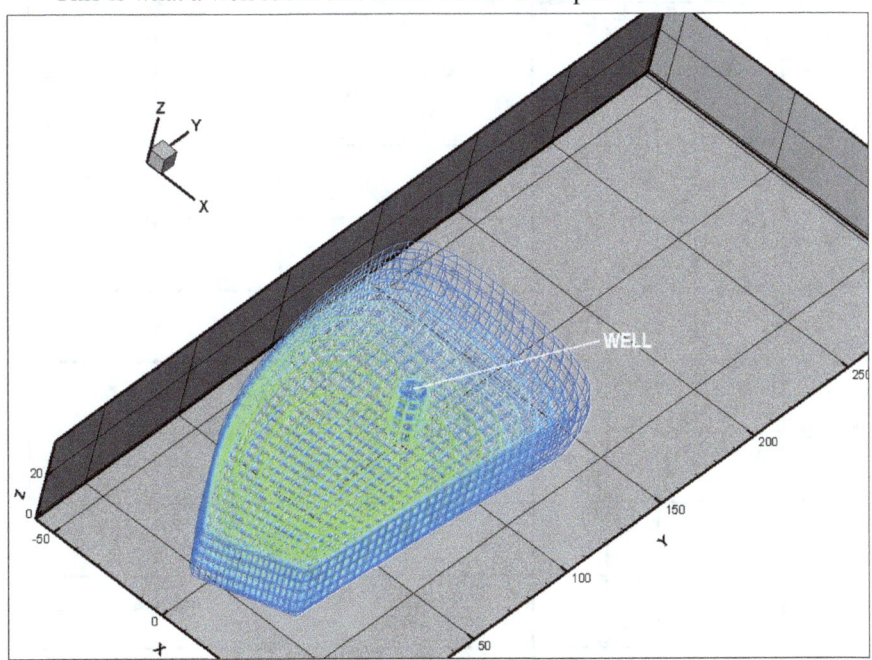

Figure 73. Well Added to CAFB Example

81

Chapter 9. Reaction and Decay

Chemical reactions and radioactive decay occur in some contaminants. Both can usually be approximated by the standard exponential decay formula for concentration over time:

$$C(t) = C_0 e^{-\frac{t}{\tau}} \quad (9.1)$$

The rate of change is easily calculated:

$$\frac{dC}{dt} = \frac{-C_0}{\tau} e^{-\frac{t}{\tau}} = \frac{-C}{\tau} \quad (9.2)$$

Equation 9.1 is shown in the following figure:

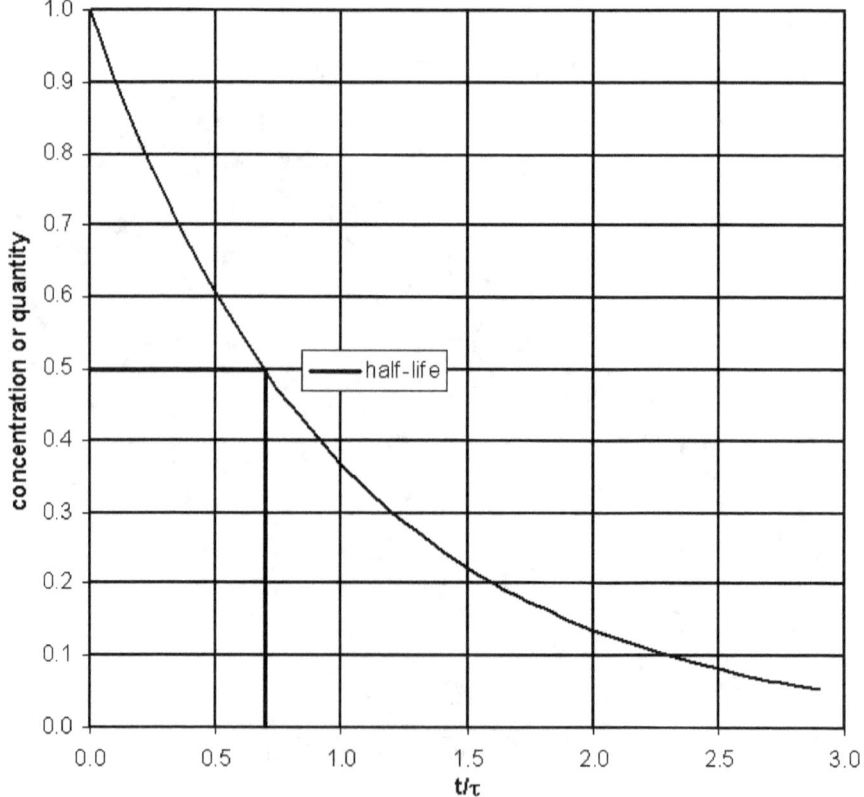

Figure 74. Exponential Decay

The "half-life" is the time required for the concentration (or quantity) to reach one half of the initial value. This happens at $\exp(-t/\tau)=0.5$ or $t/\tau=\ln(2)$. This is easily implemented in function AdvanceSolution() any one of the codes;

for example in ctmod.c we add the last term to the concentration update for each node:

```
decay=exp(-(t+dt)/tau)/exp(-t/tau);
for(z=1;z<Nlay;z++)
  {
  for(y=1;y<Nrow;y++)
    {
    for(x=1;x<Ncol;x++)
      {
      ic=node(x,y,z);
      if(!Node[ic].active)
        continue;
      Node[ic].C+=dt*dCdt[ic];
      Node[ic].C*=decay;
      }
    }
  }
```

While we might implement decay by using Equation 9.2:

```
Node[ic].C+=dt*dCdt[ic]-Node[ic].C/tau;
```

this may lead to overshoot and sometimes negative concentrations. This problem is easily avoided by simply multiplying by the ratio of exponentials as shown above. Using the WINE example and $\tau=5$ years, the corresponding half-life is 3.47 years. Concentrations with and without decay after 25 years are shown on the following page for comparison:

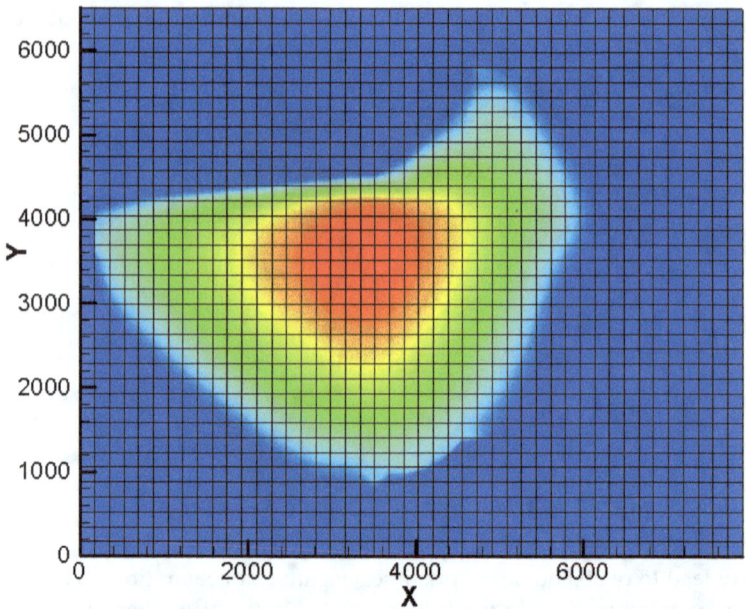
Figure 75. WINE Concentrations (log) after 25 Years without Decay

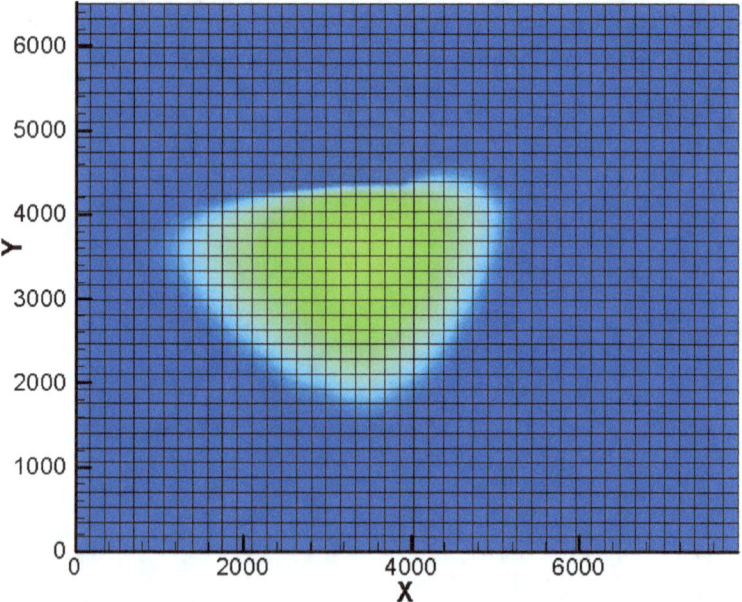
Figure 76. WINE Concentrations (log) after 25 Years with Decay

Chapter 10. AT123D Analytical Solution

AT123D is a numerical implementation of an analytical solution to the governing partial differential equation expressing the conservation of mass for the contaminant, Equation 3.8, in 1D, 2D, or 3D. The original FORTRAN code was developed by Dr. G. T. Yeh of the Oak Ridge National Laboratory.[9] A Web search will turn up documents, examples, models, and more. This useful code has been the basis of many successful remediation projects, as the literature reflects. AT123D was also used to validate my particle tracker, PTRAX.[10] It can also be used to validate other numerical methods, such as the ones accompanying this text, which can be found in the online archive in the example folders.

Because AT123D is an analytical solution, it is somewhat limited in what it can do because of the simple boundary conditions and domain shapes. An analytical model also cannot handle discontinuous properties, such as occur across soil type boundaries and rock formations, let alone karst or fractures. Still, it is a very useful tool. At one time the AT123D FORTRAN code could be downloaded free of charge from the ORNL web site, as it was developed by the Department of Energy. AT123D has been incorporated into and is available with at least one commercial package (SEVIEW).

Twenty-three years ago I translated the original FORTRAN into C and significantly optimized the code. I also replaced the crude (and slow) error function approximation with a better one from the standard reference on such.[11] I replaced the hardwired arrays with allocatable ones and modified the output to create files that can be read by TP2 or Tecplot™. I haven't made this optimized code available on the Web, as I have not been released to do so. Should you need it and can get permission from DoE to use it, I'd be glad to send it to you.

---

[9] G. T., Yeh, "AT123D: Analytical One-, Two-, and Three-Dimensional Simulation of Waste Transport in the Aquifer System," Environmental Sciences Division Publication No. 1439, Report. ORNL-5602, 1981.

[10] Benton, D. J., Young, S. C., and Williams, N. J., "Description and Verification of PTRAX - A Random Walk Model for Predicting Groundwater Solute Transport," U.S. Department of Energy, Report MMESD 8.13-005, 1995.

[11] Abramowitz, M. and I. A. Stegun, *Handbook of Mathematical Functions*, first published by the National Bureau of Standards as Technical Monograph No. 55. This valuable reference may be obtained free online as a PDF from several different web sites.

We can use AT123D to illustrate several processes. This first pair of figures shows a simple spreading, advecting plume.

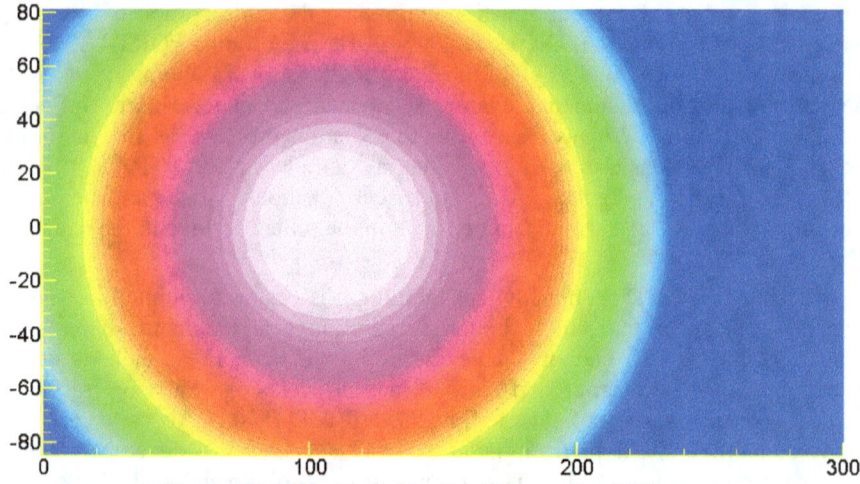
Figure 77. Diffusion/Advecting Plume after 15 Years

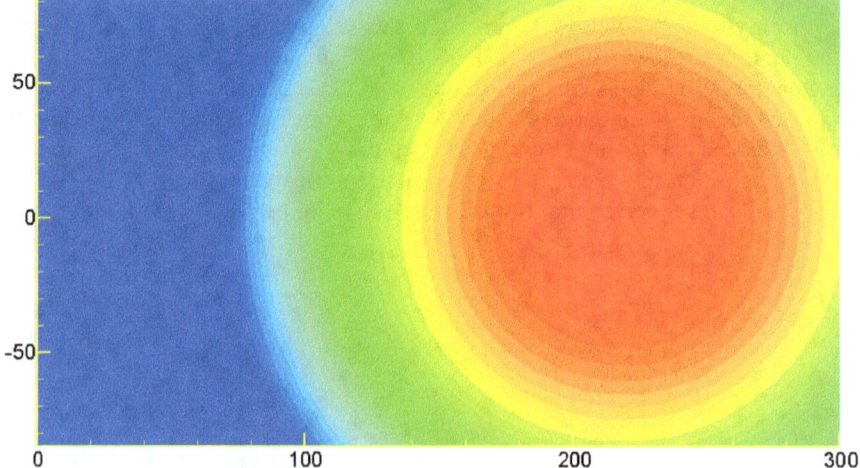
Figure 78. Diffusion/Advecting Plume after 30 Years

If we add decay and hold everything else constant, we get:

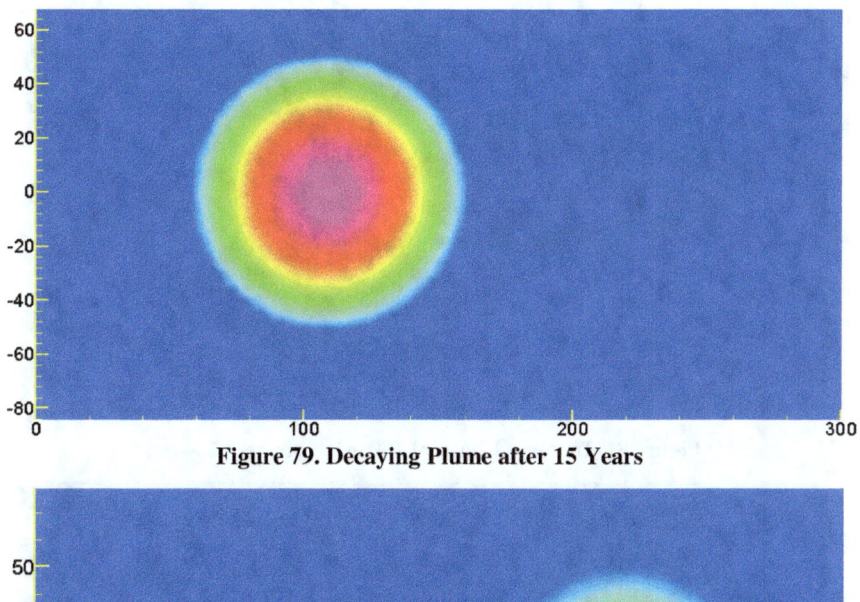

Figure 79. Decaying Plume after 15 Years

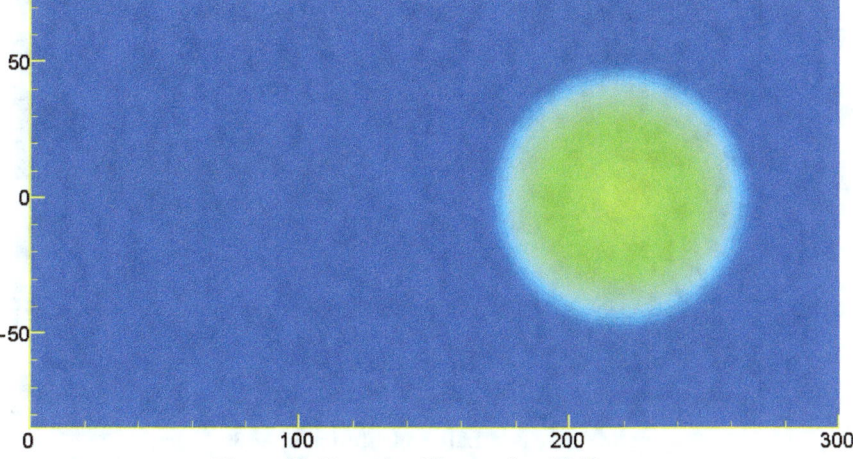

Figure 80. Decaying Plume after 30 Years

If we increase the lateral diffusion, we get:

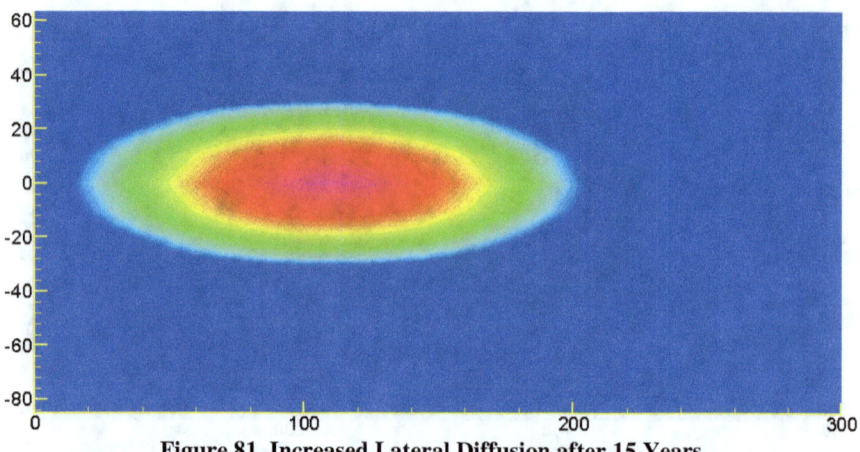

Figure 81. Increased Lateral Diffusion after 15 Years

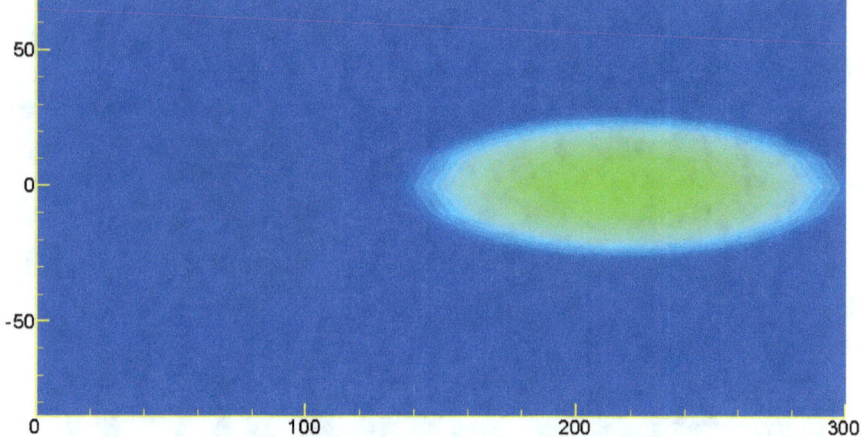

Figure 82. Increased Lateral Diffusion after 30 Years

If we add dispersion, we get:

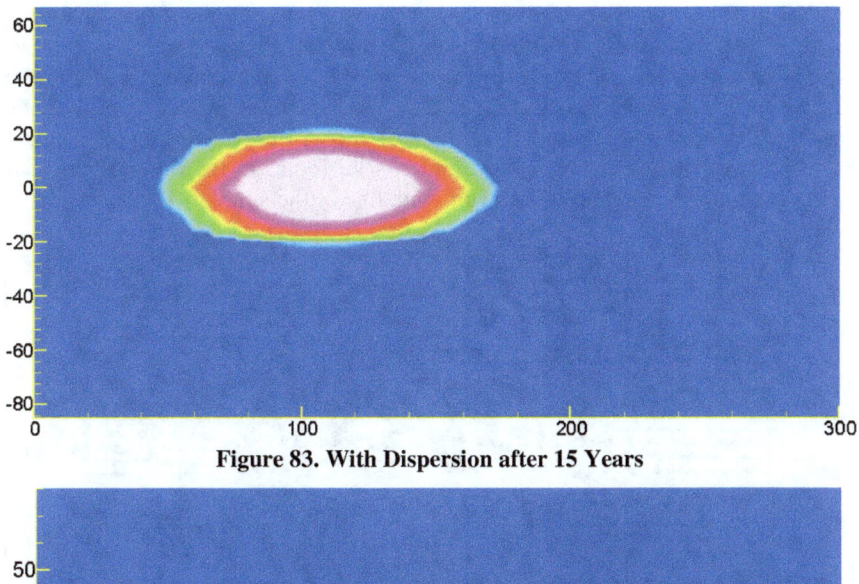

Figure 83. With Dispersion after 15 Years

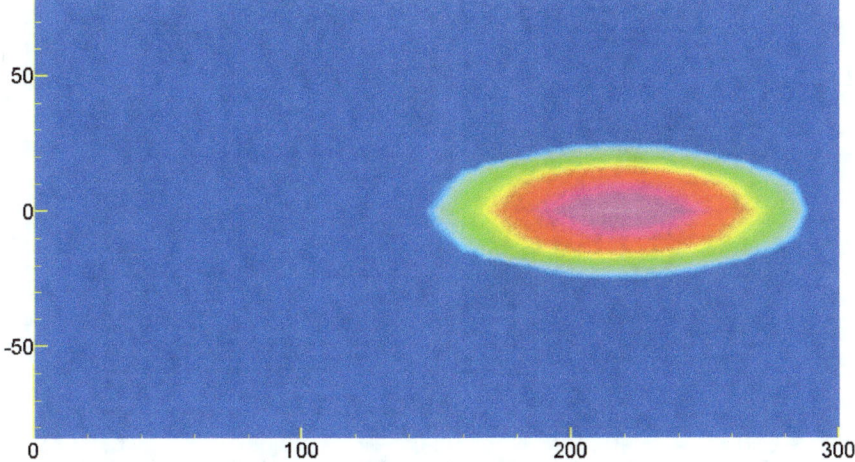

Figure 84. With Dispersion after 30 Years

Chapter 11. Two-Dimensional Lagrangian Tracking

In order to illustrate the process, we need a two-dimensional analytical velocity field. Potential flow theory is the logical choice here. We will not explain potential flow here only utilize it. There is some discussion in Appendix H. Should you need an explanation and derivation, there are numerous sources available on the Web. The code (potflow.c) we will use in this chapter comes from my book, *Complex Variables*, and can be found in the online archive in folder examples\potflow\2D. We begin with one of the simplest fields:

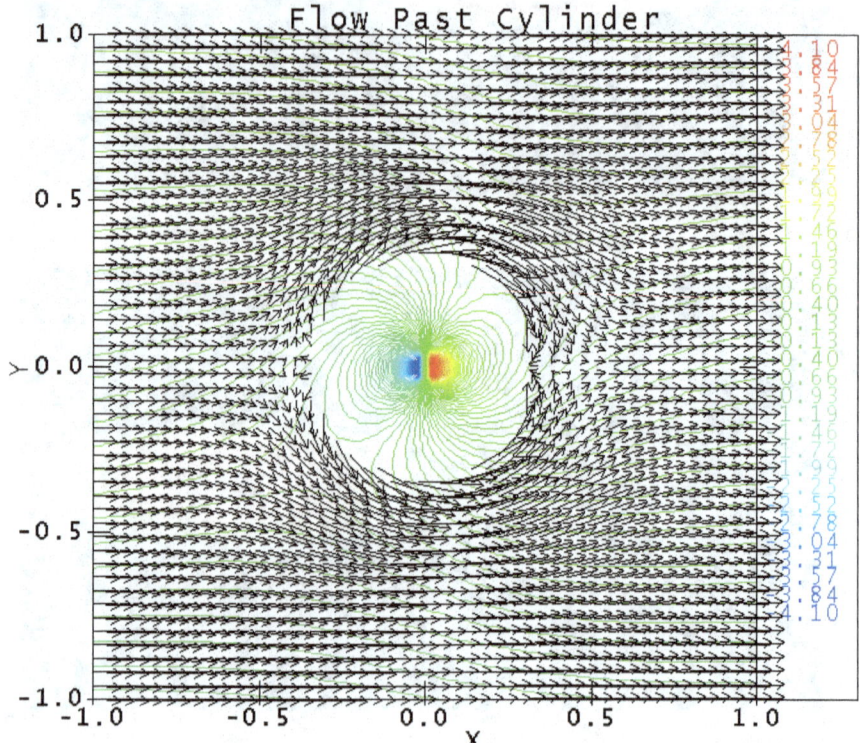

Figure 85. Potential Flow Past a Cylinder

The X and Y components of the velocity vectors (little arrows with tail length proportional to magnitude) are readily calculated at any location, as are the streamlines (colored curves). We will use the velocities, but not the streamlines. Of course, the particles should approximately follow the streamlines, else we've done something wrong. The equations of motion are trivial and arise directly from the definition of the velocity components.

$$u = \frac{\partial x}{\partial t} \quad v = \frac{\partial y}{\partial t} \quad (11.1)$$

$$x = x_0 + \int_0^t u\,dt$$
$$y = y_0 + \int_0^t v\,dt \qquad (11.2)$$

The initial position (x_0, y_0) is where we drop the particle at time, t=0. The velocity components (u and v) are functions of the position (x,y). We could use any number of integration techniques, such as Runge-Kutta, the simplest being fully-explicit forward Euler:

$$x_{t+\Delta t} = x_t + u\Delta t$$
$$y_{t+\Delta t} = y_t + v\Delta t \qquad (11.3)$$

We simply drop a few particles along the boundary at X=-1 and step through time. The code (track1.c) contains the necessary parts of the first (potflow.c) and allows you to easily select which flow field: flow over a cylinder, flow over a cylinder with circulation, flow past a doublet, flow past a half-body, flow past a Rankine body (an ellipsoid), flow past a source, flow past a sink, flow past a source and sink, stagnation flow (against a wall), and uniform (horizontal) flow. The result should be no surprise:

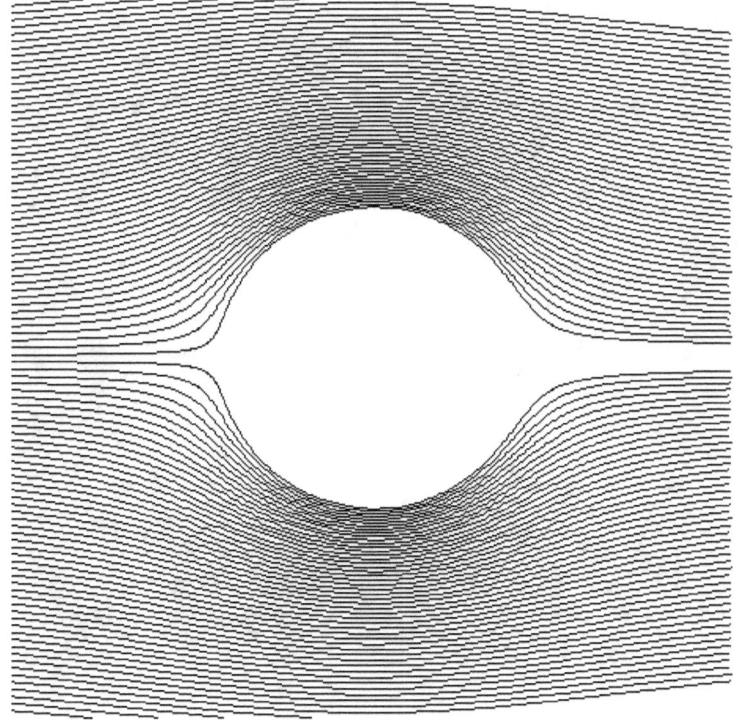

Figure 86. Streamlines over a Cylinder

The code couldn't be any simpler:
```
for(n=0;n<seeds;n++)
  {
  x=-1.;
  y=-1.+(n+1)*2./(seeds+1);
  t=0.;
  fprintf(fp,"%lG %lG %lG\n",x,y,t);
  for(i=0;i<max_steps;i++)
    {
    p=flow(x,y);
    x+=time_step*p.u;
    y+=time_step*p.v;
    t+=time_step;
    fprintf(fp,"%lG %lG %lG\n",x,y,t);
    if(x>=1.||y<-1.||y>1.)
       break;
    if(hypot(p.u,p.v)<FLT_EPSILON)
       break;
    }
```
We turn on circulation and obtain:

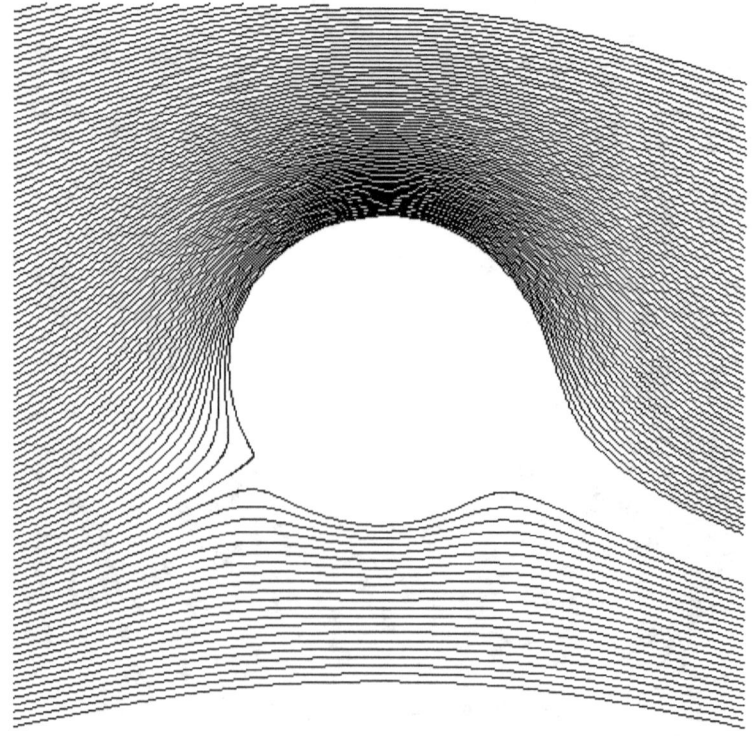

Figure 87. Streamlines over a Cylinder with Circulation

For a flow stagnating at the horizontal plane passing through the middle of the domain, we sprinkle particles along the top and bottom boundaries to obtain:

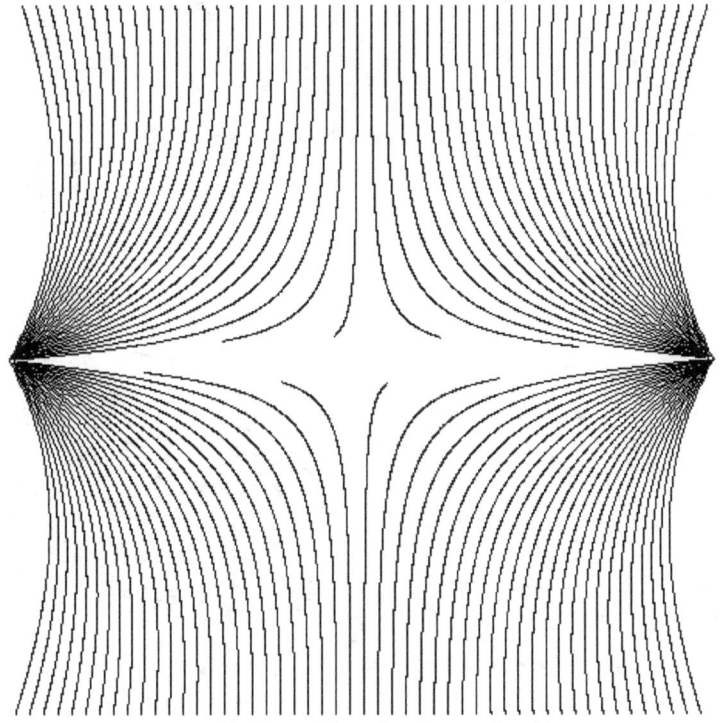

Figure 88. Streamlines for Stagnation Flow

While there's no need here, we will implement a 4th-order Runge-Kutta procedure. I cover the entire family of such methods in my book, *Differential Equations*, showing by example that there is no real advantage to higher orders or implicit variants or error estimates or step-length control embellishments beyond the basic method. While these additions are of historical and theoretical interest, they are of little practical value, as their effectiveness is entirely unpredictable. Simply decreasing the time step until the results level off will always work if anything will. That is, if this doesn't work, then don't expect anything more elaborate to work either. More details on the Runge-Kutta method can be found in Appendix J.

The main advantage of using Runge-Kutta instead of the forward Euler method is greater accuracy with a larger time step, which may or may not be computationally as efficient as using one-fourth the time step, which would still take less processing time than 4th-order Runge-Kutta. Still, I have selected the parameters to illustrate a non-trivial example. The same particles flowing over a cylinder using 4th-order Runge-Kutta are shown in this next figure. The code (track2.c) can be found in the same folder.

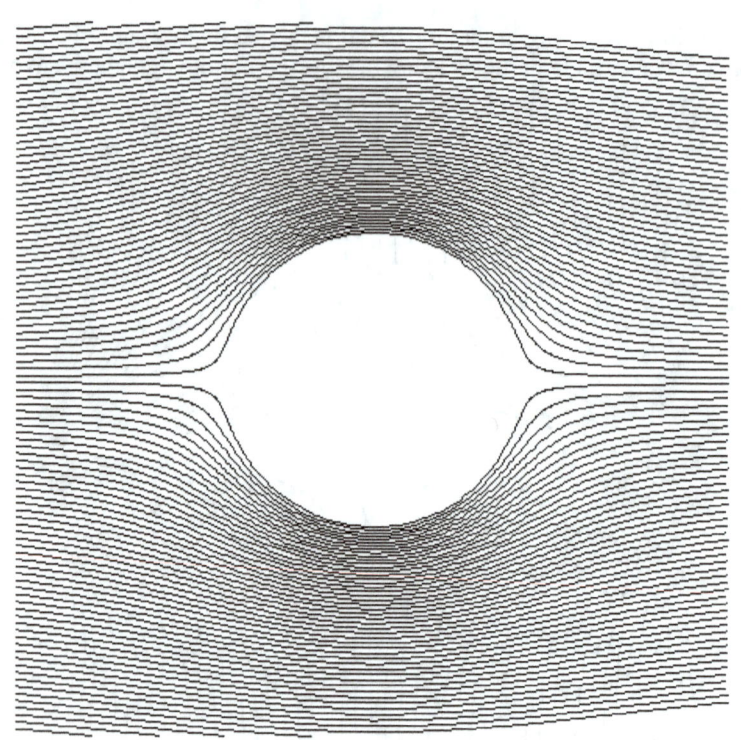

Figure 89. Streamlines Using 4th Order Runge-Kutta

The differences can be seen on the on the downstream side (back) of the cylinder.

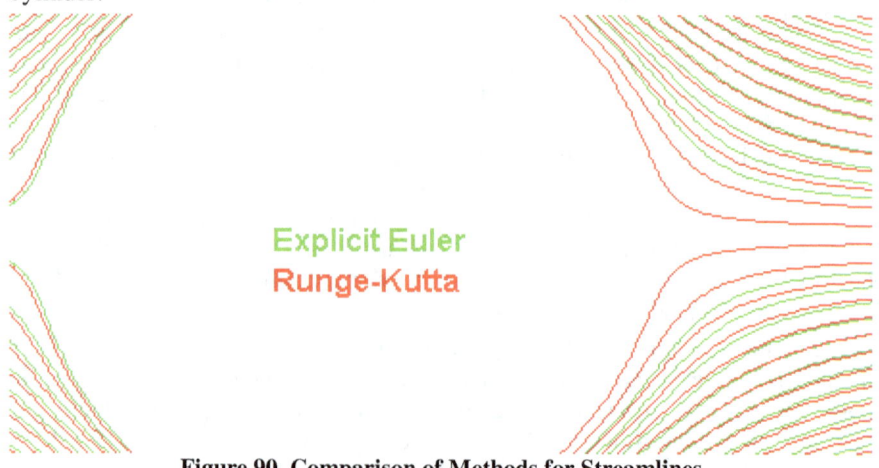

Figure 90. Comparison of Methods for Streamlines

It is important to note that the velocity components and streamlines are symmetric, as this is potential (i.e., inviscid) flow. The velocity components don't change more abruptly on the back side than the front. That's not what's happening here. What causes these two sets of particle tracks to diverge is the effects of small cumulative differences in the calculations. Even a small difference in trajectory early in the history of a particle can result in a large difference in where it ends up, much like the metaphor of a ship being slightly off course on a long journey. See how tightly the Runge-Kutta handles the circulation case:

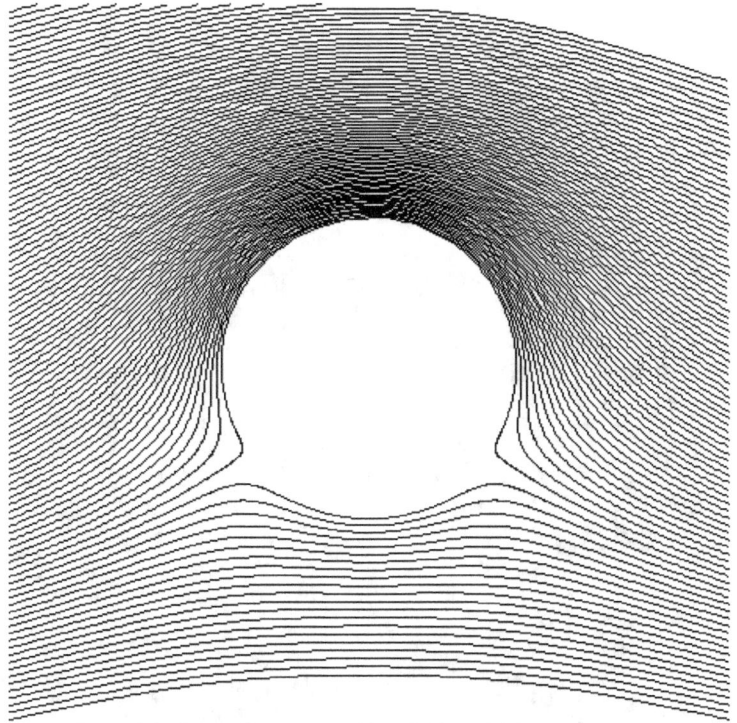

Figure 91. Streamlines with Circulation Using Runge-Kutta

Chapter 12. Three-Dimensional Lagrangian Tracking

The transition from two- to three-dimensional particle tracking is conceptually simple. The biggest implementation difference is in graphically displaying the results on a flat page. The additional formulas are:

$$z = z_0 + \int_0^t w\,dt \qquad (12.1)$$

$$z_{t+\Delta t} = z_t + w\Delta t \qquad (12.2)$$

There are fewer analytical solutions for 3D potential flow than for 2D. The flow code (potflow.c) is in the online archive in folder examples\potflow\3D. The explicit Euler (track1.c) and Runge-Kutta Lagrangian tracking codes are also in this folder. The particle tracks over a sphere are:

Figure 92. Streamlines over a Sphere

The Runge-Kutta implementation (track2.c) produces tighter particle tracks, as for the 2D examples. This next figure focuses on the zone with the most curvature in the particle paths and also fewer particles.

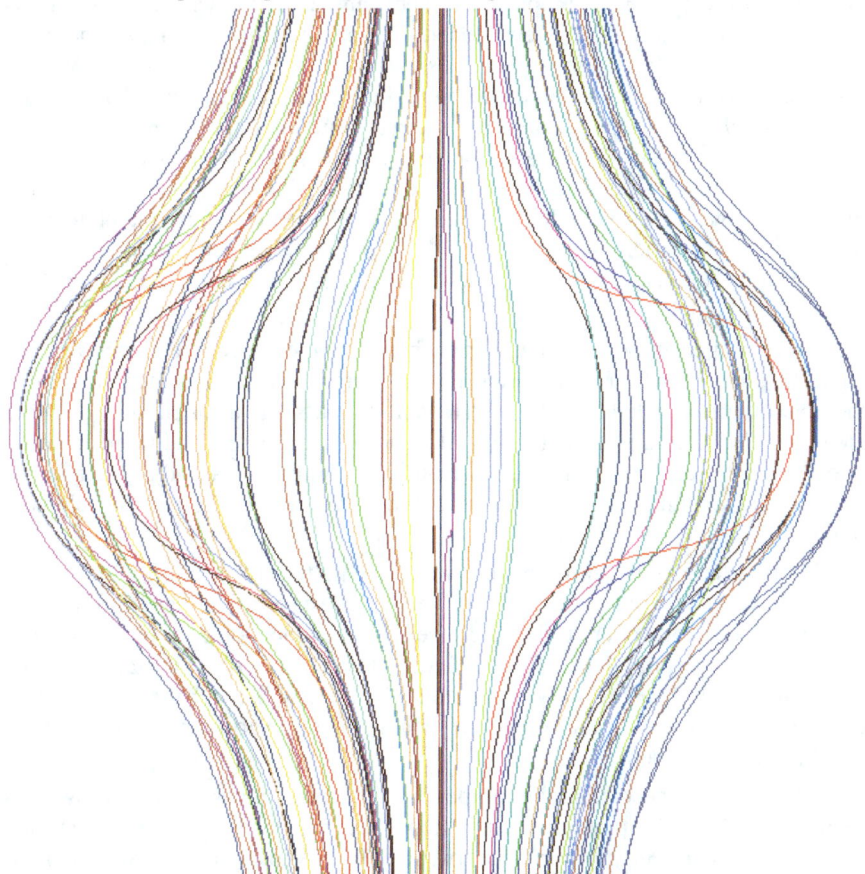

Figure 93. Select Streamlines Using Runge-Kutta

Implementation of Runge-Kutta is very simple:

```
for(i=0;i<max_steps;i++)
  {
  p=flow(y[0],y[1],y[2]);
  if(hypot3(p.u,p.v,p.w)<FLT_EPSILON)
    break;
  RungKutta4(step,&t,time_step,y,dy,3);
  r=hypot3(y[0],y[1],y[2]);
  if(r>3.*D)
    break;
  fprintf(fp,"%lG %lG %lG %lG\n",y[0],y[1],y[2],t);
  }
```

Chapter 13. Particle Tracking in Discrete Domains

As we have seen, implementation of Lagrangian particle tracking within an analytical flow field is a simple task. The number of known 2D solutions is limited and 3D solutions are few indeed. Besides, we already know where the particles will go from the streamlines. Most practical applications will be discrete domains (i.e., finite element, finite difference, or boundary element) and the flow fields will be the result of a numerical solution, rather than an analytical one. What this means to particle tracking is that the velocity will be known (usually constant) within each element discretely, not continuously over the domain. Furthermore, the velocity will most likely change from one element to the next, as a particle passes through the domain. Keeping track of which element a particle is in will be a much greater task than simply calculating the spatial coordinates.

More importantly, the boundaries between the current element and all adjacent ones, which are where the particle might eventually enter, determine the maximum step length. We don't want to pass on through one or more elements. We want a particle to reach the end of one element and then step over into the next element. Transitioning from one element to an adjacent one has zero spatial length and occurs in zero time. We simply change the element index.

To find the maximum time step, we calculate the intersection with each of the boundaries, taking the one with the smallest positive value. Since the velocity is constant over a single element, there's no point performing some complicated integration (i.e., Runge-Kutta). There's also no point taking a step any shorter than this. The only exception, which we will cover in Chapter 15, is diffusion and dispersion that might be seen as random variations of the velocity within a single element.

The reason most Lagrangian particle tracking codes don't follow this seemingly obvious approach (time steps determined by element size and velocity) is that *snapshots* of the particle field (i.e., position and perhaps concentration) are most easily generated at regular intervals. This is why most Lagrangian particle tracking codes advance all of the particles at the same time (i.e., marching forward simultaneously). In this methodology, every so often you write out a graphic and/or concentration field.

Another reason for programming this methodology is that the memory requirements don't continue to grow as the simulation advances because you only need to store the current values. You can also restart this process where it ended (if you save the files), which fits well with shared and scheduled computer resources and also lends itself to distributed processing. These reasons became superfluous with the ubiquitous availability of microprocessors with gigabytes of RAM—basically with the introduction of the Intel® Pentium® in 1995, about the time I ditched Lagrangian particle tracking in favor of Hamiltonian, which we will discuss in Chapter 14.

For the purposes of illustration (it's not worth putting a lot of effort into what we already know is an inefficient algorithm, especially when an efficient one has been available for 20+ years), we will use the Lagrangian method to track particles through a uniform rectangular grid with a contrived velocity field. This can be readily implemented in 2D and 3D, as the element-to-element transitions are easily calculated. The codes (Lagrangian2.c and Lagrangian3.c) can be found in the online archive in folder examples\Lagrangian. For variety, we can slightly modify the flow field:

```
vars flow(double x,double y)
{
static vars t;
t=FlowPastCylinderWithCirculation(D/6.,-3.*D,x,y);
t.v+=x-y;
if(t.r<1./3.)
   t.u=t.v=0.;
return(t);
}
```

We first calculate the contrived velocities and store them in arrays, as we might do when loading results from some more elaborate flow model:

```
for(k=i=0;i<ny;i++)
  {
  Y=Ym+(i+1)*(Yx-Ym)/(ny+1);
  for(j=0;j<nx;j++,k++)
    {
    X=Xm+(j+1)*(Xx-Xm)/(nx+1);
    p=flow(X,Y);
    U[k]=p.u;
    V[k]=p.v;
    fprintf(fp,"%1G %1G %1G %1G\n",X,Y,p.u,p.v);
    }
  }
```

Because the grid is uniform, the element-to-element bookkeeping is quite simple:

```
for(s=0;s<max_steps;s++)
  {
  j=(int)((Y[0]-Xm)*nx/(Xx-Xm));
  i=(int)((Y[1]-Ym)*ny/(Yx-Ym));
  if(i<0||i>=ny||j<0||j>=nx)
    break;
  k=nx*i+j;
  RungKutta4(step,&t,time_step,Y,dY,2);
  if(Y[0]<Xm||Y[0]>Xx||Y[1]<Ym||Y[1]>Yx)
    break;
  fprintf(fp,"%1G %1G %1G\n",Y[0],Y[1],t);
  }
```

The 4th-order Runge-Kutta integration again provides smooth particle tracks:

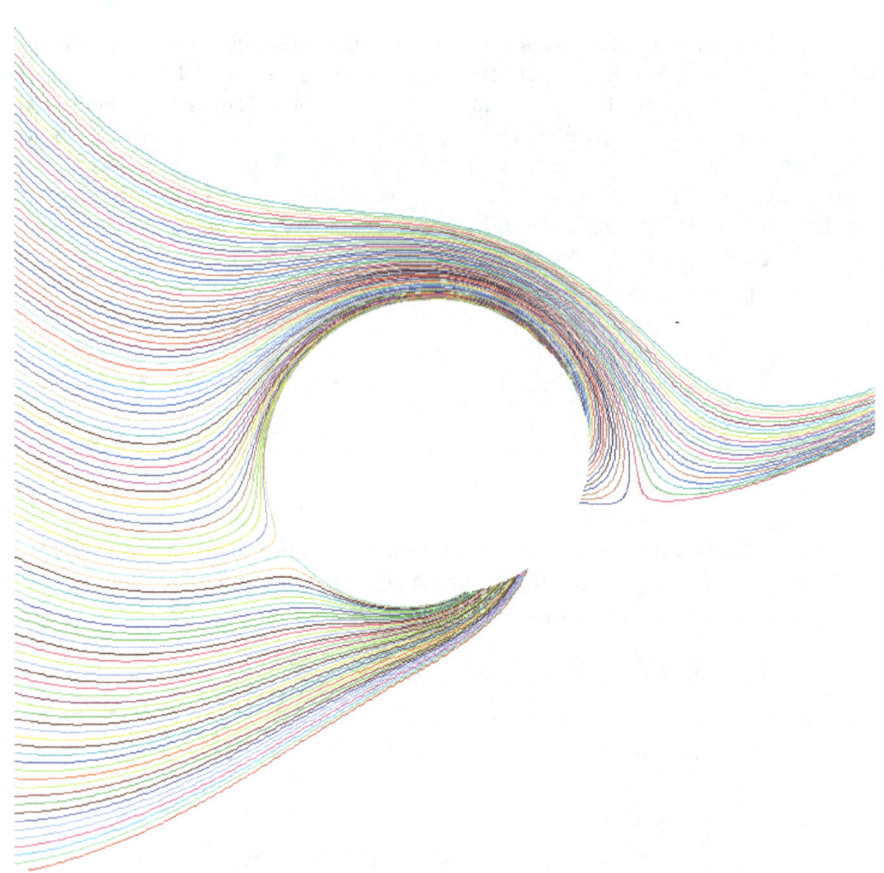

Figure 94. Select Streamlines with Circulation Using Runge-Kutta

The transition to 3D (Lagrangian3.c) is also quite simple and we can contrive a different velocity field to investigate. For illustration, we create an upward swirling flow field (i.e., a tornado):

```
vars flow(double x,double y,double z)
  {
  double a,r;
  static vars t;
  r=hypot(x,y);
  if(r>FLT_EPSILON)
     a=atan2(y,x);
  else
     a=0.;
  t.u=-sin(a);
  t.v=cos(a);
  t.w=1.;
  return(t);
  }
```

A side view of the particle tracks is:

Figure 95. Side View of Particle Tracks

Viewed from the top down (or bottom up), the tracks are circular:

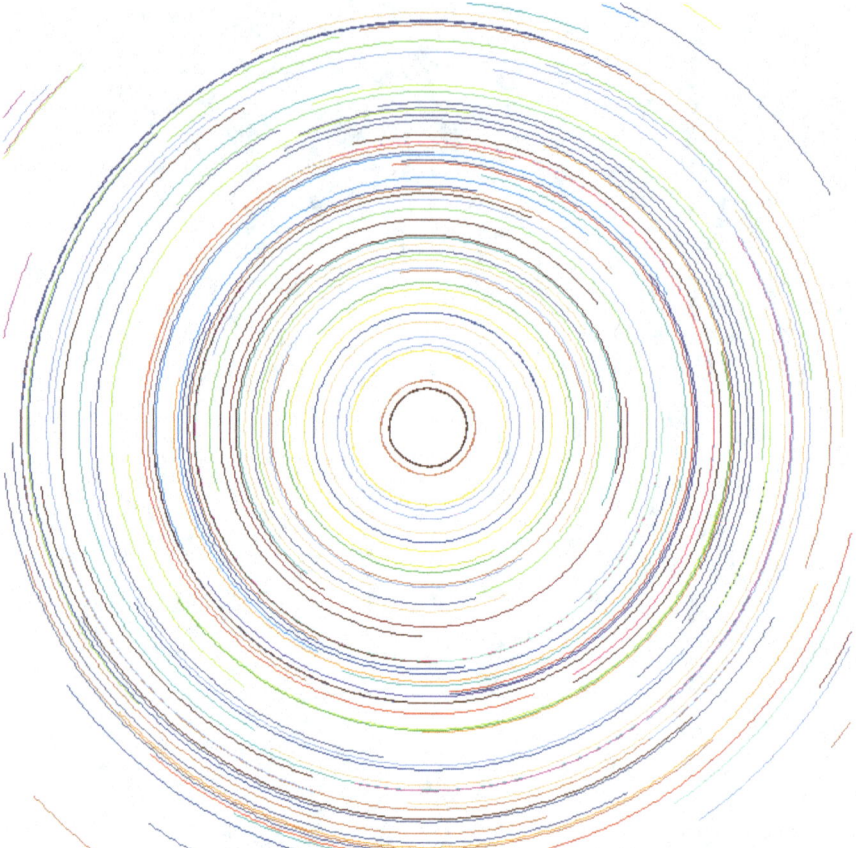

Figure 96. End View of Particle Tracks

That's it for Lagrangian Particle tracking of simplistic fields and domains. We now move on to realistic problems and practical implementations.

Chapter 14. Hamiltonian Particle Tracking

As briefly mentioned previously, what we are calling the Hamiltonian approach considers position and velocity the independent variables and time the dependant variable. Consider an incremental step in three dimensions:

$$dS = \sqrt{dX^2 + dY^2 + dZ^2} \qquad (14.1)$$

at some velocity magnitude:

$$|\vec{V}| = \sqrt{U^2 + V^2 + W^2} \qquad (14.2)$$

all within a single element. Except for diffusion and dispersion, which we will discuss in Chapter 15, the velocity is constant over the element, so there is no reason not to step completely across it to the adjacent element. The time within this element is then:

$$\Delta t = \frac{\sqrt{dX^2 + dY^2 + dZ^2}}{\sqrt{U^2 + V^2 + W^2}} \qquad (14.3)$$

<u>Implementation</u>

To facilitate calculating intersections, all quadrangles are split into two triangles, prisms are split into three tetrahedra, and bricks (hexahedra) are split into four tetrahedra. Thus, we only need calculate the intersection of a 2D vector with three line segments for each triangle or a 3D vector with three bounded triangular planes for each tetrahedron. The splitting is automatically performed inside the particle tracker to minimize preprocessing.

If snapshots or concentrations are required, these are initialized after reading loading the nodes and elements, then updated for each particle as it passes through each element. If decay occurs, such as with a radioactive component, this is calculated as needed for each particle as it passes through each element. We can even color the tracks based on decay so that they begin with red and proceed through the rainbow to blue.

Flow model results may or may not include porosity, as in groundwater flow through porous media. Porosities and hydraulic conductivities can be read in for each element or assigned by default. Units (SI, English, whatever) don't matter, as long as they are consistent (velocity=length/time). Default porosity (in the .CFG file) can be used to scale the velocities. I began work on the code, PTRAX, in May of 1995 and completed it in March of 1997 with minor features added and a few bugs fixed since then. It has been thoroughly validated against analytical solutions and field tests (see Appendix M) and has withstood the test of time (and court!). While PTRAX began with groundwater applications, it has been successfully used in surface water, atmospheric, branched pipe networks, solids, crystals, and even astronomical simulations. Contaminant animations (which are automatically generated when activated) were first a diagnostic tool, but turned out to be the most compelling feature, especially helpful in court cases.

Two-Dimensional Examples

The fastest way to create complex 2D flow fields is the boundary element method (see Appendix I). The code (PFLOW.c) is in the online archive in folder examples\bem. All you need to get started is a polygon. You can create and edit polygons with AutoCAD®, but a far more convenient tool is my polygon editor, PolyEdit, which can be free downloaded from the web site listed in the Forward. I have modified the BEM code to spit out all the files needed to run the particle tracker. You create the boundary file (see examples *.BEM), run PFLOW, and then run PTRAX. All of the files for a particular project must have the same name but different extensions.[12] The PFLOW input file is: name.BEM.

Table 5.1. File Extensions

recognized file extensions	
extension	contents
name.2DV	generic 2D elements
name.3DV	generic 3D elements
name.BEM	PFLOW input file
name.ELM	elements
name.FPR	element properties
name.NOD	nodes
name.P2D	2D tracks or polygons
name.P3D	3D tracks or polygons
name.TB2	2D tabular fields
name.TB3	3D tabular fields
name.V2D	2D velocities
name.V3D	3D velocities
name.VEF	fracture velocities
name.VEP	porous media velocities

All of these are recognized by my graphics program, TP2[13], which can be freely downloaded from the site listed in the Forward.

---

[12] If you're running Windows®, you really should go to Control Panel, Folder Options, and turn off [x] hide file extensions for known file types, as this is one of the stupidest things Bill Gates ever came up with. Of course you need to see the file extensions and know what they mean, not just the icon of the application that they might be associated with.

[13] I wrote TPLOT in 1980. A coworker took a copy of the source, which was eventually turned it into a commercial product without his or my involvement. TP2 is the second generation TPLOT and freely available. It recognizes 27 different file types and runs on any version of Windows®.

Strait of Gibraltar

The files (beginning with the polygons) can be found in the online archive in the folder examples\bem having names that begin with Gibraltar. The dimensions are nautical miles (i.e., minutes of latitude) with X=0 at Greenwich and Y=0 at the equator. The velocities are arbitrary and can be scaled as desired, since the flow is inviscid. Viscous flow with large-scale turbulence and eddies is beyond the scope of this book. Here we are concerned with particle tracking.

Boundary conditions for the BEM are defined along the polygons (there must be an external polygon and there may be none or several internal ones representing islands). See header at the top of PFLOW.c for details.

Table 5.2. Boundary Condition Types

type	boundary condition
0	U defined
1	dU/dN defined
2	dU/dT defined
3	U defined and end of this boundary
4	dU/dN defined and end of this boundary
5	dU/dT defined and end of this boundary

The top of Gibraltar.BEM is:

```
Strait of Gibraltar
*dimensions are nautical miles (one minute of latitude)
93 -1000
209.0 2206.0 0 1
189.2 2205.0 0 1
169.5 2202.4 0 1
150.9 2196.0 0 1
131.1 2196.0 0 1
111.2 2193.7 0 1
91.4 2192.0 0 1
71.5 2190.0 0 1
52.8 2183.6 0 1
34.1 2176.7 0 1
16.1 2168.1 0 1
2.2 2153.9 0 1
-15.8 2149.9 0 1
-32.5 2146.8 0 1
```

The first line is the title. Any line beginning with an asterisk (*) is a comment. The second line gives the number of boundary points and optionally the number of internal velocities to calculate. Enter a positive number if you intend to supply the points after the boundary or a negative number if you want them automatically distributed uniformly within the interior. The following lines are the boundary (x,y) the type of boundary condition and the value. The first entry is x=209.0, y=2206.0, type=0, value=1.

105

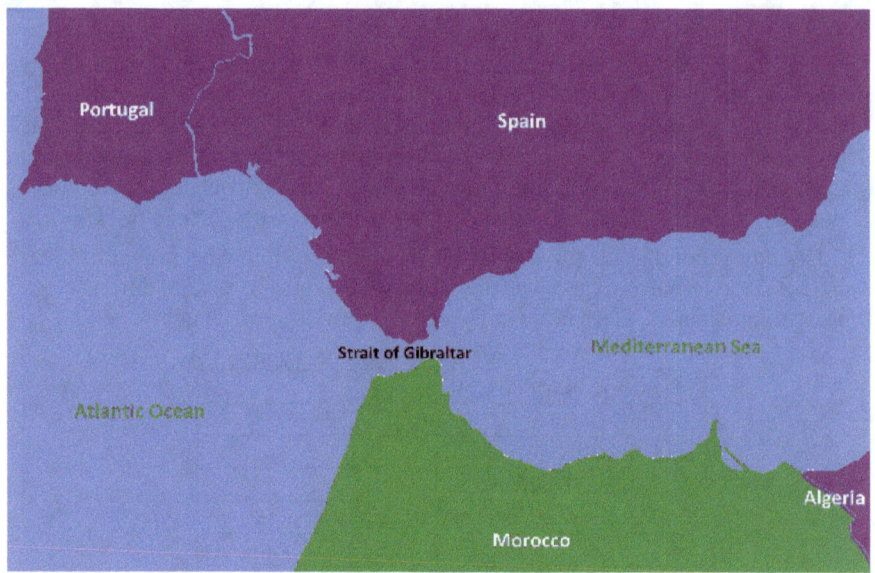

Figure 97. Strait of Gibraltar

The boundary element results from PFLOW.c are shown below. Again, PFLOW solves for the flow field and generates all of the output files in a few milliseconds. This figure shows velocity vectors on top of the potential field.

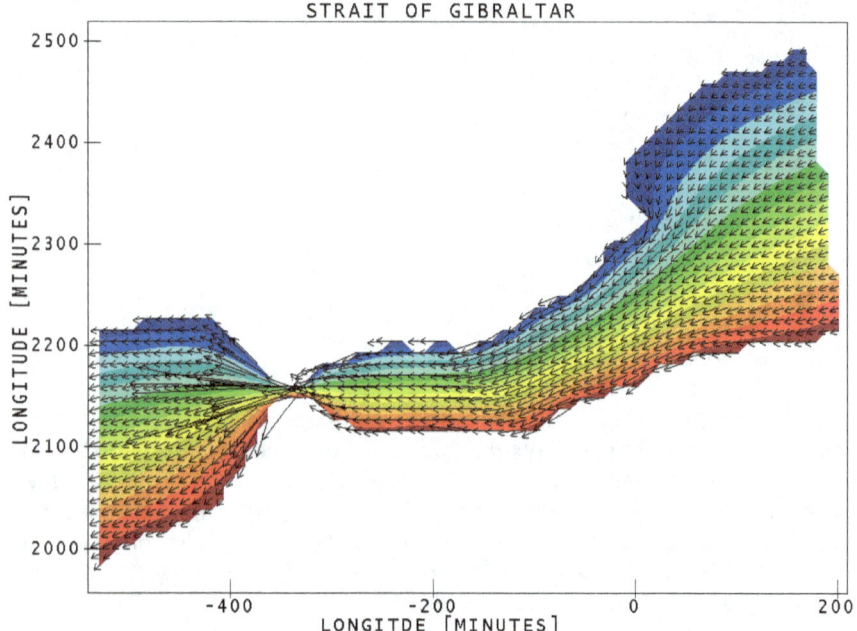

Figure 98. Potential Flow for Strait of Gibraltar

The automatically generated grid of triangles is shown below:

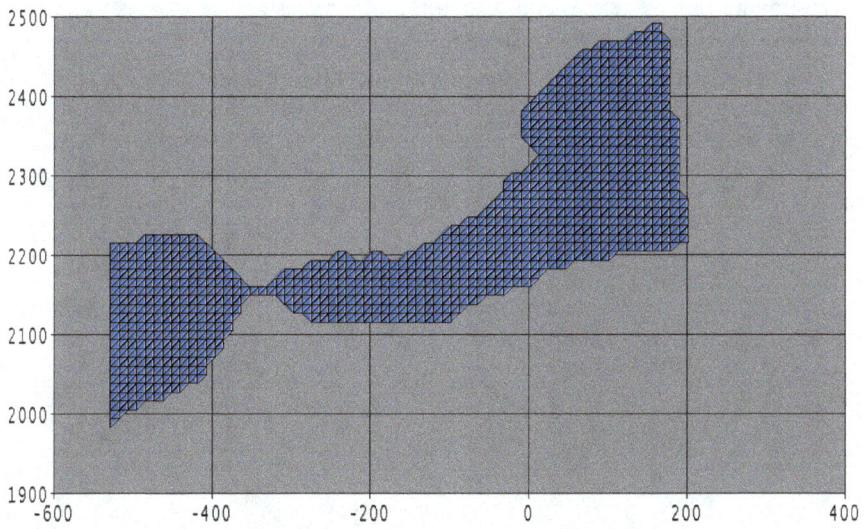

Figure 99. Triangular Grid for Gibraltar

All of the input files are then ready for PTRAX, so you just run it. The particle tracks will be displayed and a plot file will also be created.

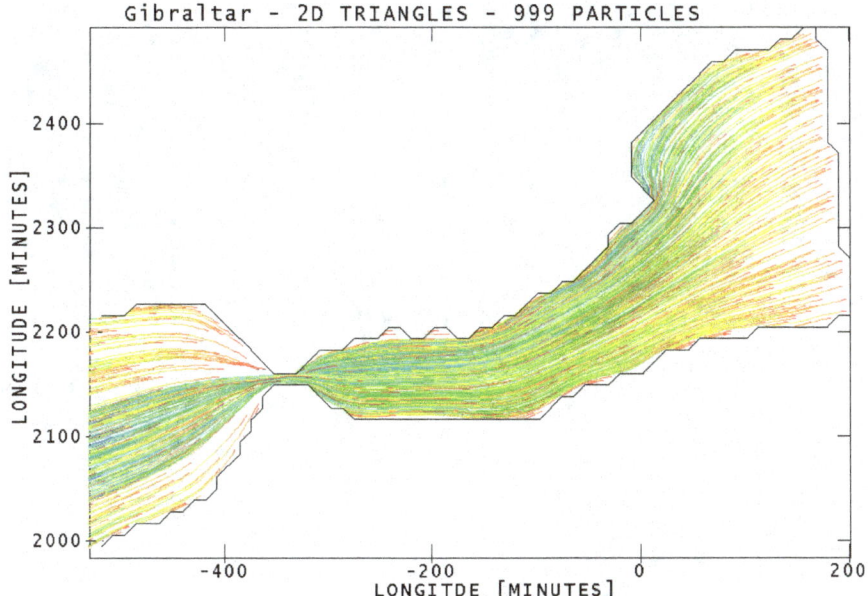

Figure 100. Streamlines for Strait of Gibraltar

The default is 999 particles randomly seeded throughout the domain. As shown in the log file (PTRAX.LOG), the time required to read the model, generate and track the particles, and create the output files is 3 seconds (on a 3GHz Intel® processor). Also by default, the particles age so that they start red and decay through the rainbow to blue.

I already had polygons for the world, including continents, islands, lakes, and countries. I simply cut out this portion using PolyEdit and pasted it onto Gibraltar.BEM, then added 0 0 or 0 1 to define where the potential changes along the north and south boundaries. This Gibraltar example took about five minutes to create and solve. It took longer than that to find a suitable map with Google®.

Strait of Hormuz

The geometry is similar, though the water bodies are quite different. The coordinates are in nautical miles and all of the files are named Hormuz.*.

Figure 101. Strait of Hormuz

The boundary element input file is also similar:
```
Strait of Hormuz
*dimensions are nautical miles (one minute of latitude)
51 -1000
3030.8 707.7 0 0
2978.1 687.6 0 0
2922.4 678.8 0 0
2866.9 668.0 0 0
2814.3 654.4 0 0
2760.5 646.8 0 0
2711.6 633.6 0 0
2655.7 628.0 0 0
```

```
2613.8 669.6 0 0
2580.1 696.0 0 0
```
The velocity vectors and potential are shown in this next figure:

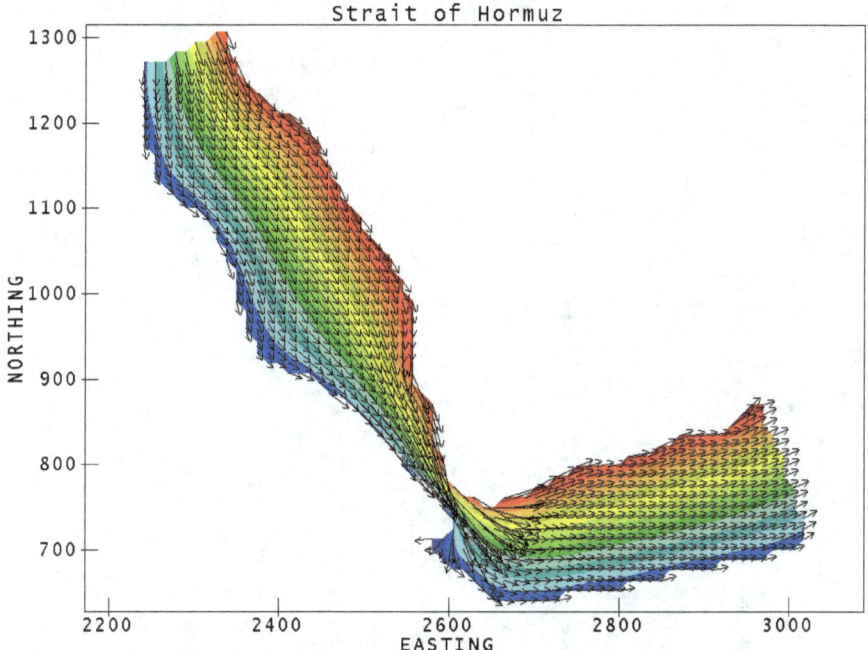

Figure 102. Potential Flow for Strait of Hormuz

The default particle tracking parameters are in Hormuz.cfg:

```
999        <- default number of particles to track
36500      <- track duration (days)
365        <- time step (days)
999        <- maximum steps along a single particle track
0.3 1 1    <- default porosity and retardation factor in
   matrix and fractures
0 0 0      <- default longitudinal, lateral, and
   transverse dispersivity in feet
1E30 1E30  <- default matrix/fracture decay half-lives
0.05 0 0   <- default velocities
0 0 0      <- matrix diffusion coefficients
0          <- user-defined synchronous time step (set to
   zero for automatic)
4          <- maximum times a particle can enter an
   element
4          <- maximum times a particle can enter a
   fracture
0 0        <- matrix & fracture tortuosity (0=none,
   1=complete)
```

The PTRAX informational display is shown below:

```
999 particles tracked in 3 seconds
PTRAX/V5.01: Particle Tracker by Dudley J. Benton
animations ..................................... enabled
fractures ...................................... enabled
input format ................................... Frac3D & ModFlow

application prefix ............................. Hormuz
reading default parameters ..................... Hormuz.CFG
default seeds .................................. 999 particles
track duration ................................. 36500 day(s)
time step ...................................... 365 day(s)
maximum steps along particle track ............. 999 steps
default porosity & retardation factor .......... 0.3,1,1
default dispersivities ......................... 0,0,0 ft
default matrix half-life ....................... 1E30 day(s)
default fracture half-life ..................... 1E30 day(s)
default velocities ............................. 0.05,0,0 ft/day
matrix diffusion coefficients .................. 0,0,0 ft²/day²
synchronous time step .......................... default
maximum times a particle can enter an element .. 4
maximum times a particle can enter a fracture .. 4
matrix,fracture tortuosities ................... 0,0
concentrations divided by porosity ............. NO
stray particles will be ignored ................ OK
trapped particles will be ignored .............. OK
include empty elements in snapshot files ....... OK
create track file .............................. OK

input model .................................... FRAC3D
FRAC3D node file: Hormuz.NDE ................... 996 nodes
range of X ..................................... 2241.13≤X≤3017.47
range of Y ..................................... 640.13≤Y≤1307.3
FRAC3D element file: Hormuz.ELM ................ 1804 elements
elements composed of 3 nodes ................... TRIANGLES
reading element nodes .......................... 1804 elements
node:element links ............................. 5412 links
element:element links .......................... 5226 links
model boundaries ............................... 186 faces
computing element areas ........................ 1804 elements
grouping elements .............................. 437 groups
FRAC3D velocity file: Hormuz.UEP ............... 1804 velocities
properties ..................................... default
transport properties ........................... default
dispersion ..................................... OFF
diffusion ...................................... OFF
scattered seeds ................................ 999 seeds
particle track file ............................ Hormuz.TRK
time to track particles ........................ 1 seconds
net performance ................................ 59940 seeds/minute
tracks ended at boundaries ..................... 945 seeds
tracks ended due to maximum time ............... 54 seeds
track plot command file ........................ Hormuz.TP2
total elapsed time ............................. 3 seconds
3 notes + 0 warnings
for summary see log file, PTRAX.LOG
```

Figure 103. Output of PTRAX for Strait of Hormuz

It took all of 3 seconds to read in the files and track 999 particles.

110

The PTRAX on-screen particle tracks are shown in this next figure:

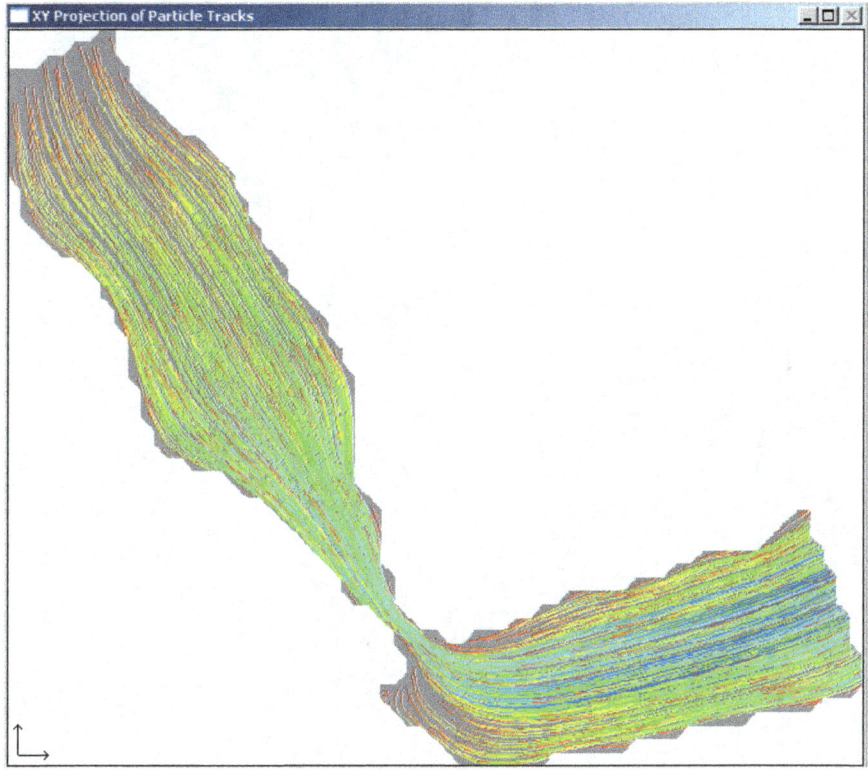

Figure 104. Particle Tracks for Strait of Hormuz

Dardanelles

The Dardanelles, also known from Classical Antiquity as the Hellespont, is a narrow, natural strait and internationally significant waterway in northwestern Turkey that forms part of the continental boundary between Europe and Asia, separating Asian Turkey from European Turkey. One of the world's narrowest straits used for international navigation, the Dardanelles connects the Sea of Marmara with the Aegean and Mediterranean Seas, while also allowing passage to the Black Sea by extension via the Bosphorus. The boundary element input file begins with the following:

```
Dardanelles
*the strait that divides Europe from Asia Minor (i.e.,
   Turkey)
*dimensions are nautical miles (one minute of latitude)
56 -1000
1770.7 2441.1 0 2
1760.8 2440.6 0 2
1751.0 2439.5 0 2
1741.3 2437.9 0 2
```

Figure 105. Dardanelles

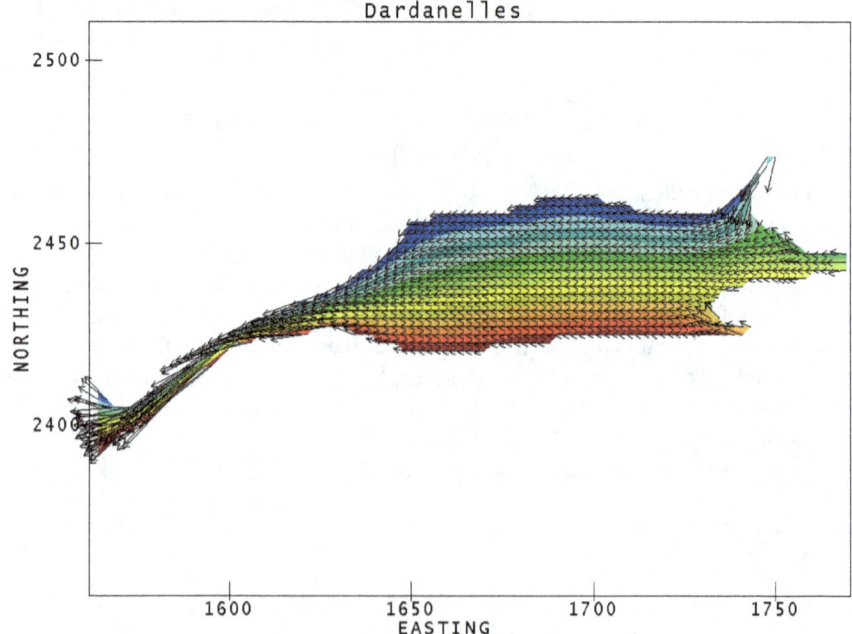

Figure 106. Potential Flow for Dardanelles

The default particle tracks are:

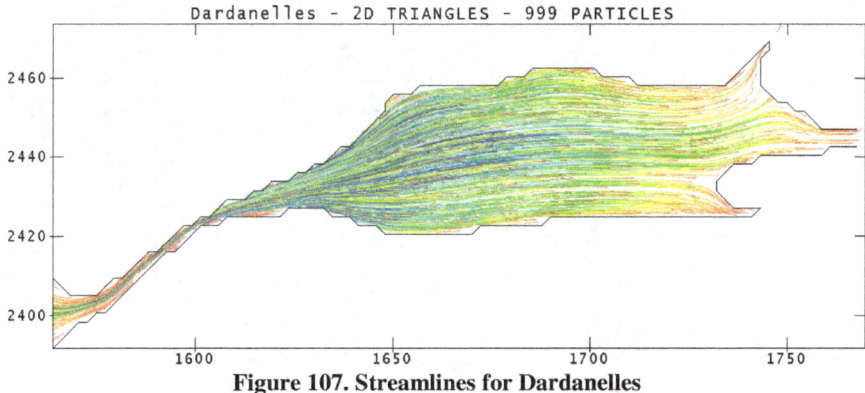
Figure 107. Streamlines for Dardanelles

Islands

The boundary element method can also handle islands. Note the "end of this boundary" designations in Table 5.2. There is no limit to the number of polygons used to describe the domain, although the BEM does eventually require solution of simultaneous linear equations and this is definitely limited. The number of equations is roughly twice the number of boundary points.

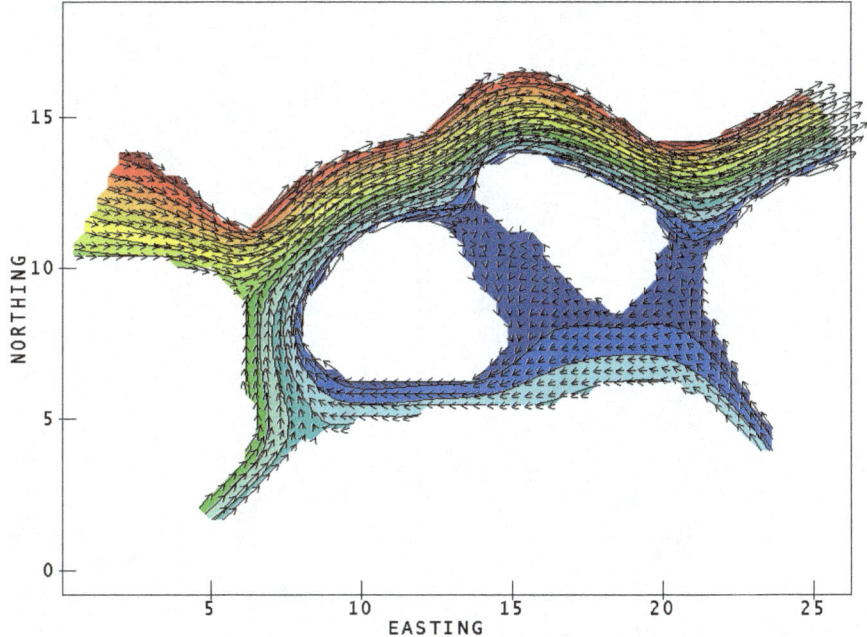
Figure 108. Potential Flow for Lake with Two Islands

The default particle tracks are:

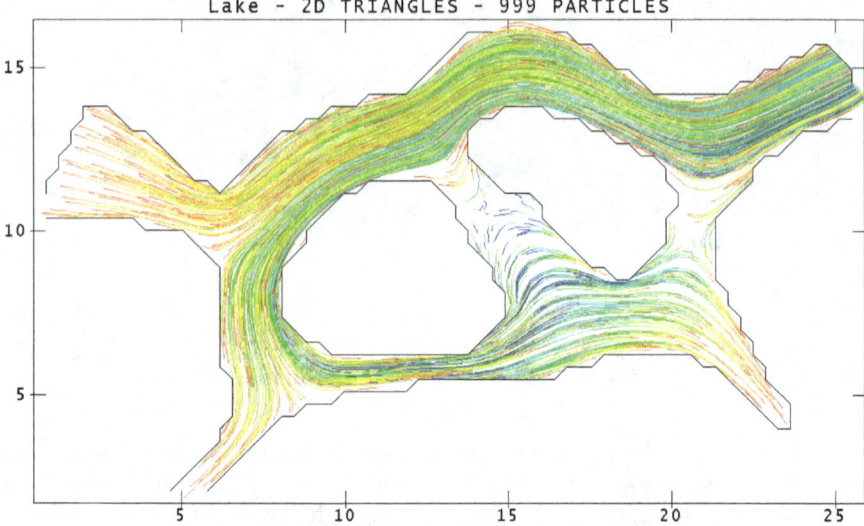

Figure 109. Streamlines for Lake with Two Islands

Chapter 15. Diffusion and Dispersion

In the context of particle tracking, diffusion is random movement of the particles through the domain over time, irrespective of the velocity field or even in the absence of any velocity. This is not just any random movement. There are specific quantitative expectations. Diffusion is well-defined analytically and diffusion coefficients, derived from experiment, have been published for many years. If enough particles are tracked, the result must match analytical solutions where these are available. How many particles are enough? The validation runs described in Appendix M were done using 800,000 and the agreement is excellent. The following example tracked 50,000 particles, which was ample to illustrate the process.

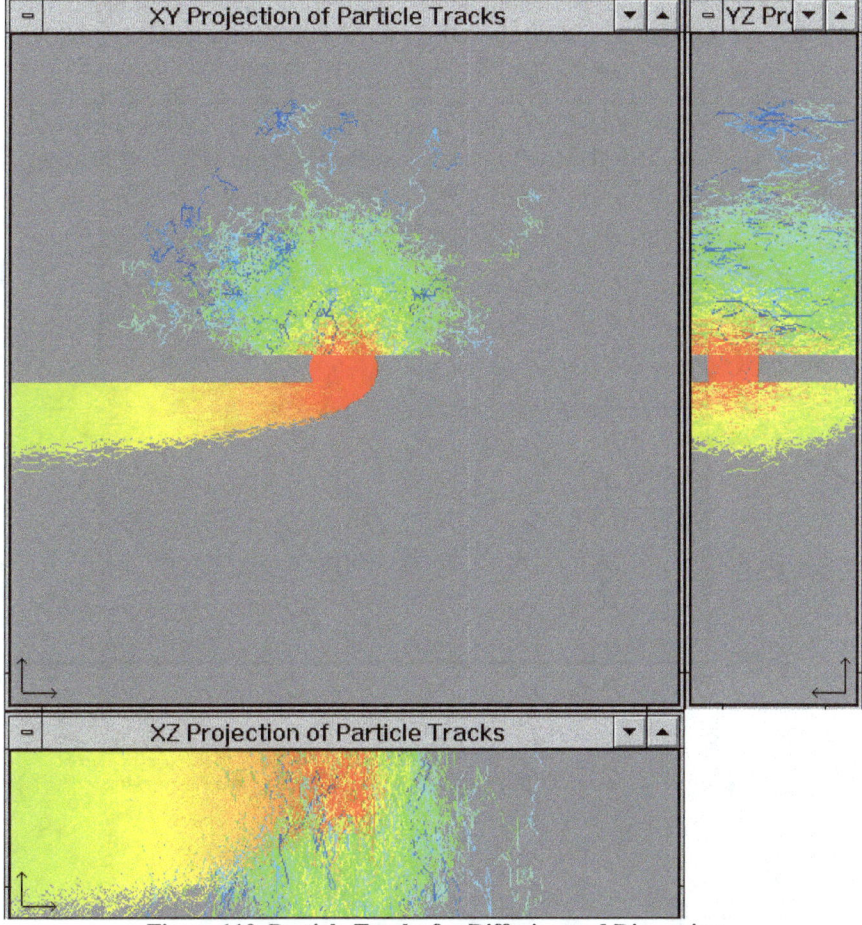

Figure 110. Particle Tracks for Diffusion and Dispersion

Dispersion is very similar to diffusion, in that it is random movement, but dependant on (i.e., proportional to) local velocity. Experimentally measured dispersion coefficients have also been published and so there are also definite expectations for dispersion. If a particle tracker doesn't meet these expectations, then it's just making up numbers and holds forth no real promise of predictive capability, which is essential for effective remediation and design. The preceding figure illustrates the difference between diffusion and dispersion, both within a single model.

The red circle (actually a vertical cylinder in this 3D model) in the middle of the domain is where the xxx particles were seeded. The white horizontal slot is a section of elements with no dispersion or diffusion. These particles are *stuck* and never move. Elements in the upper portion have zero velocity and non-zero diffusion coefficients. Elements in the lower portion have uniform velocities to the left, non-zero dispersion coefficients, and zero diffusion coefficients. The particles age so that they start out red, proceed through the rainbow, and end blue. The ones in the center along the stagnant slot remain red because they are trapped in the elements and aren't tracked. PTRAX generates four-dimensional results (three spatial dimensions plus time) in the form of concentration fields that can be sliced and also contoured using Tecplot® or TP2. The plan view contours (looking from the top down) are:

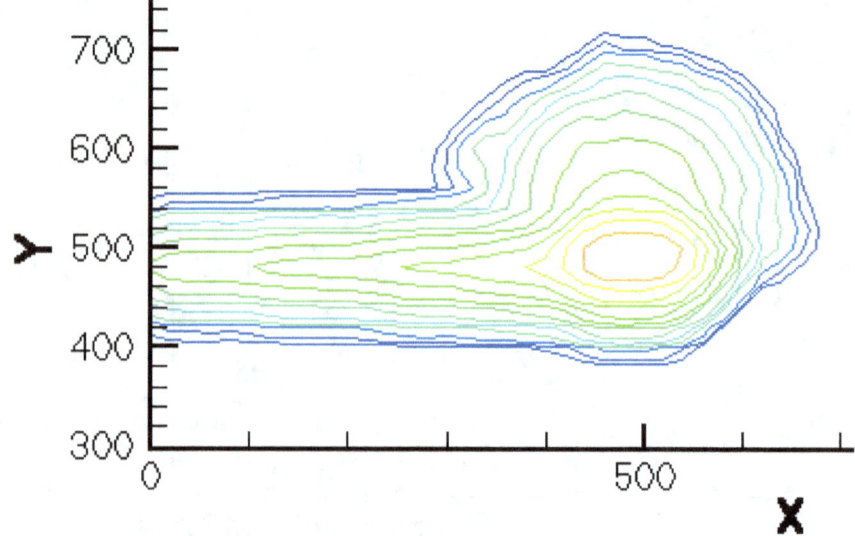

Figure 111. Concentration Contours for Diffusion and Dispersion

Three-dimensional contours are more accurately described as *shells* of constant value.

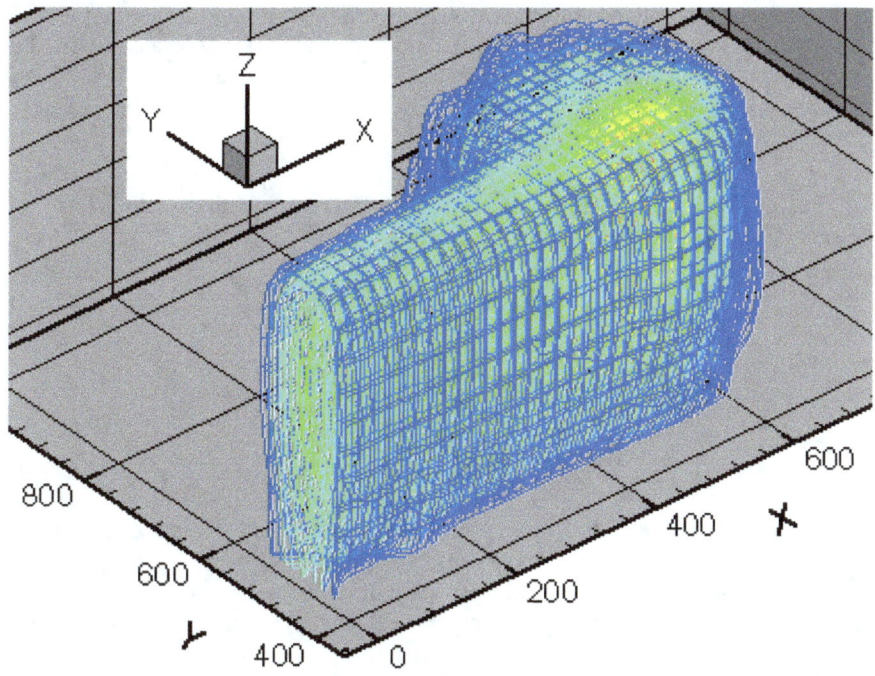

Figure 112. 3D Concentration Shells

Random Walk

In order for particle tracking results to be meaningful, they must match analytical results. The stepping process used is called a *random walk*, and is often compared to a drunken sailor staggering back to the ship after a long night at the pub. While the final equation seems so very simple, the process of arriving at it in 1995 was anything but simple. Many formulations were tried, tedious simulations performed, results compared, and approaches discarded before arrive at one that was adequately consistent. The random step length associated with a dispersion length is defined by the following equation:

$$\Delta S = R \sqrt{2\alpha |\vec{V}_m| \Delta T} \qquad (15.1)$$

where ΔS is the random step length, R is a normalized random number (i.e., having a mean of 0 and a standard deviation of 1), α is the dispersion length, $|V_M|$ is the magnitude of the mean (i.e., non-random) velocity, and ΔT is the time step. Diffusion is handled in this same way, except that the dispersion length times the magnitude of the vector velocity is replaced by the diffusion coefficient, D.

For dispersion (or diffusion) in several directions, multiple random numbers (i.e., Rs) and directionally associated dispersion lengths (i.e., α_X, α_Y, α_Z) are combined to form the random steps (i.e., ΔS_X, ΔS_Y, ΔS_Z). For a particle traversing a cell, there is an effective random velocity associated with the random step length and implied time step.

$$\left|\vec{V}_R\right| = \frac{\Delta S}{\Delta T} \qquad (15.2)$$

For dispersion in several directions, the effective velocity components can be represented by a mean and random part:

$$U_T = U_M + U_R = U_M + \frac{\Delta S_X}{\Delta T} \qquad (15.3)$$

$$V_T = V_M + V_R = V_M + \frac{\Delta S_Y}{\Delta T} \qquad (15.4)$$

$$W_T = W_M + W_R = W_M + \frac{\Delta S_Z}{\Delta T} \qquad (15.5)$$

where U, V, and W are the velocity components in the X, Y, and Z directions, respectively. If dispersion lengths are specified along the longitudinal, horizontal-transverse, and vertical-transverse directions, the corresponding steps along the principle axes are computed using standard trigonometric relationships.

<u>Statistical Requirements</u>

For a statistically large sample (i.e., many particles), the net influence of the random walk on the ensemble of particles must exhibit several properties:

1) The spreading (over that without dispersion) in the direction associated with each α is proportional to the square-root of α and ΔT.

2) The net displacement of the particles (compared to that without dispersion) is zero.

3) The net movement of the mass-weighted centroid of the particles is the same with or without dispersion.

Given these properties and the relationships between the random step length, mean and random velocity components, and time steps, the following requirements can be deduced:

$$\sum_{i=1}^{n} \Delta S_i \approx 0 \qquad (15.6)$$

$$\sum_{i=1}^{n} \frac{\Delta S_i}{\Delta T_i} \approx 0 \qquad (15.7)$$

These summations must hold for a single particle as well as for the ensemble, and they must hold in each dimension. In order to simultaneously satisfy these pairs of relationships, the time steps must be equal (i.e., if they are equal, then ΔT can be brought outside the summation).

Because these statistical relationships require equal time steps, an immediate problem arises, regardless of whether conventional Lagrangian particle tracking or the Hamiltonian method is used. Efficient implementation of a Lagrangian method requires a dynamically adjusted step length. Implementation of the Hamiltonian scheme results in time steps varying over orders of magnitude, as a particle may pass through a cell near a vertex and cover a small distance in a correspondingly small time.

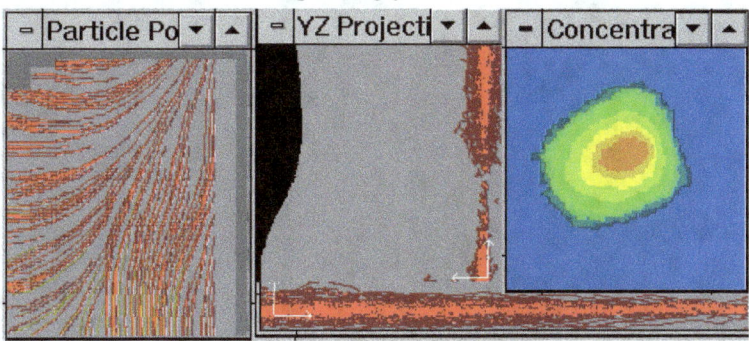

Figure 113. Typical PTRAX Output

If a constant time step is required, then the smallest required time step becomes a limiting factor and results in impractical runtimes. It is for this reason that the random walk is not often used in large particle tracking applications or such applications are run on super computers. Rather than using such a brute force approach, PTRAX steps around this problem by *joggling* the particles and *quantizing* the time steps. The resulting algorithm is very fast and also accurate, as shown in Appendix M.

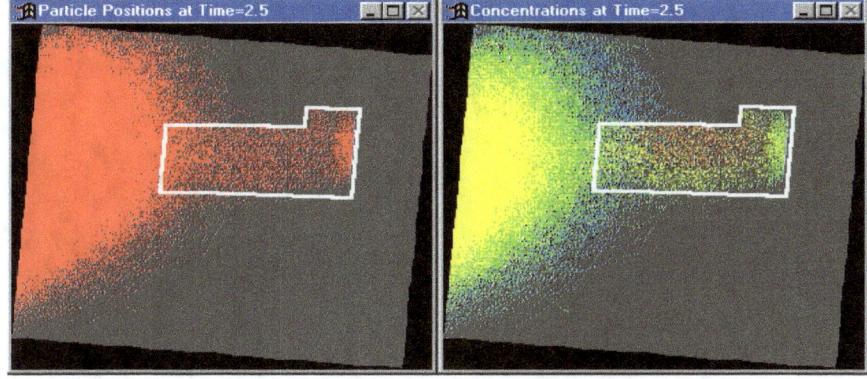

Figure 114. Particles and Concentrations

Pure Diffusion

Pure diffusion in 3D results in spreading of the substance, as approximated by the particles, in this case 50,000.

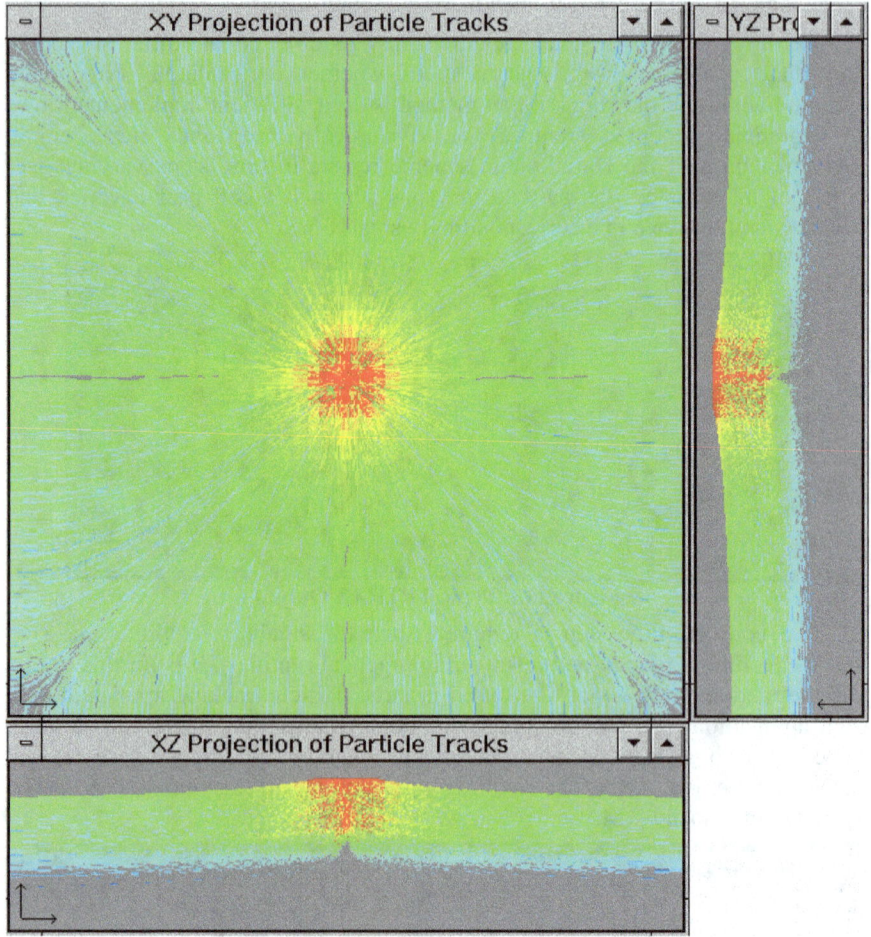

Figure 115. Particle Tracks for Pure Diffusion

Modified Diffusion

When dispersion is added, even in the presence of uniform velocity, the tracks take on a more random pattern, which shows that PTRAX can handle both types of displacements using a random walk algorithm.

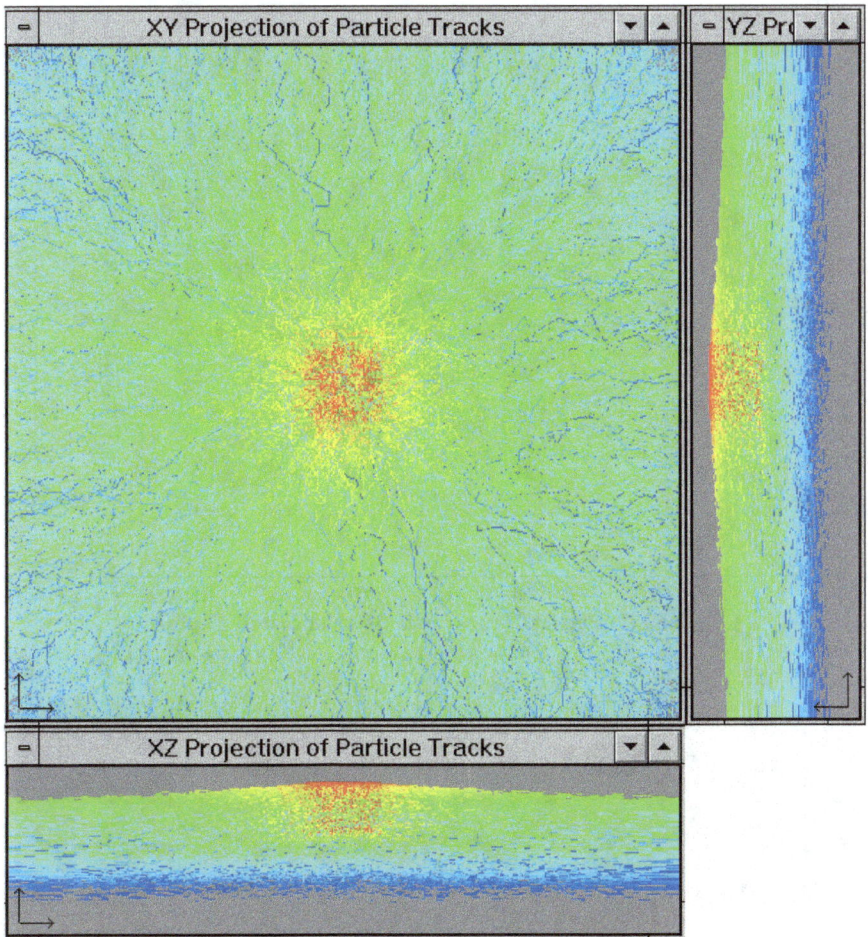

Figure 116. Particle Tracks for Dispersion

Capture Wells and Walls

In order to accurately evaluate potential remediation strategies, it is necessary to implement wells and walls. Wells are straight-forward enough, but walls are usually deep, narrow trenches lined with special cloth and filled with gravel, like a French drain. Both are commonly installed to inject and/or withdraw water from the ground and process it to remove contaminants. PTRAX handles wells that inject (red) and withdraw (cyan) as well as walls. The robust algorithms handle any sort of boundary, not just smooth domains.

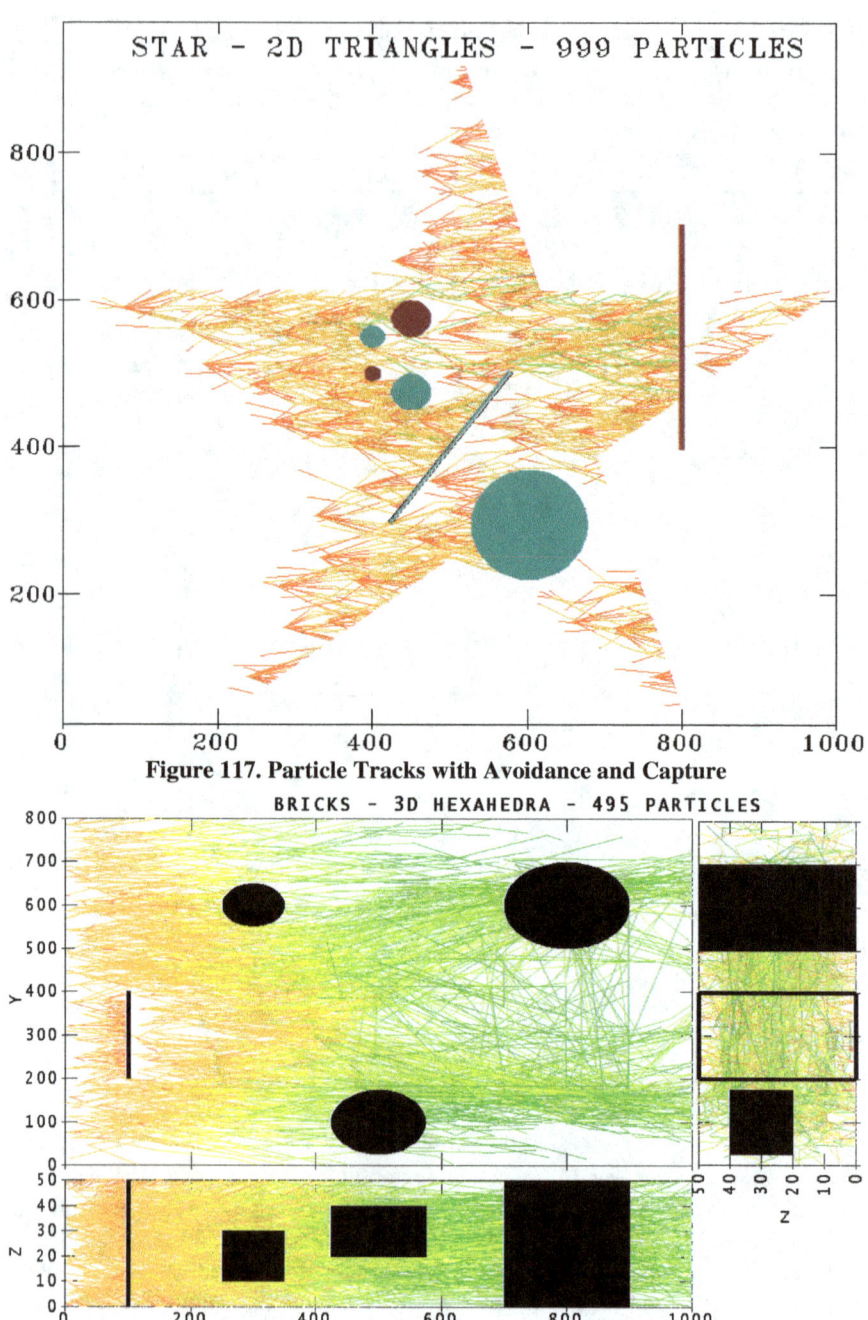

Figure 117. Particle Tracks with Avoidance and Capture

Figure 118. Particle Tracks with Walls and Wells

Once the programming logic was refined, any number of complex 3D obstructions can be handled, as illustrated in the previous figure.

2D Dispersion Example

The following example of dispersion in a two-dimensional surface water flow can be found in the online archive in folder examples\bem in files with names beginning with "Reservoir."

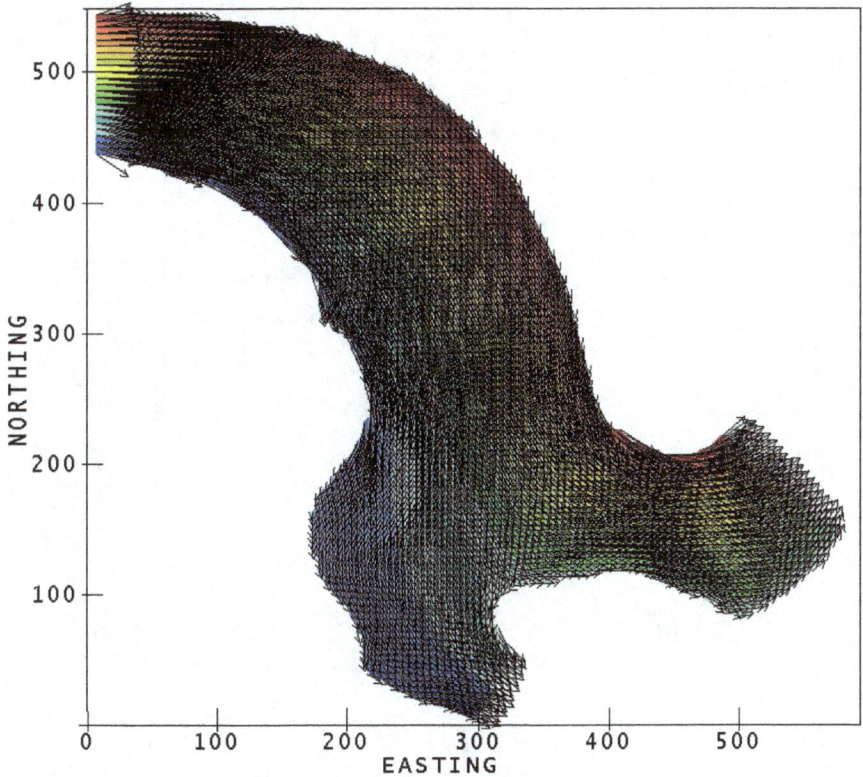

Figure 119. Potential Flow for a Branch

The particle tracks with dispersion are:

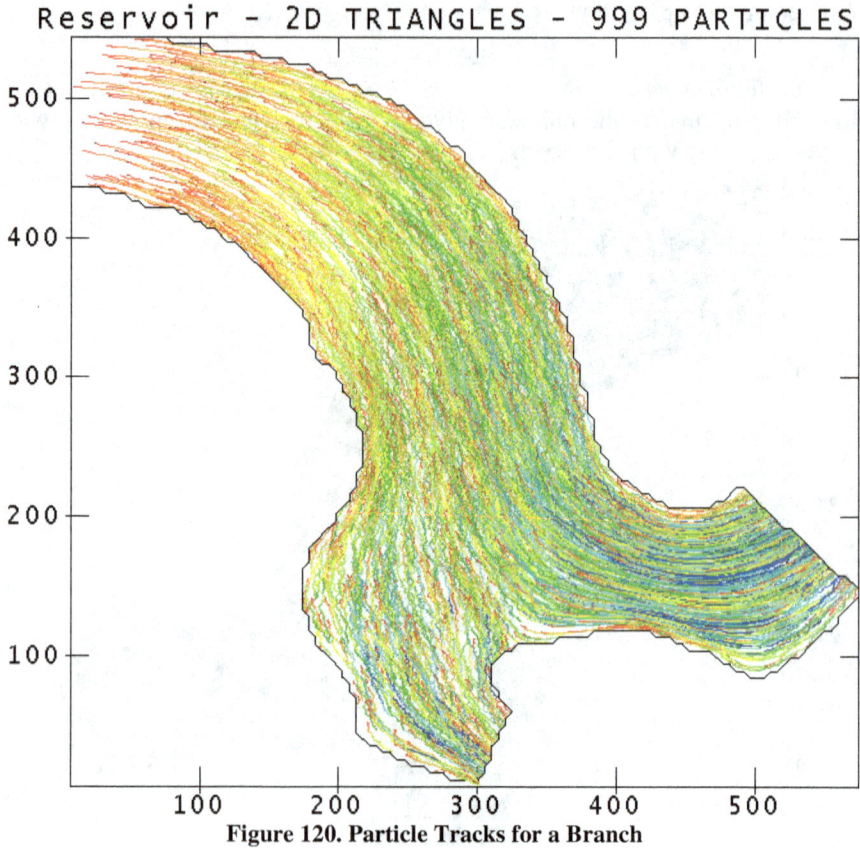

Figure 120. Particle Tracks for a Branch

3D Dispersion Example

This first figure shows the particle tracks without dispersion:

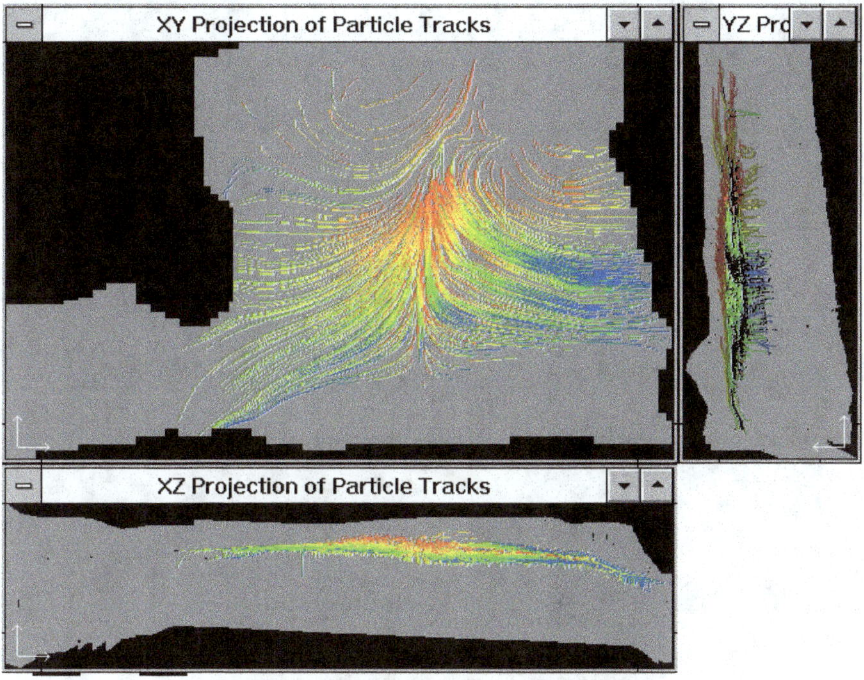

Figure 121. Particle Tracks for 3D Diffusion

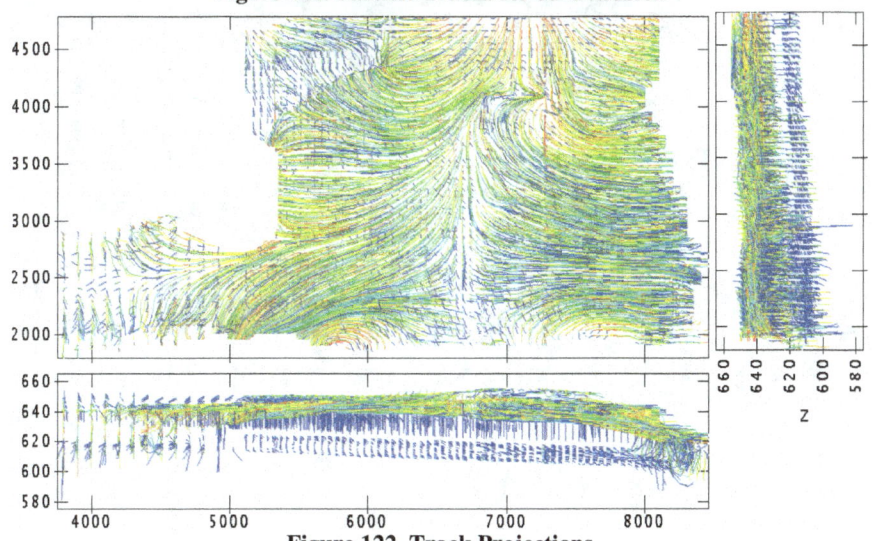

Figure 122. Track Projections

The second figure shows the particle tracks with dispersion:

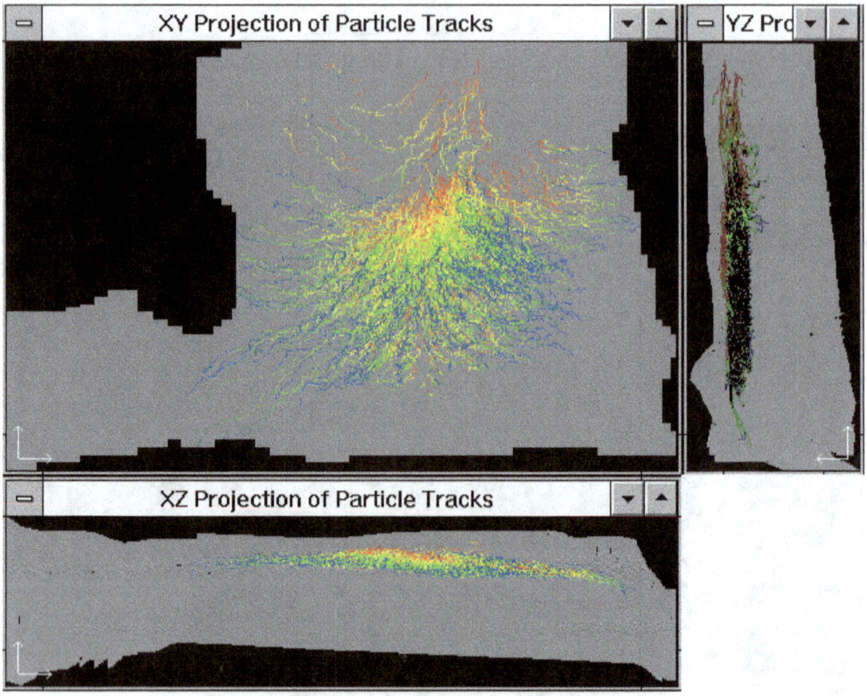

Figure 123. Particle Tracks for 3D Dispersion

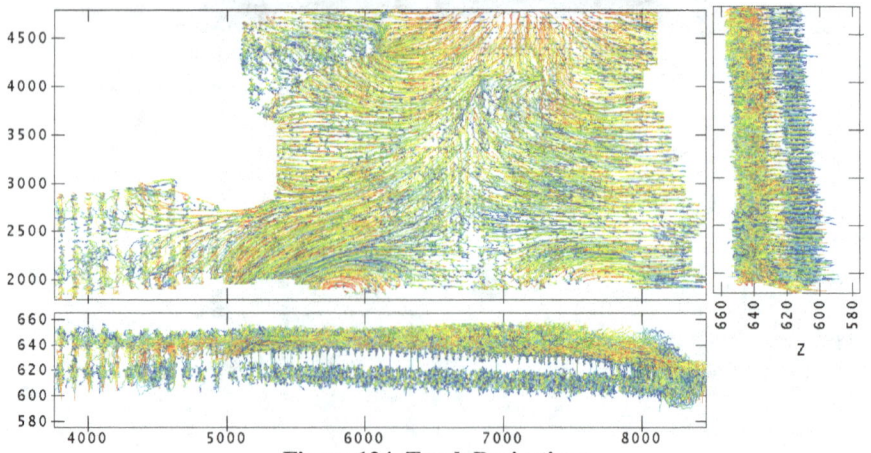

Figure 124. Track Projections

A different 3D domain without diffusion or dispersion:

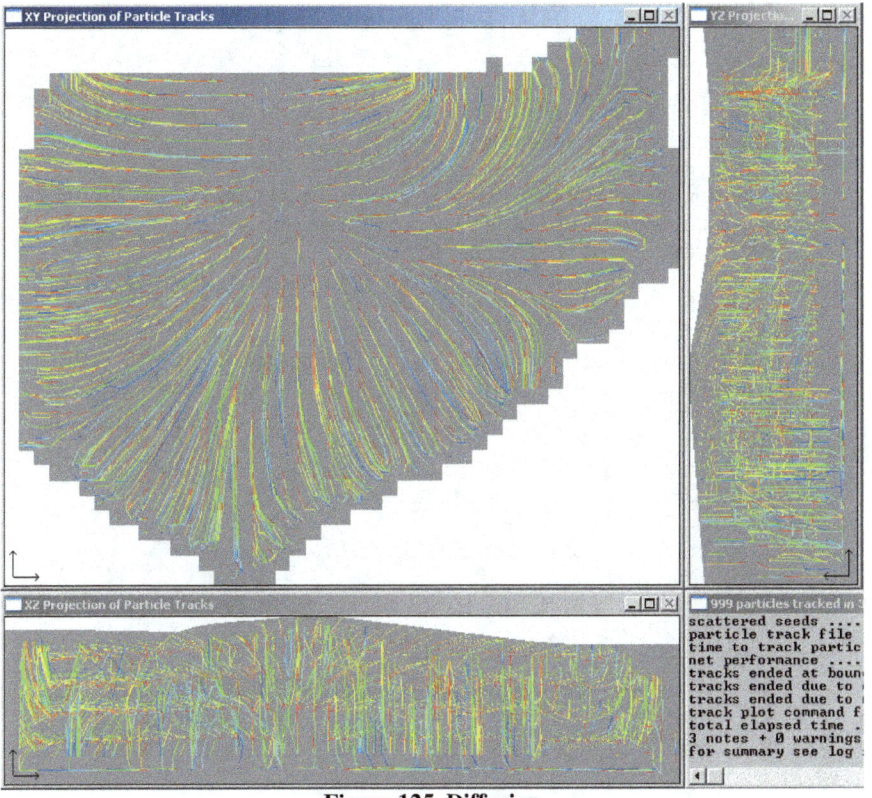

Figure 125. Diffusion

Typical information in the log file:
```
PTRAX/V5.01: Particle Tracker by Dudley J. Benton
animations:    enabled
fractures:     enabled
input format: Frac3D or ModFlow
parameter 2: track
parameter 3: wait
parameter 4: ok
application prefix: wine
checking for existing output files
  none found
reading default parameters from file: wine.CFG
default parameters
  default seeds: 999
  track duration: 500 days (1.4 years)
  time step: 100 days (0.3 years)
  maximum steps along particle track: 999
  default porosity: 0.3
```

```
default retardation factors: 0.5.1
default dispersivities: 3,3,0.3 ft
default matrix half-life: 1E+030 day(s)
default fracture half-life: 1E+030 day(s)
default velocities: 0.05,0,0 ft/day
matrix diffusion coefficients: 1,1,0.01 ft²/day²
synchronous time step default
maximum times a particle can enter an element: 99
maximum times a particle can enter a fracture: 99
matrix tortuosity (0=none, 1=complete): 0
Note: matrix tortuosity OFF
fracture tortuosity (0=none, 1=complete): 0
Note: fracture tortuosity OFF
reading MODFLOW files
Note: concentrations will NOT be divided by porosity
stray particles (outside domain) will be ignored
trapped particles will be ignored
include empty elements in snapshot files
create track file
MODFLOW basic input file: wine.BAS
  Title1: PREFIX: WINE
  Title2: CREATED BY BUILD3D/V1.71
  grid: 45x37x5
  total cells: 8325
  active cells: 5860
  resulting nodes: 10488
  element type: HEXAHEDRA
MODFLOW block-centered flow file: wine.BCF
  06X67889.85
  06Y66500.16
  145.916Z6235.105
linking domain
  node:element links
  66600 links
  element:element links
  45800 internal faces
  4150 external faces (boundaries)
  3810 external elements
  4152 external nodes
  computing element volumes
grouping elements
  largest element: 175.33x175.68x24.485
  27x22x2=1188 groups
  group size: 303.456x309.531x89.195
  sorting groups
  indexing groups
  there are 546 active and 642 empty groups
  the smallest group is 56, containing 5 members
  the largest group is 1, containing 20 members
  the active groups contain an average of 15 members
```

```
MODFLOW binary flow file: wine.CBB
characteristic parameters
  length threshold = 1.0223 feet
  volume threshold = 0.00457439 ft^3
  time threshold = 0.0630275 day(s)
  velocity threshold = 1.62199E-005 ft/day
  mean velocity = 16.2199 ft/day
  mean time to traverse element = 5.92921 day(s)
  synchronous time step = (automatic)
random walk
  dispersion ON
  diffusion ON
scattered seeds: 999
  particles tracked: 999
  time to track particles: 2 seconds
  average particles tracked per minute: 25409
  average steps per particle: 126
  average particle track: 887.037
  average particle life: 6911.78
  average particle speed: 0.128337
  total  particle movement: Sp=886150
  random particle movement: Rx/Sp=-0.00255429
  random particle movement: Ry/Sp=0.00155633
  random particle movement: Rz/Sp=-0.00166988
  random time steps: 2.08967 ñ 5.93962 day(s)
  tracks ended at boundaries: 621
  tracks ended due to circulation: 20
  tracks ended due to maximum steps: 63
  tracks ended due to maximum time: 295
Summary of Particle Tracking by Centroid of Mass
```

snap	year	mass	%mass	Xcentroid	Ycentroid	Zcentroid	Xradius	Yradius	Zradius
1	0.0	2.83E-05	100	3560.7	3602.9	183.3	2056.4	1557.5	21.8
2	0.3	1.44E-05	50.8509	3499.2	3196.1	173.6	2299.1	1987.8	18.5
3	0.5	1.21E-05	42.7427	3482.7	2948.1	172.3	2315.0	2077.1	19.0
4	0.8	1.14E-05	40.1401	3516.8	2902.4	170.9	2341.6	2094.8	18.7
5	1.1	1.10E-05	38.8388	3517.2	2890.5	170.5	2344.4	2109.3	18.7
6	1.4	1.07E-05	37.8378	3477.3	2863.4	171.0	2340.7	2109.7	19.4

The configuration file contains pertinent modeling information:

```
999        <- default number of particles to track
500        <- track duration
100        <- time step
999        <- maximum steps along a single particle track
0.3 0.5    <- default porosity and retardation factor
3 3 0.3    <- default longitudinal, lateral, and
   transverse dispersivity in feet
1E+030     <- default decay half-life
0.05 0 0   <- default velocities
1 1 0.01   <- default matrix diffusion factors
```

```
0          <- user-defined synchronous time step (set to
zero for automatic)
99         <- maximum times a particle can enter an
element
99         <- maximum times a particle can enter a
fracture
0          <- scattering factor (0=no scattering,
1=complete scattering)
```

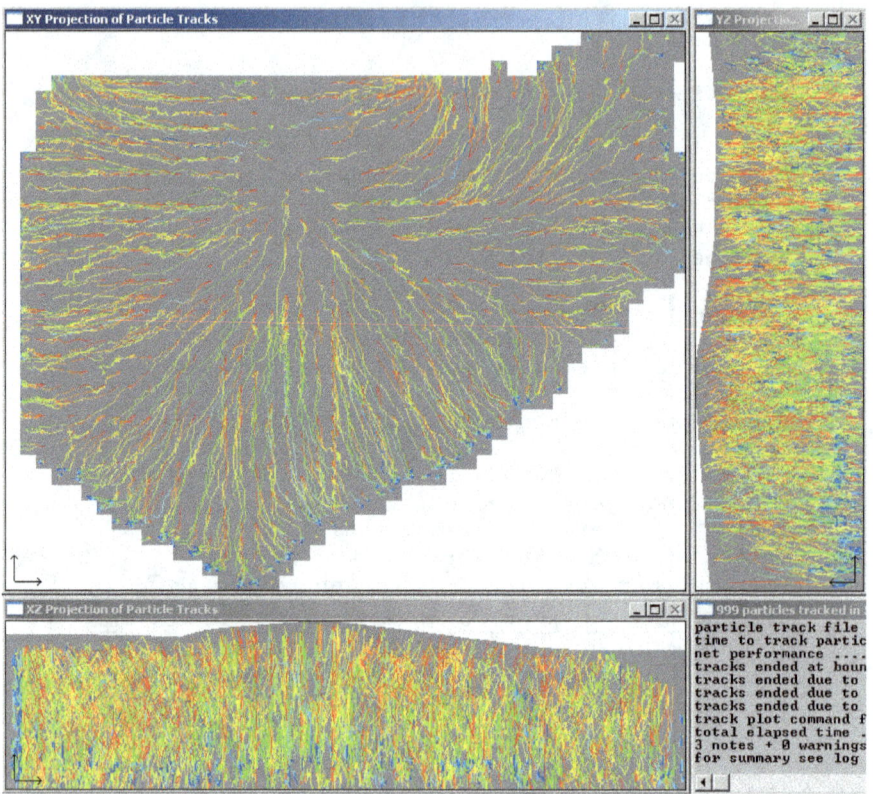

Figure 126. Diffusion + Dispersion

Chapter 16. Flow in Fractures

Fractures are like short-circuits in porous media. Fractured rock is often called *karst*, especially when limestone-based. In modeling, fractures are a 2D network imposed on top of a 3D domain, typically having much higher velocities, as would be the case in actual Earth formations. Particles can enter factures, pass quickly along them, and then return to the porous media at another location, sometimes far away. I devised a series of test cases to demonstrate that the coding works.

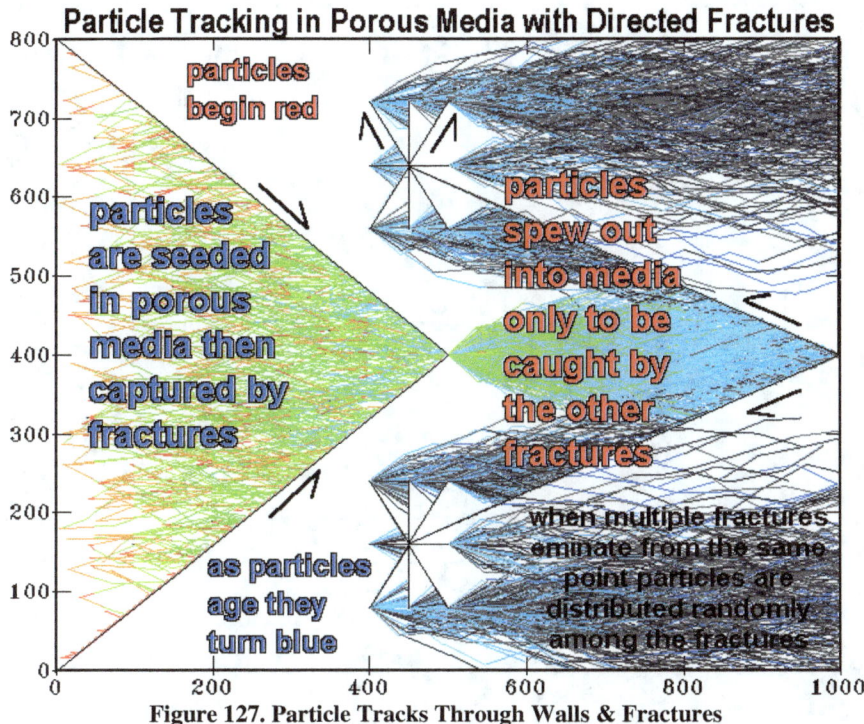

Figure 127. Particle Tracks Through Walls & Fractures

In the preceding example, particles are seeded on the left and flow toward the right. When they encounter the first two long diagonal fractures, the particles are trapped and quickly flow to the point, where they spew back out into the porous media. They continue flowing right and begin to spread out, then are caught by the second two long diagonal fractures. These second pair of factures flow back to the left, so the particles run back to the two stars, which consist of eight fractures. When multiple fractures emanate from the same point, the particles are randomly distributed between the possible paths, flow along these, and eventually back into the porous media, where they are again swept to the right. Some escape the right boundary, while others are caught by the fractures again and loop back through again. The particles age along the way from red to blue so that you can tell which ones have been around more than once. The algorithm is robust enough to traverse a 3D maze:

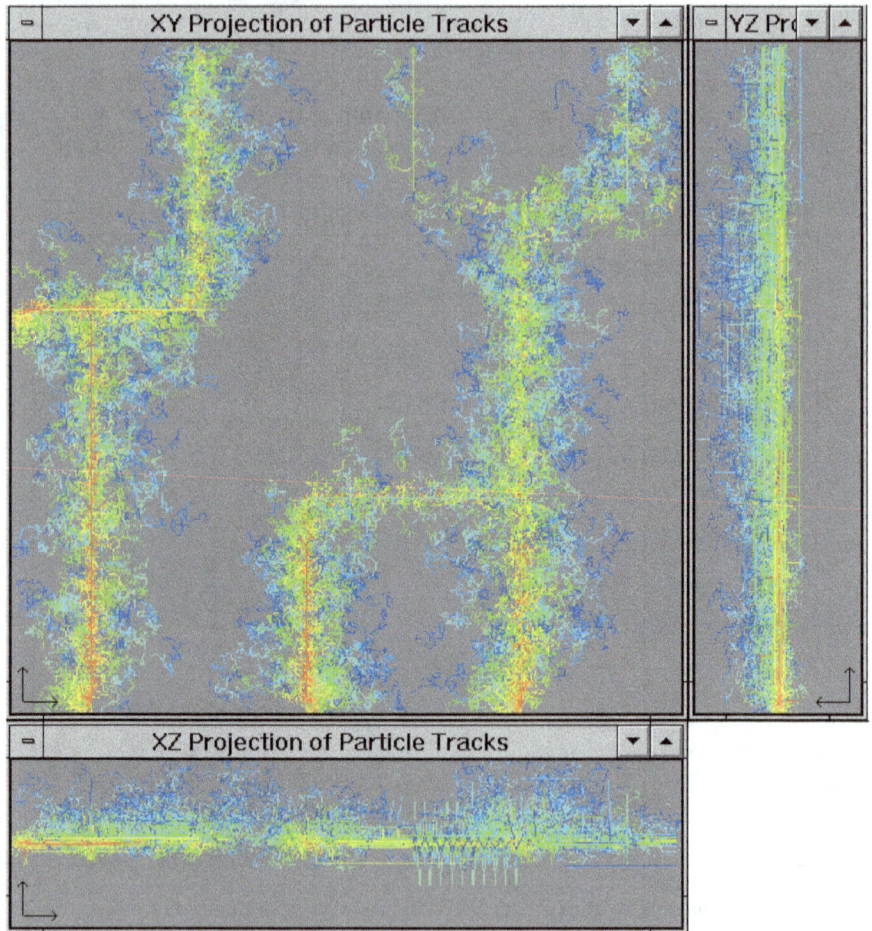

Figure 128. Diffusion + Dispersion with Fractures

This next figure shows the concentrations:

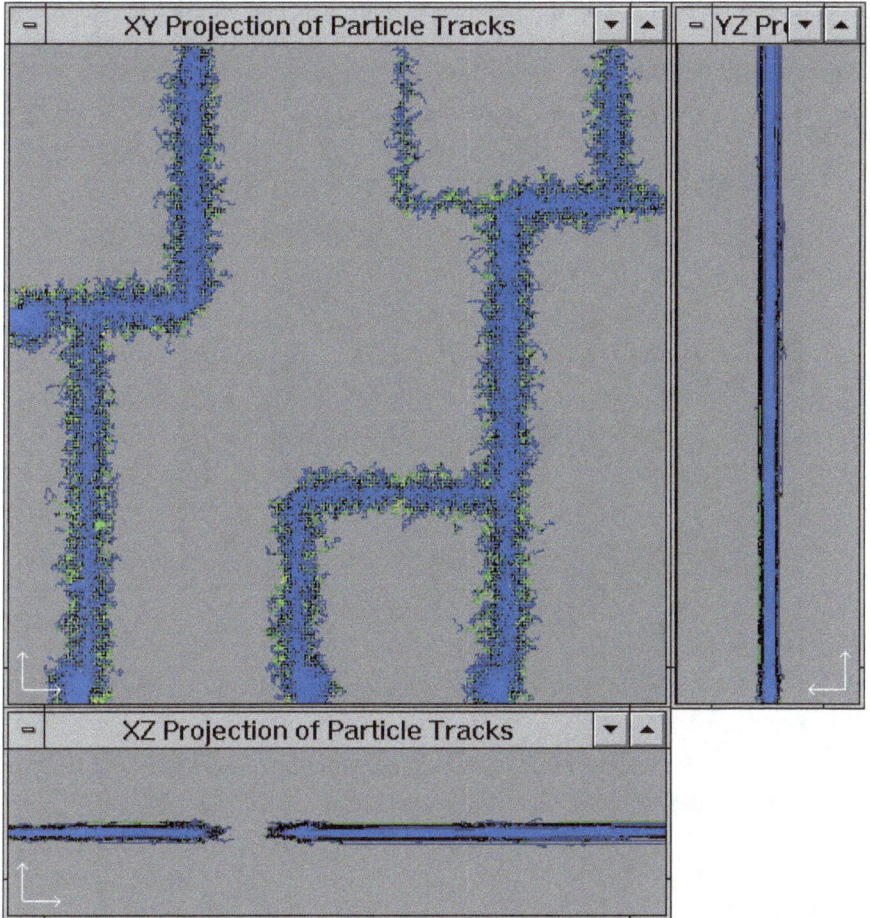

Figure 129. Resulting Concentrations

This representation loses a lot, being static. The original (which I still have) is animated, as is the analytical solution on the next page. The files for both of these examples can be found in the online archive in folder examples\fractures.

This next figure is as close as we could get with an analytical solution. Needless to say, setting up the particle tracking was much easier than building the analytical solution, which we accomplished with finite elements and superimposed domains.

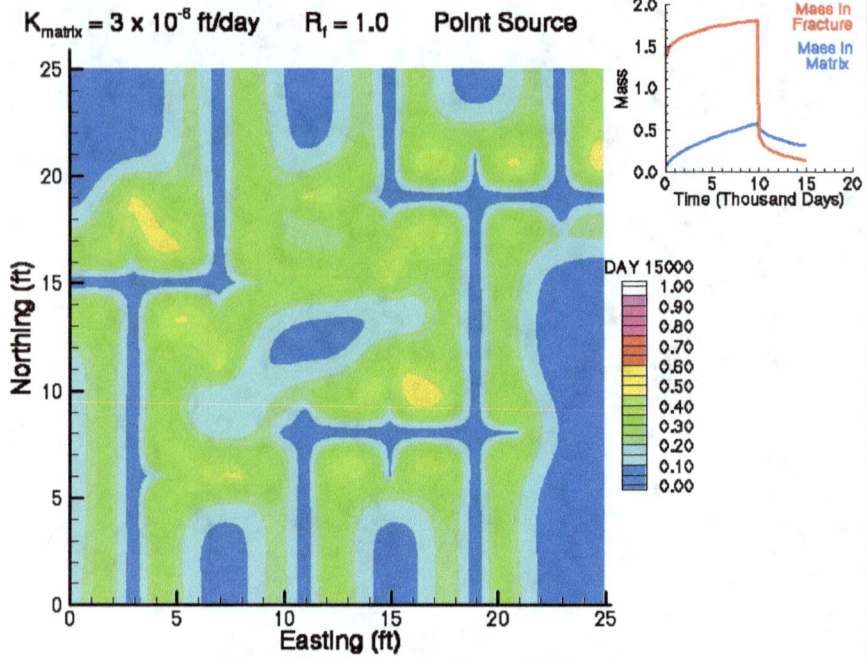

Figure 130. Concentrations with Fractures

Chapter 17. Contaminant Plumes

Contaminant plumes were the motivation for developing the software in the first place. Somebody put something in the ground that they shouldn't have and now it must be tracked, contained, and removed, which is the purpose of remediation. Contaminant plumes are often the starting point for particle tracking and I have devoted years to constructing these. Whether in the ground or a body of surface water, you will never be able to measure it all. Only point measurements are available plus sometimes a total spill amount. A volume of variable concentrations must be inferred from the point measurements. This example is typical of the information available, except that there were far more data points than are often available. The thick magenta curve represents the estimated boundary. The red curves represent surface features, including roads and a parking lot. The data points are colored based on the log of the concentration of the contaminant.

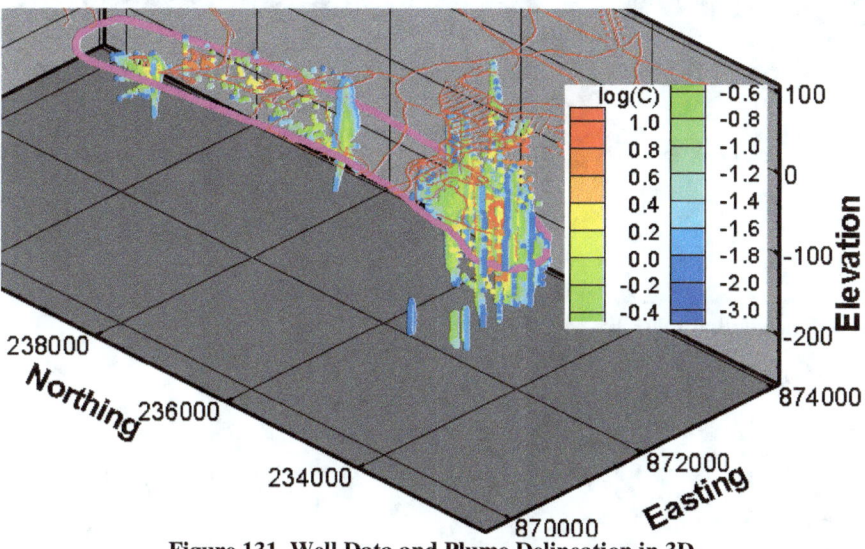

Figure 131. Well Data and Plume Delineation in 3D

In groundwater, data points are most often a vertical sequence, as these are withdrawn from wells. The contaminant plume must be defined for the entire initially affected zone. This requires specialized software and interpolation strategies, including: linear, inverse distance, and kriging.[14]

---

[14] Used mainly in geostatistics, kriging or Gaussian process regression is a method of interpolation in which the continuously estimated values are approximated by a Gaussian relationship. In some cases, kriging gives the best linear unbiased prediction of the continuous values. Interpolating methods based on other criteria (such as smoothness) may not yield the most likely results and can often produce unwanted artifacts. This method is also known as Wiener–Kolmogorov prediction, after Norbert Wiener and Andrey Kolmogorov.

Both TP2 and Tecplot® have such capabilities. The blue mesh indicates the three-dimensional extent of the plume at the level of log(C)=-3. The white specks indicate were most of the particles will be seeded. Generating appropriate particle seeds will be covered in the next chapter.

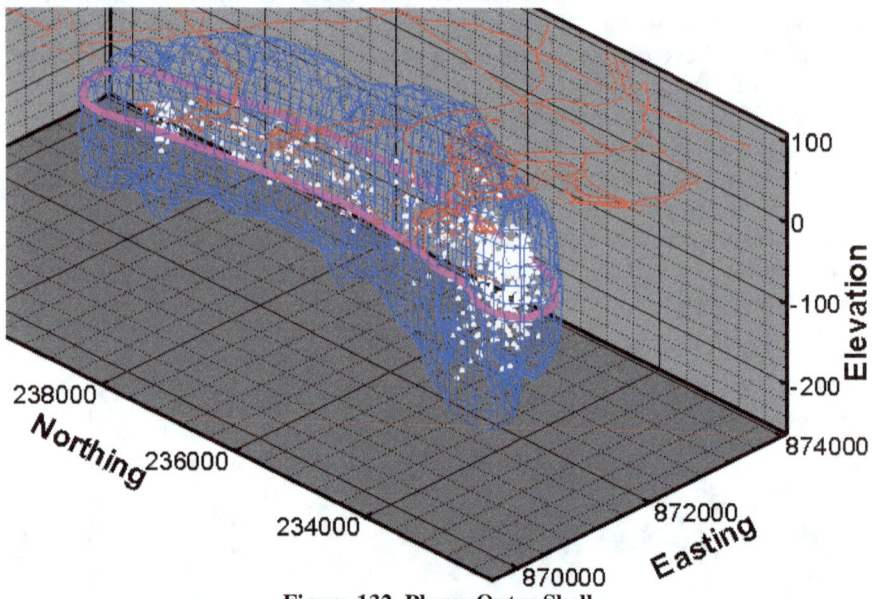

Figure 132. Plume Outer Shell

The shell at log(C)=-2 is shown in this next figure:

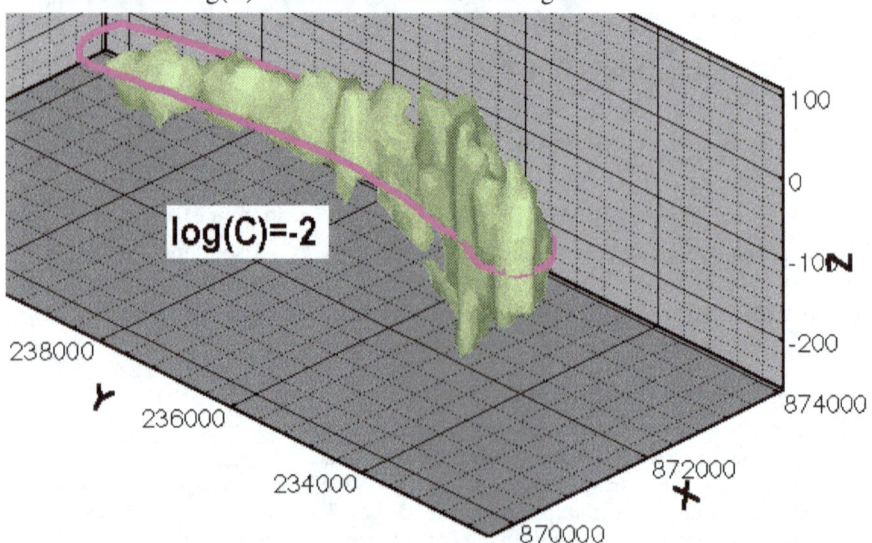

Figure 133. 3D Concentration Shells

A horizontal slicing plane of concentrations is shown in this next figure:

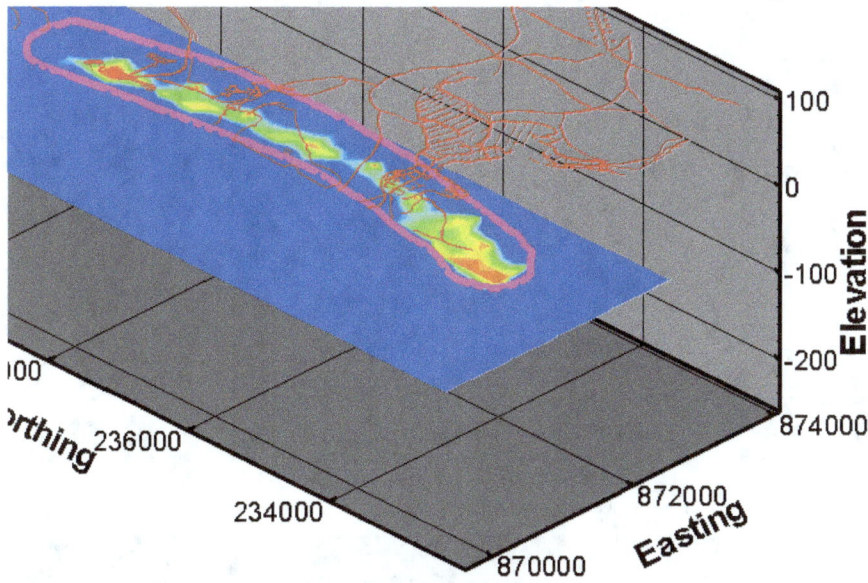

Figure 134. Horizontal Slices through Plume

A vertical slicing plane of concentrations is shown in this next figure:

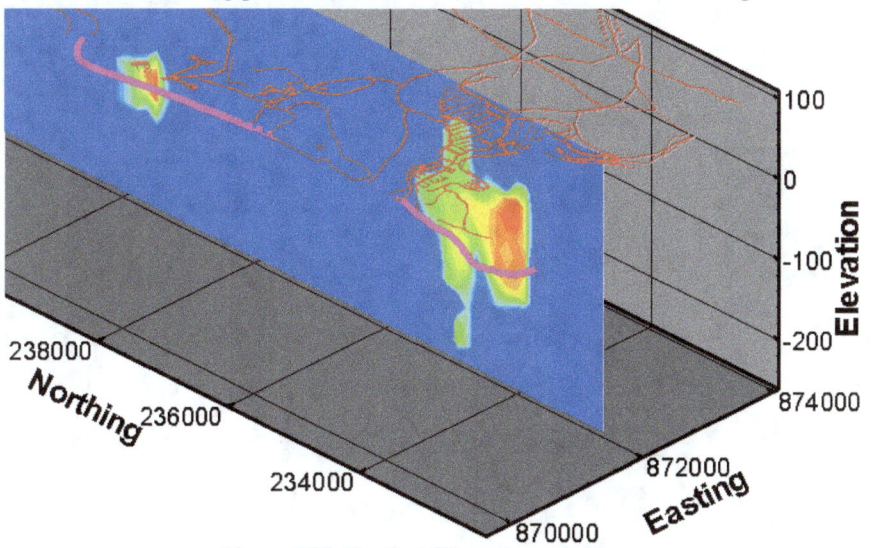

Figure 135. Vertical Slices through Plume

The most interestingly shaped plume we ever worked with was affectionately called the *starship*:

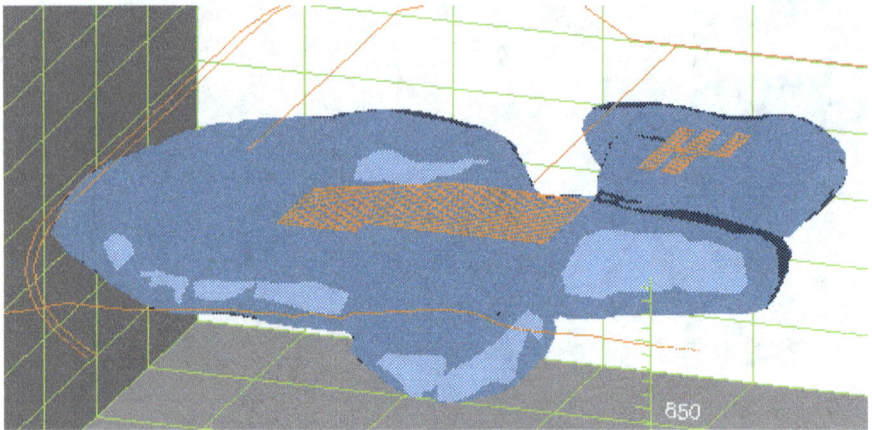

Figure 136. Outer Plume Shell Side View

A view from the top:

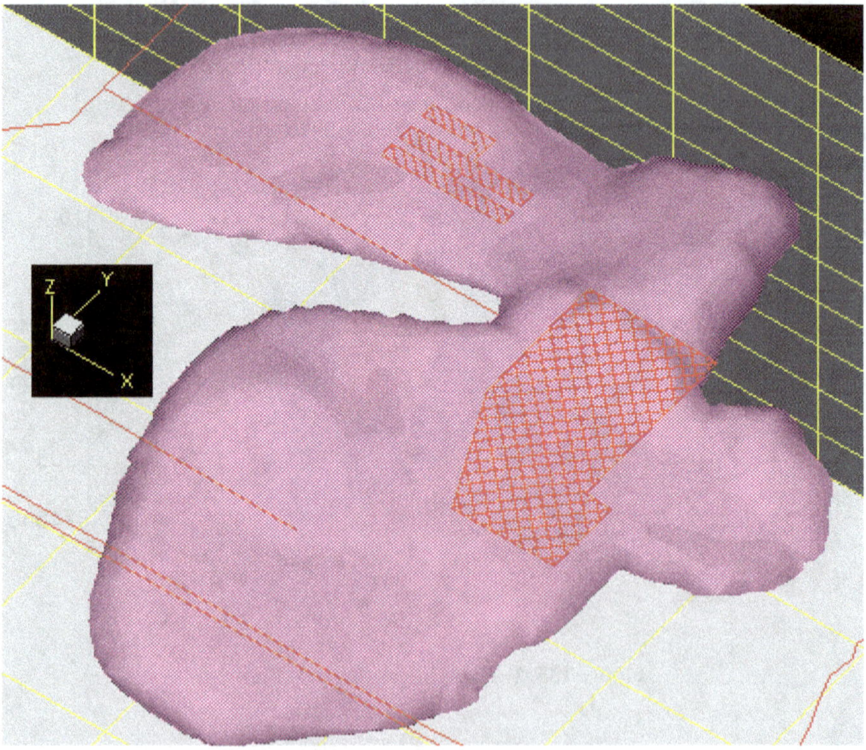

Figure 137. Outer Plume Shell Top View

Chapter 18. Particle Seeds

Creating a *swarm* of particles that accurately represent an initial concentration field in three-dimensions is a complex task. PTRAX can handle particle seeds having different mass and/or concentration and can keep these separate, even coloring them differently, as shown in the following figure:

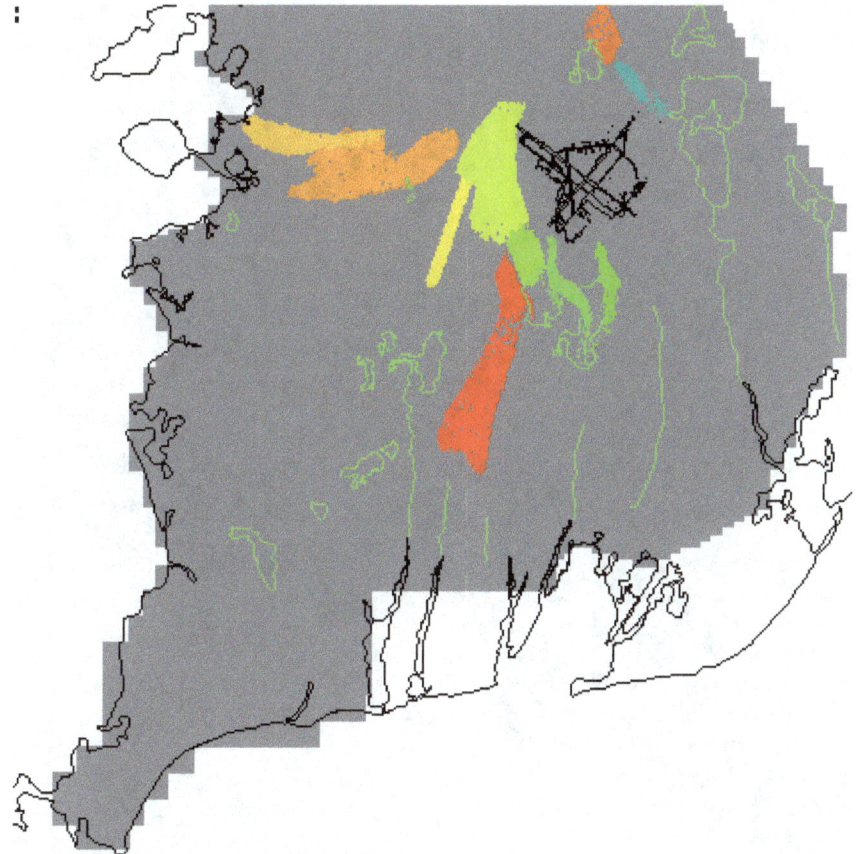

Figure 138. Multiple Plumes

This is the first frame in an animation showing how the then current seven distinct contaminant plumes can be expected to evolve over time, should no steps be taken to remediate.

This next figure shows where wells were positioned to capture the contaminants. The circles are pump-and-treat wells and the plusses are monitoring wells to make sure the contaminant didn't get past the extraction wells. The calculated arrival of particles over time was also used to design the treatment systems, as different chemical substances and concentrations were expected.

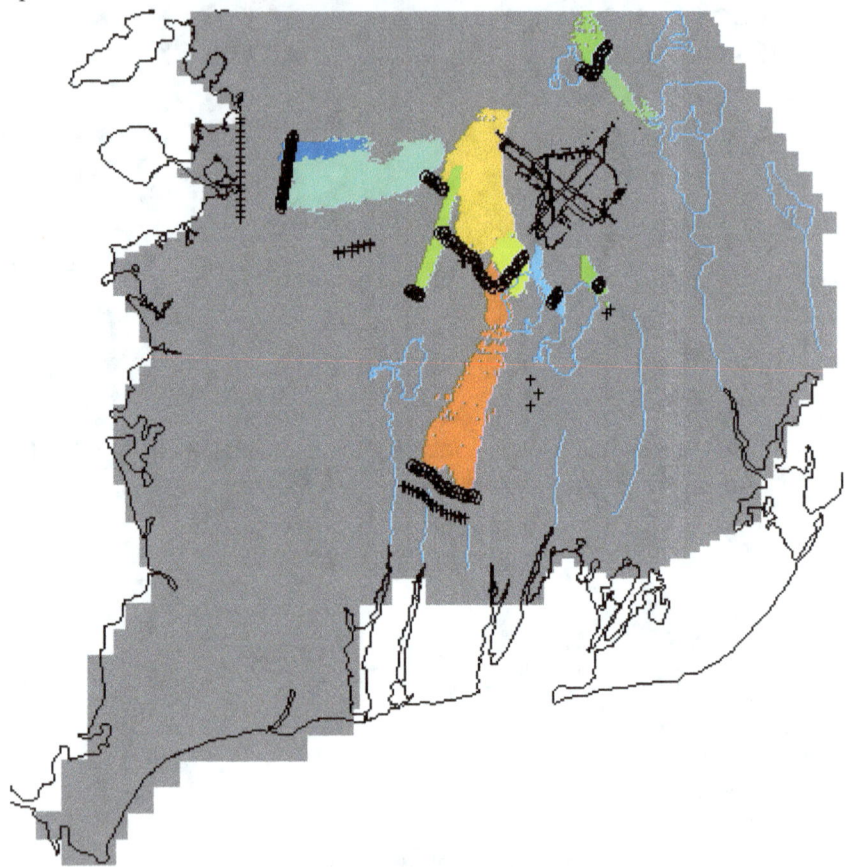

Figure 139. Multiple Plumes with Wells & Streams

The first step in creating the particle seeds is defining the three-dimensional field describing the initial concentrations. This can be done by TP2 or Tecplot® resulting in a name.TB3 file. The format of this file is Nx, followed by the X values, Ny followed by the Y values, Nz followed by the Z values, and finally Nc=Nx*Ny*Nz followed by the concentrations. Both TP2 and Tecplot® can display 3D fields in several ways, including shells, as illustrated in the previous chapter, and slices along planes, as illustrated in this next figure:

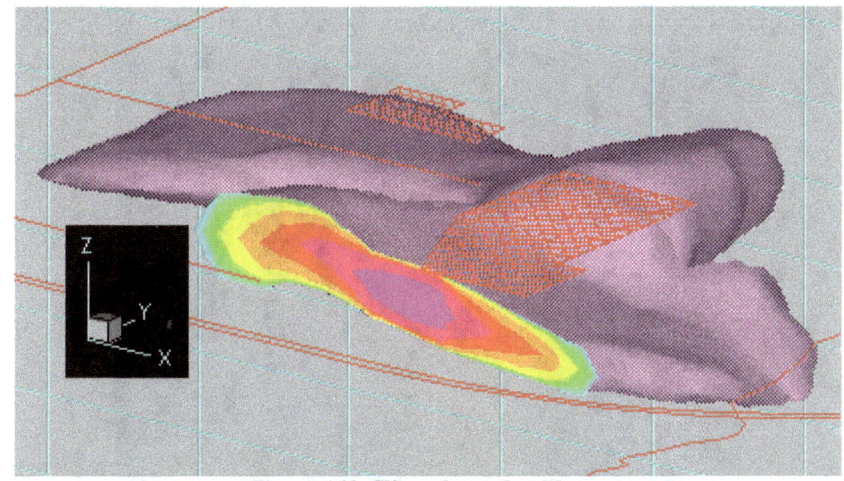

Figure 140. Slices through a Plume

And a different-shaped plume:

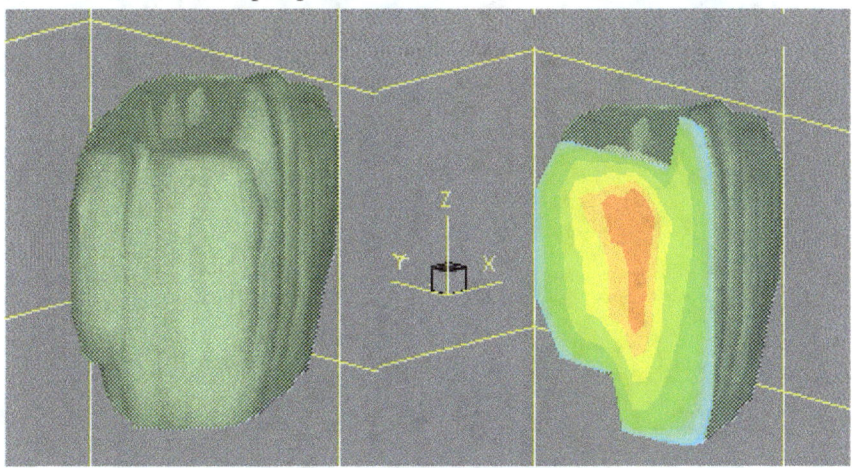

Figure 141. "Apple" Plume Sliced

You will find 5 example three-dimensional concentration files in the online archive in folder examples\seeds. I have many more. These were all created with TP2 using inverse distance interpolation and also some smoothing, which is under the Tools menu.

The crosshairs in this next figure indicate the slicing planes in 3D:

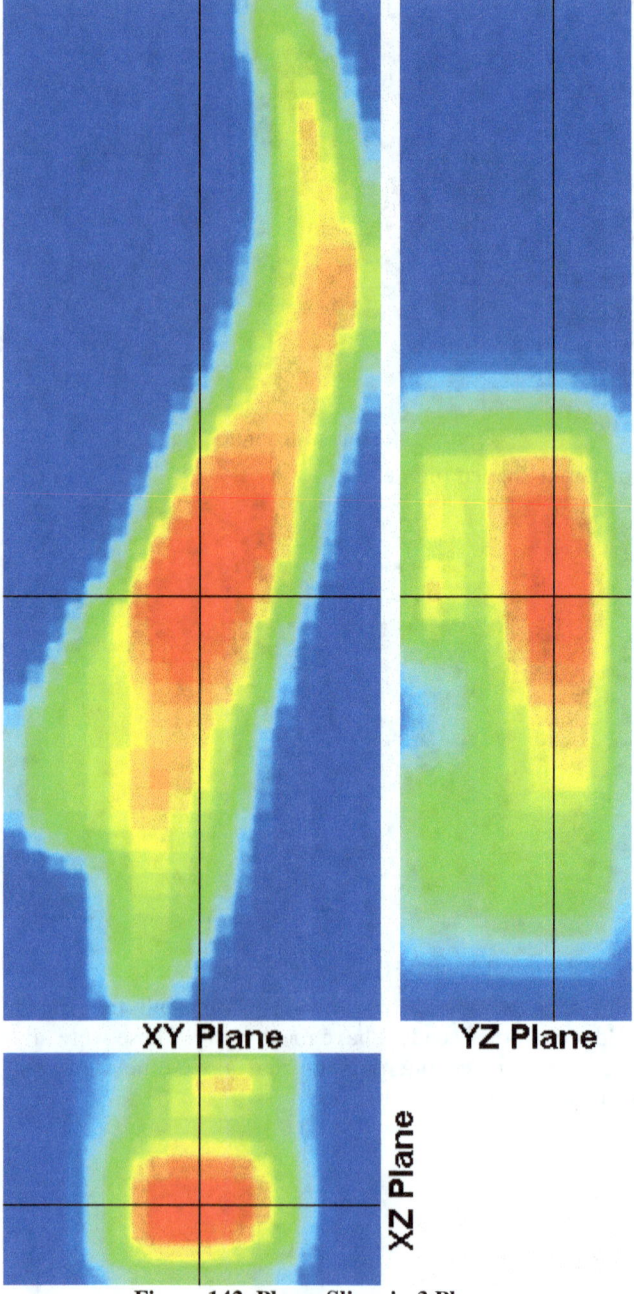

Figure 142. Plume Slices in 3 Planes

The seed generation program is seed3d.c, which outputs something like the following:

```
SEED3D/V1.13: create seed file from a .TB3 file
prefix: ash
seeds = 10000
extent of plume: ash.p2d
polygon: 222 points
concentration file: ash.TB3
seed file: ash.SED
field: 19x50x12=11400
concentration: -1.30103≤log(C)≤1.87935 µg/l
threshold = 10 µg/l
plan view area = 1221.94 acre (4.94502 km²)
plume volume = 31927.5 acre-ft (39380.8 Ml)
average thickness = 26.1285 ft (7.96395 m)
total plume mass = 915.705 kg
average concentration = 23.2526 µg/l
9679 actual seeds
total seed mass = 915.705 kg
TP2 command file: ash.TP2
seed mass: 3.23389E+006≤M≤3.77226E+006
```

It read the field from name.TB3 and the boundary from name.P2D and produces the seeds, which look something like:

```
* Note: mass is in æg-ft^3/liter, when
*       divided by volume in cubic feet,
*       this becomes æg/l or ppb.
*   X       Y       Z      Mass      Time
857532 220962 -78.2251 3.37734E+006 0
857864 220970 -74.3877 3.37734E+006 0
857675 221093 -69.7951 3.37734E+006 0
857839 220809 -69.3116 3.37734E+006 0
857824 220942 -76.7429 3.37734E+006 0
857537 220776 -75.9333 3.37734E+006 0
857592 220805 -67.5764 3.37734E+006 0
857715 220787 -67.3602 3.41488E+006 0
857535 220889 -60.3033 3.41488E+006 0
857766 220977 -59.2923 3.41488E+006 0
857599 221001 -61.3863 3.41488E+006 0
857676 220762 -59.2853 3.41488E+006 0
857854 221056 -60.4611 3.41488E+006 0
857655 221085 -57.6919 3.41488E+006 0
857925 221499 -73.5912 3.48598E+006 0
857590 221316 -77.6624 3.48598E+006 0
857886 221217 -70.3733 3.48598E+006 0
857879 221527 -67.4268 3.48598E+006 0
857783 221289 -77.2488 3.48598E+006 0
857653 221465 -80.4957 3.48598E+006 0
857686 221170 -71.7453 3.48598E+006 0
```

The seeds are sprinkled throughout the domain, as shown in this next figure:

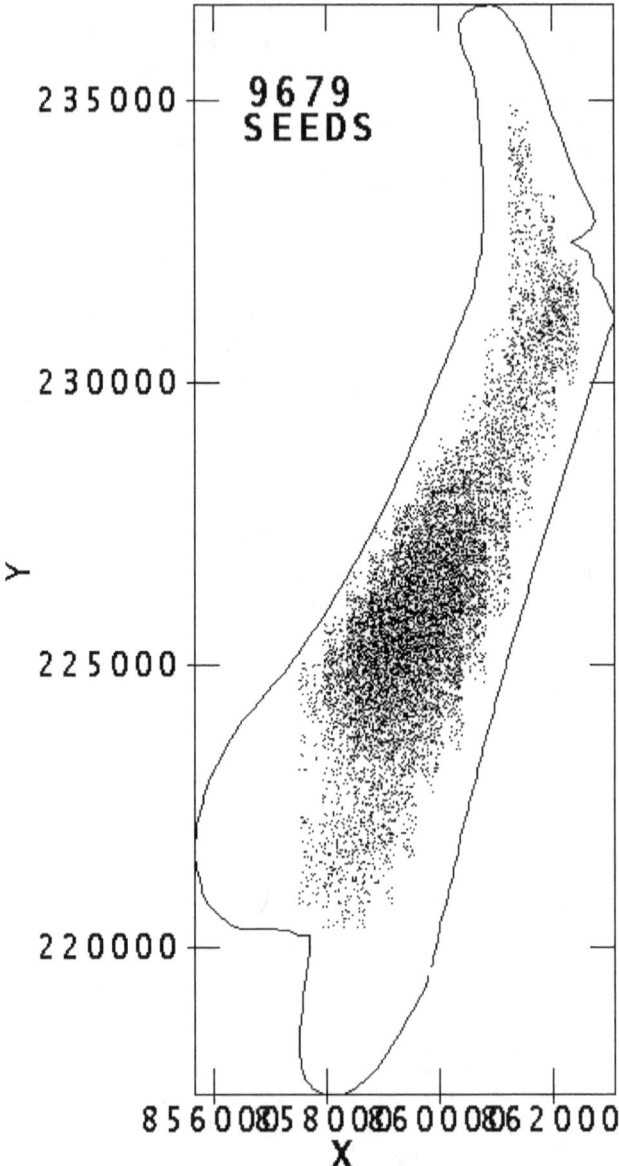

Figure 143. Location of Seeds (Planar View)

The actual number of seeds will be determined to most accurately represent the field. An approximate number of 10,000 was specified in this case. You can see from this figure that most of the seeds are placed in the highest concentration (i.e., red) zone from the previous slice graphic.

The core of seed3d.c is the following code:
```
for(x=1;x<Nx-1;x++)
  {
  X1=(Xn[x-1]+Xn[x])/2;
  X2=(Xn[x]+Xn[x+1])/2;
  dX=X2-X1;
  for(y=1;y<Ny-1;y++)
    {
    if(!InsidePolygon(Xp,Yp,Np,Xn[x],Yn[y]))
      continue;
    Y1=(Yn[y-1]+Yn[y])/2;
    Y2=(Yn[y]+Yn[y+1])/2;
    dY=Y2-Y1;
    for(z=1;z<Nz-1;z++)
      {
      Z1=(Zn[z-1]+Zn[z])/2;
      Z2=(Zn[z]+Zn[z+1])/2;
      dZ=Z2-Z1;
      dV=dX*dY*dZ;
      n=(Ny*z+y)*Nx+x;
      C=Cn[n];
      if(C<Cthr)
        continue;
      M=pow(10,C)*dV;
      m=max(1,(int)(ns*M/Mtot));
      Ms+=M;
      Ns+=m;
      M/=m;
      Mm=min(Mm,M);
      Mx=max(Mx,M);
      for(i=0;i<m;i++)
        {
        X=X1+(X2-X1)*drand();
        Y=Y1+(Y2-Y1)*drand();
        Z=Z1+(Z2-Z1)*drand();
        fprintf(fo,"%lG %lG %lG %lG 0\n",X,Y,Z,M);
        }
      }
    fprintf(stderr,"%li seeds\r",Ns);
    }
  }
```

Note that the TB3 file is assumed to contain the log of the concentrations. If not, you can modify the code accordingly. Also, the TB3 file is presumed to contain rectangular elements, though not necessarily the same length on each side, especially in the vertical.

Chapter 19. Animations

PTRAX automatically creates animations of the particle tracks. As this is a Windows® application, the GUI is continuously updated as the particles are tracked, which is per particle over time. What you really want is all of the particles over time. As the Hamiltonian approach tracks each particle sequentially, rather than simultaneously, as in the Lagrangian approach, the animations must also be created while the particles are being tracked and then written out after all of the calculations are complete. Compare the 1st and 24th frames (0 and 30 years) in this example shown previously:

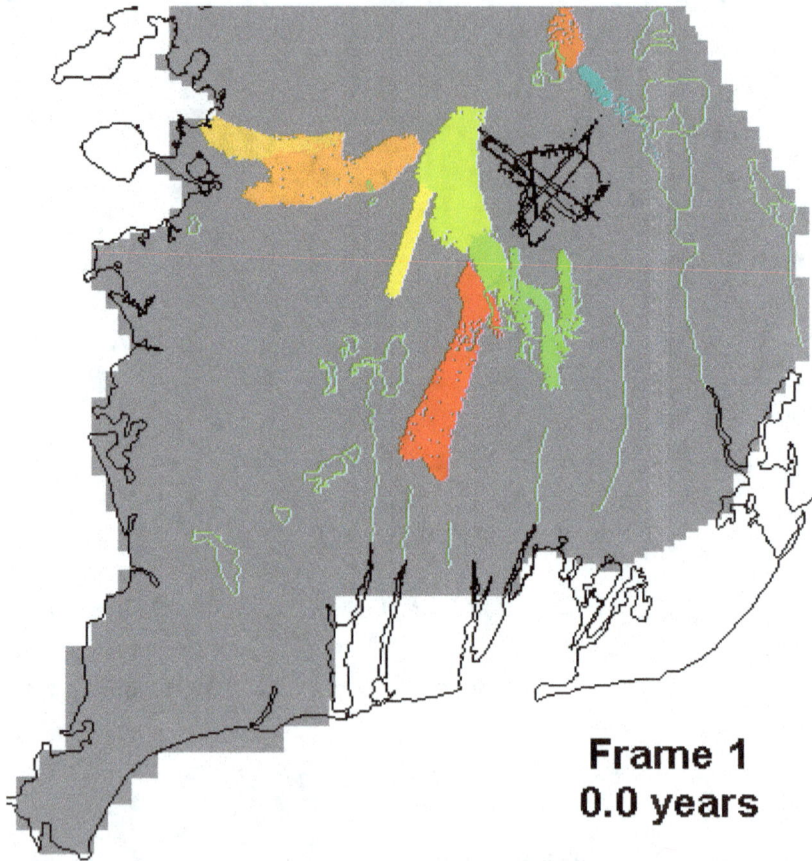

Figure 144. Remediation Model at Startup

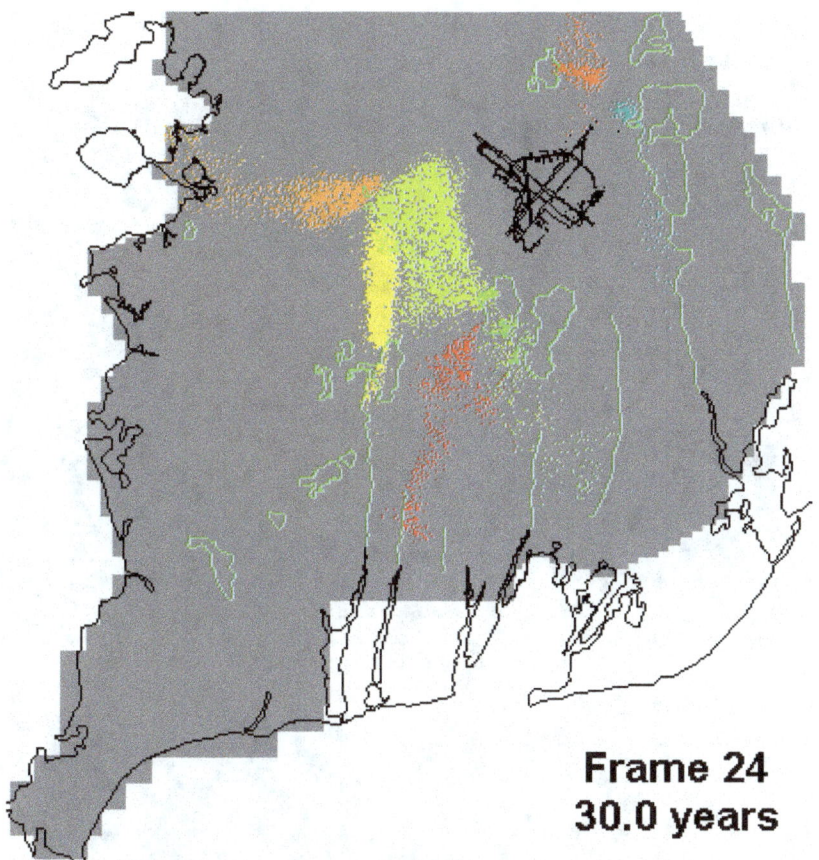

Figure 145. Remediation Model after 30 Years

Notice that the particles have dispersed and also moved away from the original contamination zones. This particular site has the highest elevation in the center with groundwater flow outward, mostly to the west, south, and east, with very little flow to the north.

This next simulation achieved only partial capture of the contaminant. Compare the 1st and 20th frames (0 and 30 years).

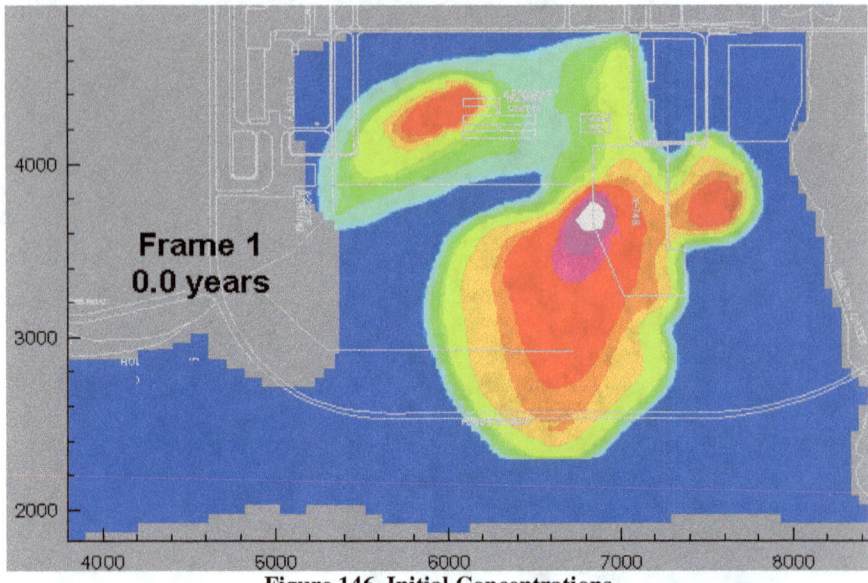

Figure 146. Initial Concentrations

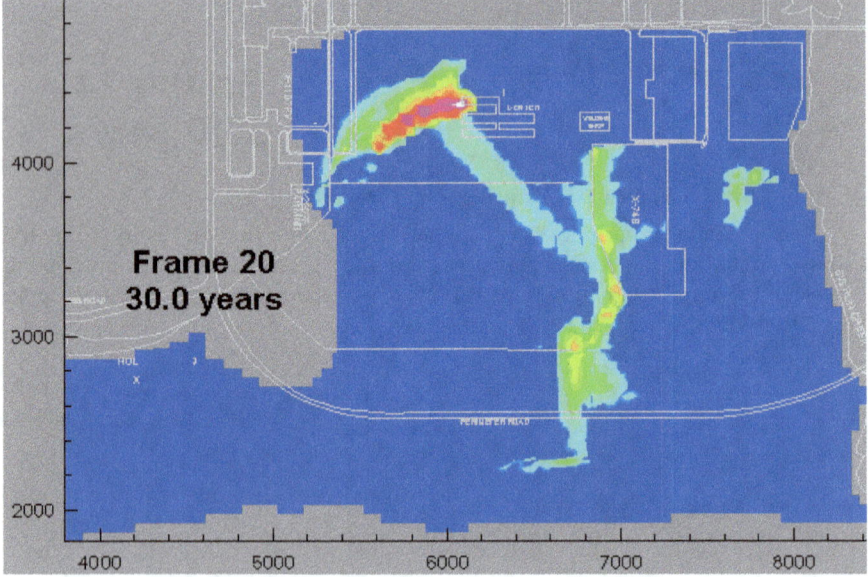

Figure 147. Expected Concentrations after 30 Years

This is a good illustration of how modeling can shape remediation strategies. It's a lot cheaper to run a model twenty times than drill a bunch of wells and wait 30 years to find out what might or might not happen!

Chapter 20. Concentration Mappings

PTRAX automatically creates concentration snapshots at regular intervals as specified. These may or may not be set within model boundaries. Both TP2 and Tecplot® have the facility to overlay and embed images in fields to enhance the graphical presentation. The previous spaceship plume is shown in these next three figures, which are estimates of taking no remedial action.

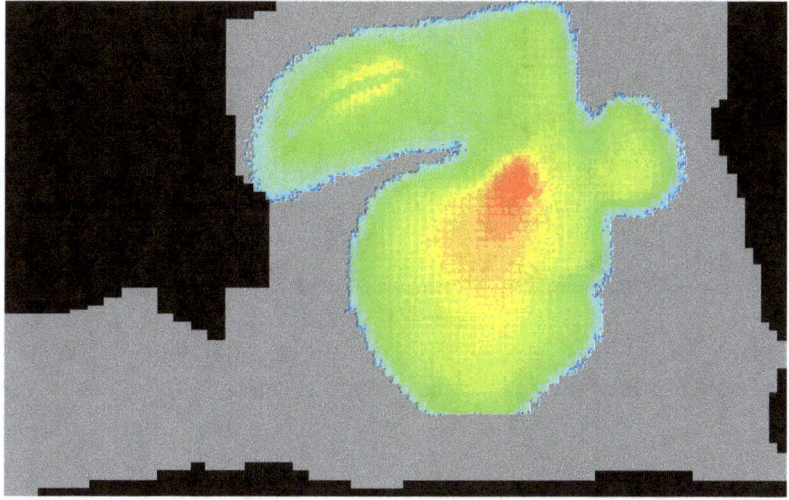

Frame 1 – 0.0 Years

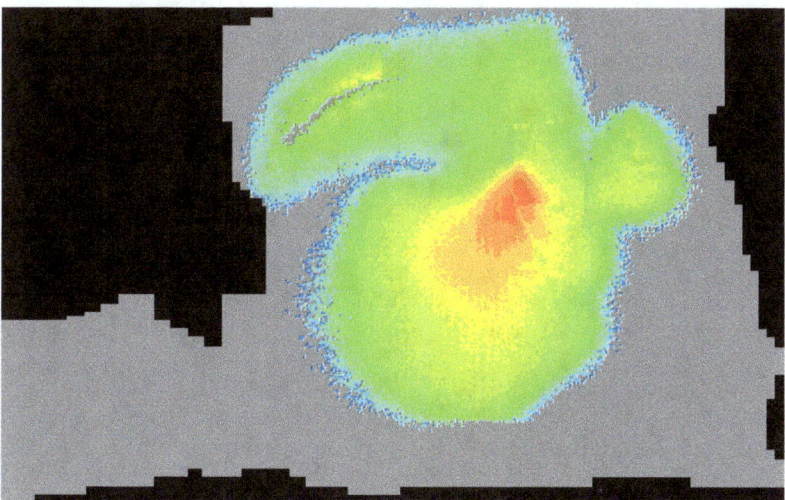

Frame 19 – 14.2 Years

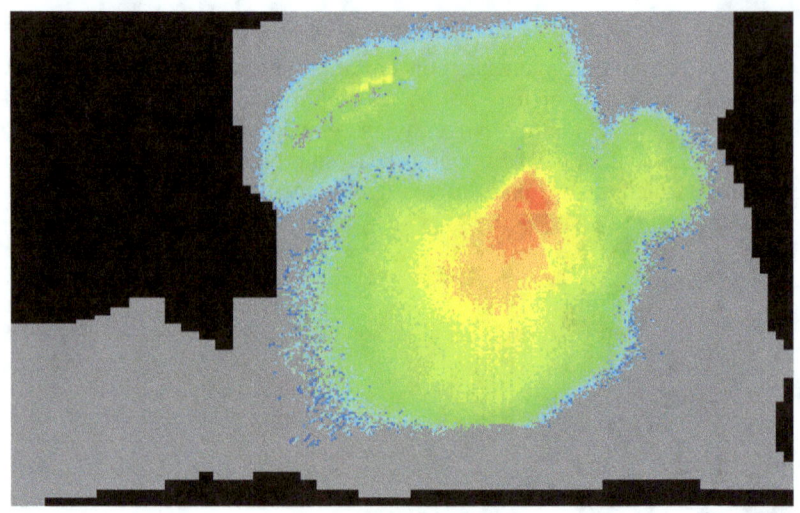

Frame 39 – 30 Years

PTRAX writes two- or three-dimensional table files (TB2 or TB3) for each concentration snapshot as well as images, such as these.

Chapter 21. Reverse Particle Tracking

Implementing reverse particle (i.e., back-) tracking is a real no-brainer. It's just a switch (BACK). When you read in the velocity field, simply reverse all the components. Adding this capability led us to one of the most interesting stories that I dare not share completely, considering the parties involved. Starting from what is and running backward in time to what might have started the current mess led is to find a vacant lot in a residential neighborhood where someone had apparently drained a tanker truck load of solvent that was supposed to be incinerated. So, build a model, make sure it agrees well with available data and then turn back the clock to see who left the mess and where!

The forward tracks for the example on page ii are:

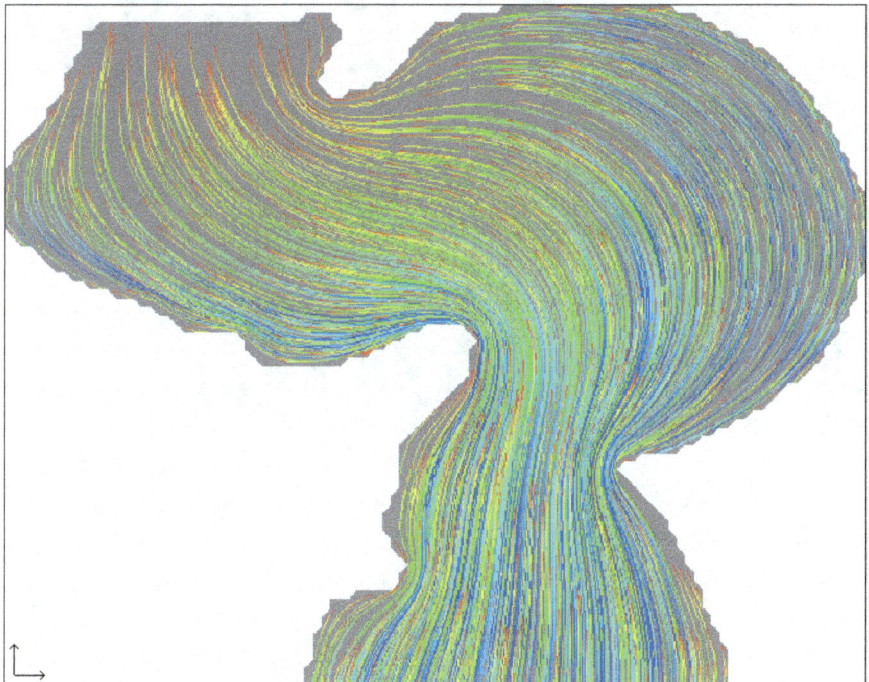

Figure 148. Forward Particle Tracks

Just add BACK to the command line or in the batch file (BEND.BAT), as in PTRAX BEND TRACK BACK WAIT OK, to track the same particles in reverse:

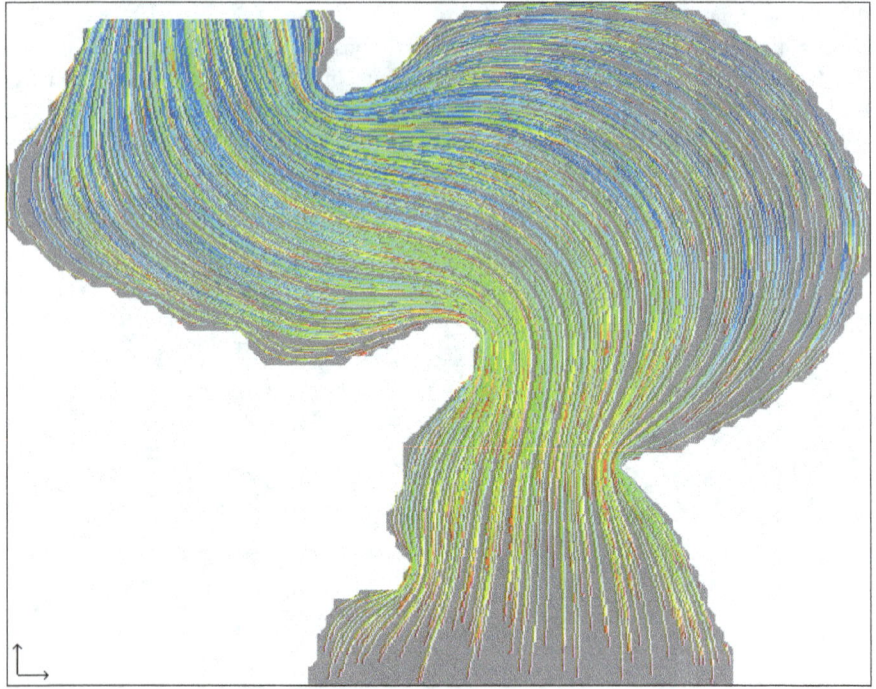

Figure 149. Backward Particle Tracks

Chapter 22. Sources and Sinks

Wells can be used to inject and withdraw so that they are either sources or sinks. The flow model must be able to handle either. PTRAX can inject or remove particles. This is typical of such a system:

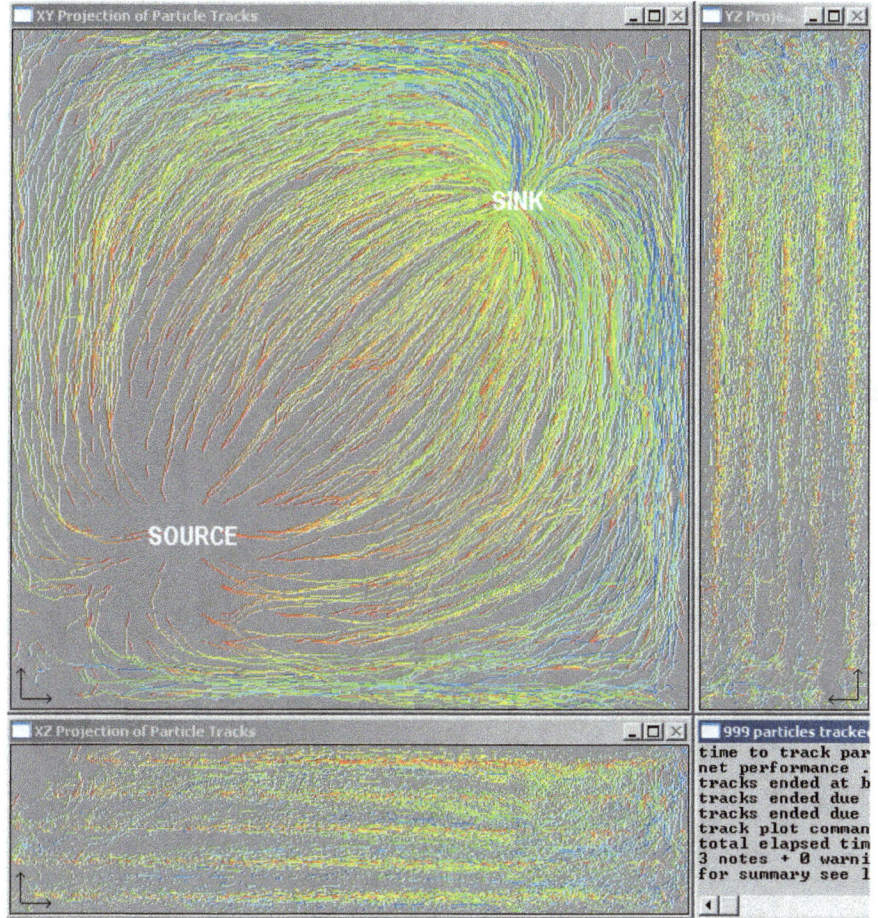

Figure 150. Source & Sink

```
PTRAX/V5.01: Particle Tracker by Dudley J. Benton
animations:    enabled
fractures:     enabled
input format: Frac3D or ModFlow
parameter 2: TRACK
parameter 3: WAIT
parameter 4: OK
application prefix: SLAB
checking for existing output files
  none found
```

```
reading default parameters from file: SLAB.CFG
default parameters
  default seeds: 999
  track duration: 10950 days (30.0 years)
  time step: 1825 days (5.0 years)
  maximum steps along particle track: 999
  default porosity: 0.3
  default retardation factors: 1.1
  default dispersivities: 3,3,0.3 ft
  default matrix half-life: 1E+030 day(s)
  default fracture half-life: 1E+030 day(s)
  default velocities: 0.05,0,0 ft/day
  matrix diffusion coefficients: 0,0,0 ft²/day²
  synchronous time step default
  maximum times a particle can enter an element: 4
  maximum times a particle can enter a fracture: 4
  matrix tortuosity (0=none, 1=complete): 0
  Note: matrix tortuosity OFF
  fracture tortuosity (0=none, 1=complete): 0
  Note: fracture tortuosity OFF
reading MODFLOW files
Note: concentrations will NOT be divided by porosity
stray particles (outside domain) will be ignored
trapped particles will be ignored
include empty elements in snapshot files
create track file
MODFLOW basic input file: SLAB.BAS
  Title1: PREFIX: SLAB
  Title2: CREATED BY BUILD3D/V1.71
  grid: 37x25x5
  total cells: 4625
  active cells: 4625
  resulting nodes: 5928
  element type: HEXAHEDRA
MODFLOW block-centered flow file: SLAB.BCF
  06X6999.999
  06Y61000
  06Z6100
linking domain
  node:element links
  37000 links
  element:element links
  25280 internal faces
  2470 external faces (boundaries)
  2210 external elements
  2472 external nodes
  computing element volumes
grouping elements
  largest element: 27.027x40x20
  22x14x2=616 groups
```

```
group size: 47.619x76.9231x100
sorting groups
indexing groups
there are 273 active and 343 empty groups
the smallest group is 135, containing 5 members
the largest group is 1, containing 20 members
the active groups contain an average of 17 members
MODFLOW binary flow file: SLAB.CBB
characteristic parameters
length threshold = 0.141774 feet
volume threshold = 9.99999E-005 ft^3
time threshold = 0.566894 day(s)
velocity threshold = 2.5009E-007 ft/day
mean velocity = 0.25009 ft/day
mean time to traverse element = 138.208 day(s)
synchronous time step = (automatic)
random walk
dispersion ON
diffusion OFF
scattered seeds: 999
particles tracked: 999
time to track particles: 1 seconds
average particles tracked per minute: 59940
average steps per particle: 27
average particle track: 425.621
average particle life: 1.12827E+006
average particle speed: 0.000377232
total    particle movement: Sp=425195
random particle movement: Rx/Sp=-0.00884701
random particle movement: Ry/Sp=-0.00415675
random particle movement: Rz/Sp=1.75661E-005
random time steps: 2681.7 ñ 395.558 day(s)
tracks ended at boundaries: 86
tracks ended due to circulation: 782
tracks ended due to maximum time: 131
```
Summary of Particle Tracking by Centroid of Mass

snap	year	mass	%mass	Xcentroid	Ycentroid	Zcentroid	Xradius	Yradius	Zradius
1	0	2.83E-05	100.0	512.5	518.3	50.8	282.9	280.7	28.3
2	5	2.77E-05	98.0	621.3	626.5	47.9	262.6	264.0	26.7
3	10	2.72E-05	96.1	672.9	673.0	46.3	232.3	241.4	25.7
4	15	2.66E-05	94.2	695.8	696.7	45.8	206.0	219.0	25.2
5	20	2.63E-05	92.9	709.9	711.8	45.4	187.4	194.7	24.9
6	25	2.61E-05	92.2	721.1	717.2	45.0	172.1	182.0	24.9
7	30	2.59E-05	91.5	727.3	720.2	44.7	160.2	170.8	24.9

```
plot command file: SLAB.TP2
Summary of Particle Tracking by Mass
```

snap	year	matrix	boundary	decay	trapped	pending	limbo
1	0	2.83E-05	0	0	0	0	0

2	5	2.77E-05	5.66E-07	0	0	0	0
3	10	2.72E-05	1.10E-06	0	0	0	0
4	15	2.66E-05	1.64E-06	0	0	0	0
5	20	2.63E-05	2.01E-06	0	0	0	0
6	25	2.61E-05	2.21E-06	0	0	0	0
7	30	2.59E-05	2.41E-06	0	0	0	0

```
Summary of Particle Tracking by Net Travel Distance
```

snap	year	%mass	dist	95%
1	0	100	1.77E-14	7.37E-14
2	5	97.998	196.354	231.939
3	10	96.0961	285.547	320.917
4	15	94.1942	326.068	356.601
5	20	92.8929	349.725	382.424
6	25	92.1922	365.715	407.887
7	30	91.4915	374.907	424.546

```
total elapsed time 3 seconds
3 notes + 0 warnings
```

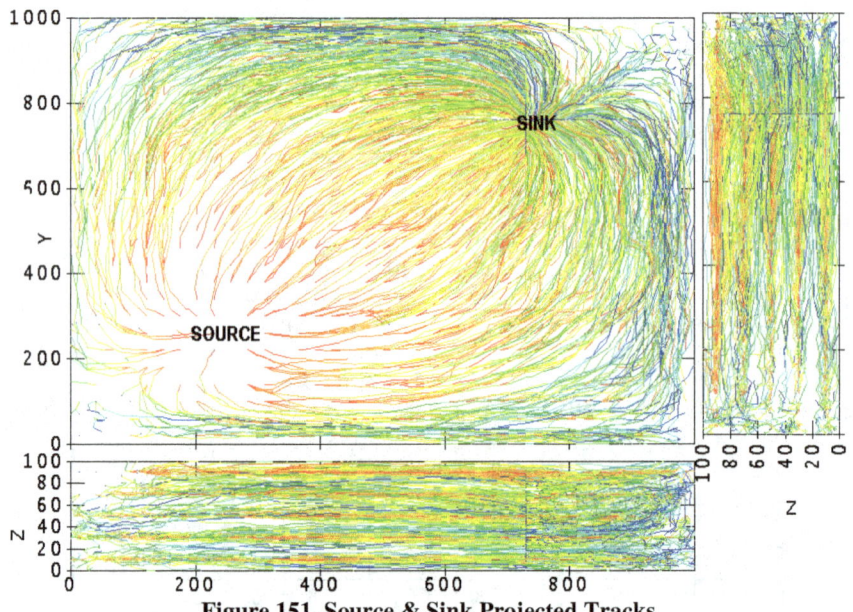

Figure 151. Source & Sink Projected Tracks

Chapter 23. Mosquito Tracking

Track mosquitoes? Not really... but sort of. One of my coworkers from the early days of PTRAX is Ken Black, a man of many talents. He volunteers with an international humanitarian organization that is concerned with one of the world's largest health crises: malaria in Zambia. This is a major concern of the World Health Organization. Even the Gates Foundation has funded various related projects, although not the one I'm about to describe.

Years ago Ken and I worked together on a project tracking runoff from surface water transporting contaminants. We began with GIS data, created 3D models, and modified several utilities (TP2 and PTRAX) to process the massive amounts of data. When the mosquito thing came up, Ken thought we might do something similar and so we did. The model is created on the fly from high-resolution topographic contours. It takes several hours to load and link three million elements and then track one million particles. The point is, where the particles end up are the most likely spots for stagnant water and mosquito breeding. Several of the areas identified by the model coincide with historical outbreaks of malaria. Here's an overview of the country:

Figure 152. Zambia Relief

This figure shows the gradient of the surface elevation (TP2 will generate this automatically from the surface):

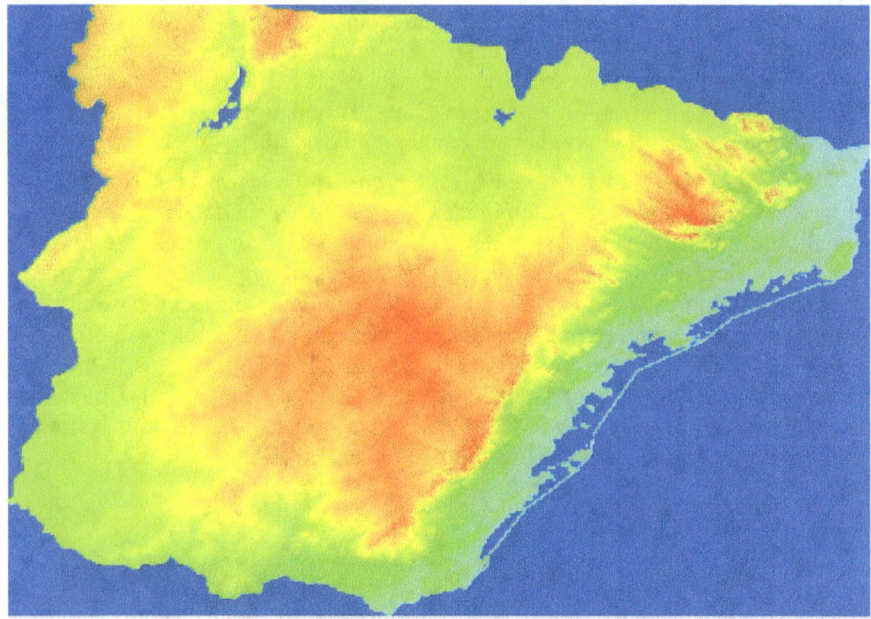

Figure 153. Zambia Elevations

This figure shows the 3D velocity vectors (TP2 also generates these from the gradients):

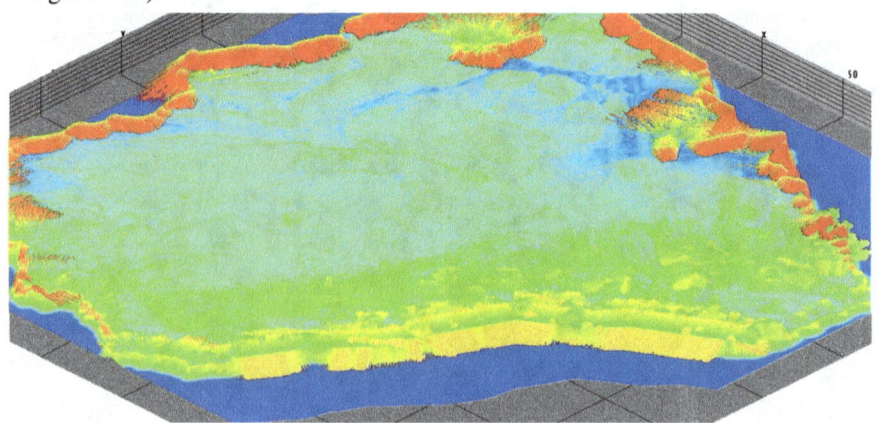

Figure 154. Surface Runoff Vectors

This next figure shows the particle tracks and stagnant water sites:

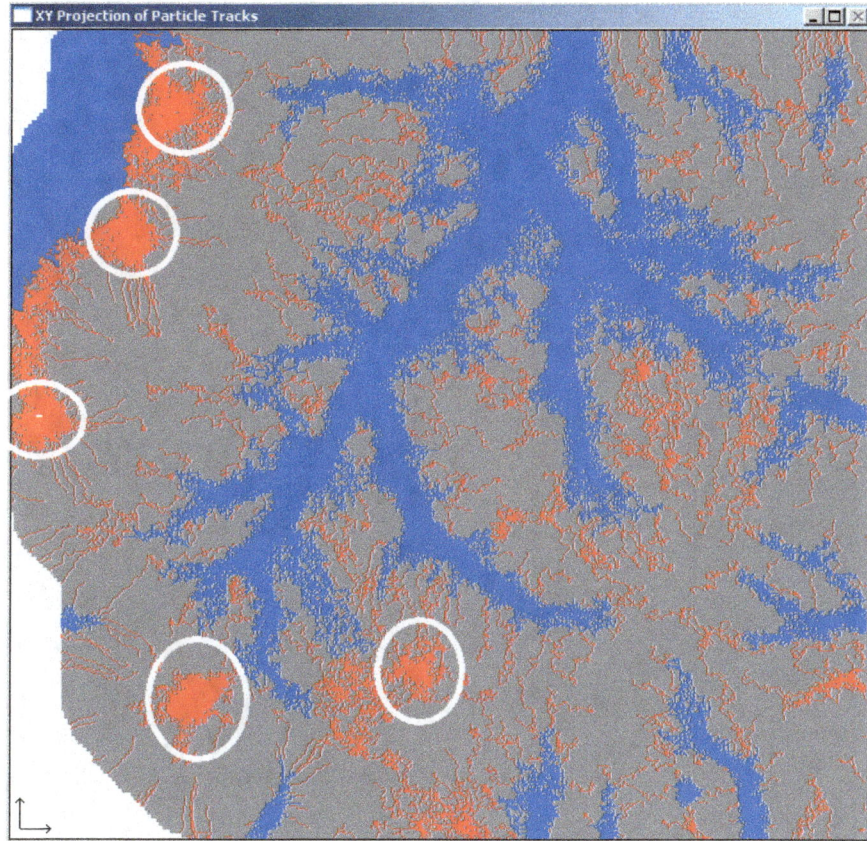

Figure 155. Particle Final Positions

Another view of the particle tracks:

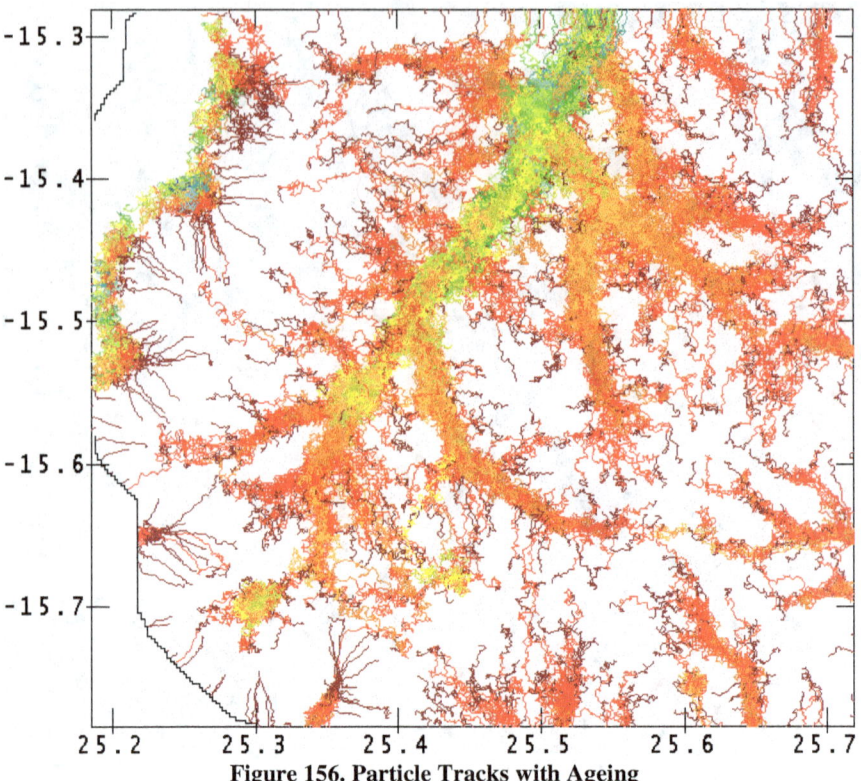

Figure 156. Particle Tracks with Ageing

Chapter 24. Tracking Particles Inside Pipes

The purpose of this model was to simulate the injection of fluorescent dye into the intake of three pumps. The three pipes converge into a single larger one. There was only one dye injection pump and only one sampling device so that all three pumps could not be tested simultaneously. By injecting the dye three times and measuring the differences, the relative performance of the three pumps could be compared. The basic arrangement is:

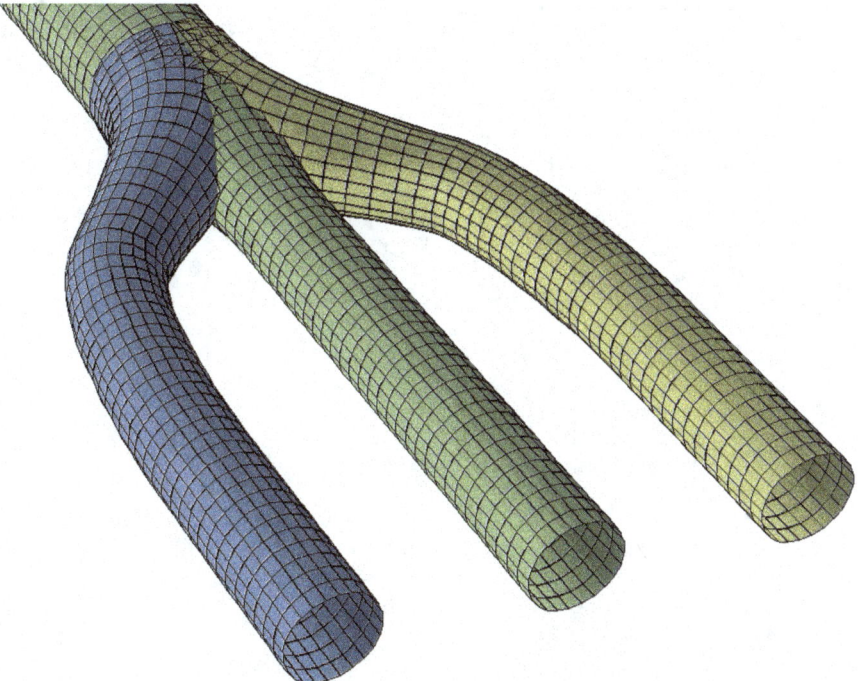

Figure 157. Three Merging Pipes

In order to get a significant statistical sample, 3,150,000 particles were found necessary, which took 14.0 minutes on a 3GHz Intel® processor.

```
PTRAX/V5.00avmf: Particle Tracker by Dudley J. Benton
compilation:   08/15/114:1734
animations:    enabled
fractures:     enabled
input format: Frac3D or ModFlow
available memory: 471518 KB
parameter 2: FOTO
parameter 3: WALL
parameter 4: OK
parameter 5: WAIT
application prefix: BLOCKS
file=BLOCKS.NDE    date=08/15/114:1824
file=BLOCKS.ELM    date=08/14/114:1909
```

```
file=BLOCKS.VEP    date=08/16/114:0104
file=BLOCKS.SED    date=09/10/119:2315
file=BLOCKS.WAL    date=08/14/114:2236
checking for existing output files
  none found
reading default parameters from file: BLOCKS.CFG
default parameters
  default seeds: 999
  track duration: 450 days (1.2 years)
  time step: 5 days (0.0 years)
  maximum steps along particle track: 999
  default porosity: 1
  default retardation factors: 1.1
  default dispersivities: 0,0,0 ft
  default matrix half-life: 1E+030 day(s)
  default fracture half-life: 1E+030 day(s)
  default velocities: 0,0,0 ft/day
  matrix diffusion coefficients: 0.0001,0.0001,0.0001
    ft²/day²
  synchronous time step default
  maximum times a particle can enter an element: 9
  maximum times a particle can enter a fracture: 9
  matrix tortuosity (0=none, 1=complete): 0
  Note: matrix tortuosity OFF
  fracture tortuosity (0=none, 1=complete): 0
  Note: fracture tortuosity OFF
reading FRAC3D files
Note: concentrations will NOT be divided by porosity
stray particles (outside domain) will be ignored
trapped particles will be ignored
include empty elements in snapshot files
create wall history file
FRAC3D node file: BLOCKS.NDE
  36 nodes
  -186X618
  06Y6320
  -56Z65
FRAC3D element file: BLOCKS.ELM
  8 elements
  element type: HEXAHEDRA
linking domain
  node:element links
  64 links
  element:element links
  14 internal faces
  34 external faces (boundaries)
  8 external elements
  36 external nodes
  computing element volumes
  Note: elements with clockwise orientation: 6
```

```
grouping elements
  largest element: 15x160x10
  2x2x2=8 groups
  group size: 36x320x10
  sorting groups
  indexing groups
  there are 1 active and 7 empty groups
  the smallest group is 1, containing 8 members
  the largest group is 1, containing 8 members
  the active groups contain an average of 8 members
FRAC3D velocity file: BLOCKS.VEP
  velocities: 8 (element-based)
using default properties
using default transport properties
characteristic parameters
  length threshold = 0.0322174 feet
  volume threshold = 1.152E-007 ft^3
  time threshold = 0.00644348 day(s)
  velocity threshold = 5E-006 ft/day
  mean velocity = 5 ft/day
  mean time to traverse element = 3.71769 day(s)
  synchronous time step = (automatic)
random walk
  dispersion OFF
  diffusion ON
wall file: BLOCKS.WAL
walls have the following dimensions
wall           area
  1            200
  2            200
wall:element links 2
seed file: BLOCKS.SED   3150000 seeds
  particles tracked: 3150000
  time to track particles: 13.9 minutes
  average particles tracked per minute: 226001
  average steps per particle: 51
  average particle track: 256.494
  average particle life: 176.015
  average particle speed: 1.45723
  total   particle movement: Sp=8.07957E+008
  random particle movement: Rx/Sp=-4.71776E-007
  random particle movement: Ry/Sp=-3.49909E-006
  random particle movement: Rz/Sp=3.60261E-007
  random time steps: 0.682491 ñ 0.022903 day(s)
  tracks ended at boundaries: 1194416
  tracks ended at capture walls: 1945092
  tracks ended due to circulation: 10492
Summary of Particle Tracking by Centroid of Mass
sorting wall histories
  wall capture file: BLOCKS.CAW
```

```
Summary of Particle Tracking by Mass
Summary of Particle Tracking by Net Travel Distance
snap   year  %mass         dist         ñ95%
available memory: 401242 KB
total elapsed time 14.0 minutes
```

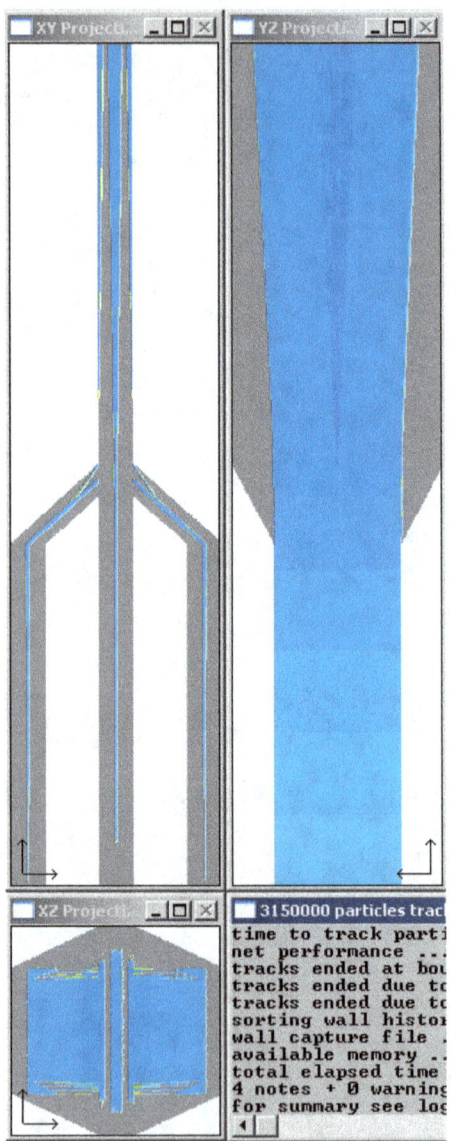

Figure 158. Particle Tracking

The analytical solution is:

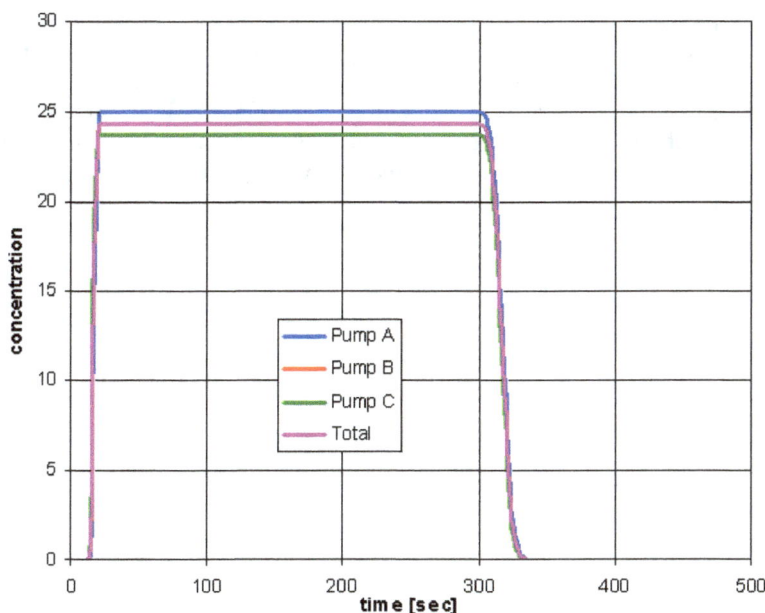

Figure 159. Results for Three Million Particles

The simulated solution based on tracking three million particles:

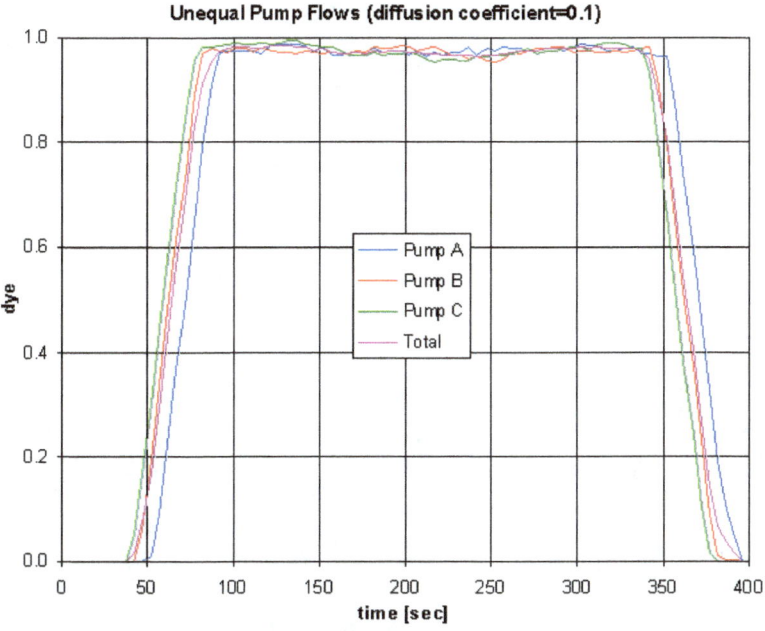

Figure 160. Calculated Dye Concentrations

The particle tracking solution is more realistic and closer to the measured results, including delayed rise on the left and the spread between the blue and green curves on the right side between 350 and 400 seconds. The analytical solution is amorphous (smeared), while the particles take a finite amount of time to disperse through the media. The velocities were scaled so that seconds became years, which was easier than modifying the code to accept the shorter time units.

Chapter 25. Simple Airborne Contaminant

The simplest plume I've ever been involved with modeling was ozone generated by some industrial processes, but mostly urban automotive exhaust around Birmingham, Alabama. The Bureau of Environmental Health had five sensors scattered around Jefferson and Shelby Counties, mostly along the corridor between I58 and I20. They provided some data and I developed software that would generate an animation of the estimated plume. A typical frame of this animation is shown below:

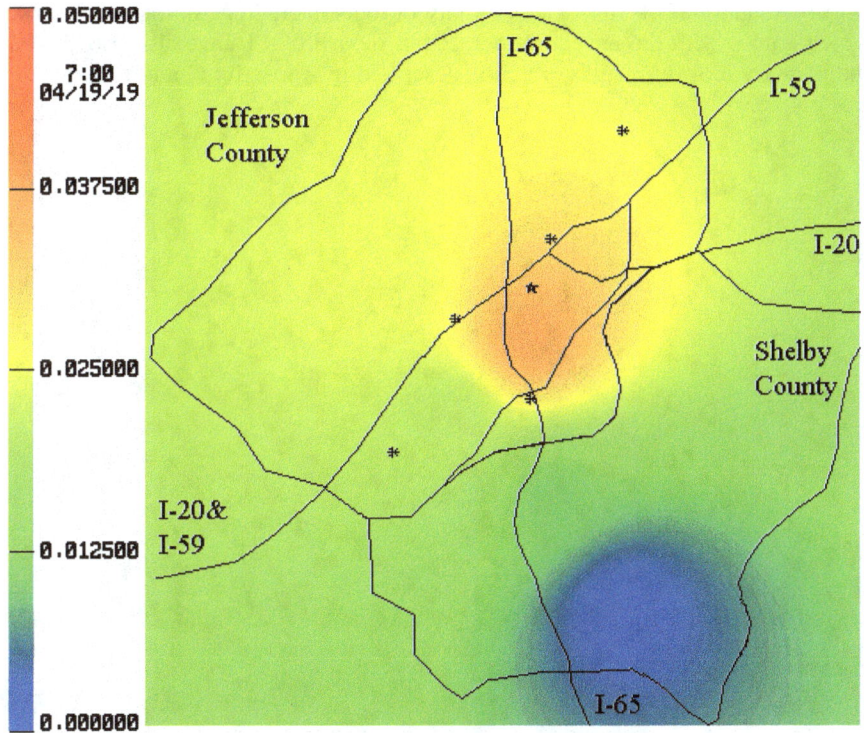

Figure 161. Airborne Concentrations

As this is very sparse data (only 5 sensors), there is no point applying elaborate processing. A simple inverse-distance weighting method for spatial interpolation of concentration is adequate. This can be expressed as:

$$C = \frac{\sum \dfrac{C_I}{R_I^{2.5}}}{\sum \dfrac{1}{R_I^{2.5}}} \tag{25.1}$$

A power of 2.5 on the distance is often applied. It is also necessary to control the concentrations in the far field (i.e., ~∞). This is easily done by adding a dozen phantom points at some distance outside the area of interest. See Appendix D for more on the inverse distance method. A more typical (and detailed) 2D contaminant plume is shown on page ii. Concentration in ppm is shown in the preceding figure, but log(C) is shown in the figure on page ii. Sometimes a log scale is more revealing, although when it comes to quantifying the contaminant, it is important to work with actual concentrations and not logs.

The code (ozone.c) produces a sequence of images (BMP files) representing the calculated spatial concentrations over time. These can be combined to form a single animation (GIF file) with a variety of tools, including Animation Shop®, which comes with the excellent tool, Paint Shop Pro®. I have also provided a rudimentary tool (bmp2gif), which is described in Appendix G and is available free online.

Chapter 26. Point Source Releases

The next level of detail we will consider are point source plumes. These are always approximate, as even a fissionable critical mass doesn't begin as an infinitesimal point. Point source plumes are the most common calculations for airborne contaminants. Interest in these calculations began in the early days of nuclear power when radioactive releases were feared imminent. When I worked at TVA I was part of a response team that would be called up in the event of a radioactive release. We had several such models ready, one for each nuclear plant and each release scenario.

We will begin by considering axisymmetric transient diffusion, as this is most often the assumption in these models. In one dimension this can be expressed by the following partial differential equation:

$$\frac{\partial C}{\partial t} = D \frac{\partial^2 C}{\partial x^2} \qquad (26.1)$$

For our purposes here, the most logical boundary conditions are an initial contamination zone (*a*) and concentration (*C₀*) or total mass (*m₀*). The analytical solution in terms of the complementary error function is:

$$C(x,t) = \frac{C_0}{2}\left[erfc\left(\frac{x-a}{\sqrt{4Dt}}\right) - erfc\left(\frac{x+a}{\sqrt{4Dt}}\right) \right] \qquad (26.2)$$

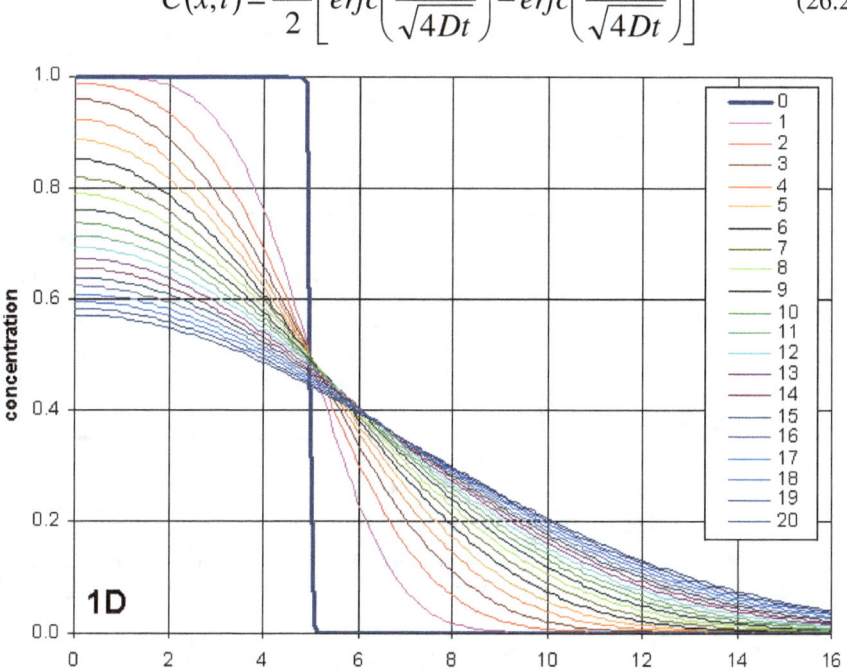

Figure 162. Diffusion in 1D for Semi-Infinite Source

This solution is illustrated for several values of distance and time in the preceding figure. The area under each of the curves is the same so that the total mass is constant over time as it spreads in one direction.

$$m_0 = \int_0^\infty \frac{C_0}{2}\left[erfc\left(\frac{x-a}{\sqrt{4Dt}}\right) - erfc\left(\frac{x+a}{\sqrt{4Dt}}\right)\right]dx = C_0 a \quad (26.3)$$

We can easily add a constant horizontal velocity (*v*) to this solution by simply substituting *x=x-vt/a*, as in:

$$C(x,t) = \frac{C_0}{2}\left[erfc\left(\frac{x-\frac{vt}{a}-a}{\sqrt{4Dt}}\right) - erfc\left(\frac{x-\frac{vt}{a}+a}{\sqrt{4Dt}}\right)\right] \quad (26.4)$$

The resulting profiles move to the right, as illustrated below:

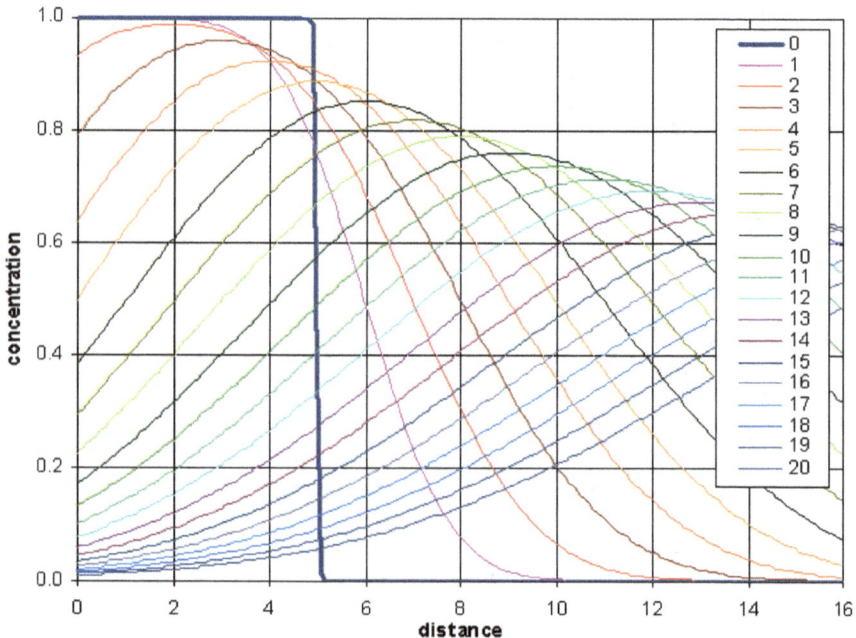

Figure 163. Diffusion in 1D for Finite Source

While this is interesting, no contaminant spreads in only one direction unless it's contained inside a pipe. We next consider two-dimensional cylindrical coordinates. Equation 26.1 becomes:

$$\frac{\partial C}{\partial t} = \frac{D}{r}\frac{\partial(rC)}{\partial r^2} \quad (26.5)$$

The analytical solution to Equation 26.5 is an infinite series of Bessel functions, which is impractical for point source plume modeling, especially when such calculations were first undertaken in the late 1970s and early 1980s. We can obtain an approximate solution with realistic behavior by substituting $x=r^2/a$ into Equations 26.2 or 26.4. The resulting profiles are shown below:

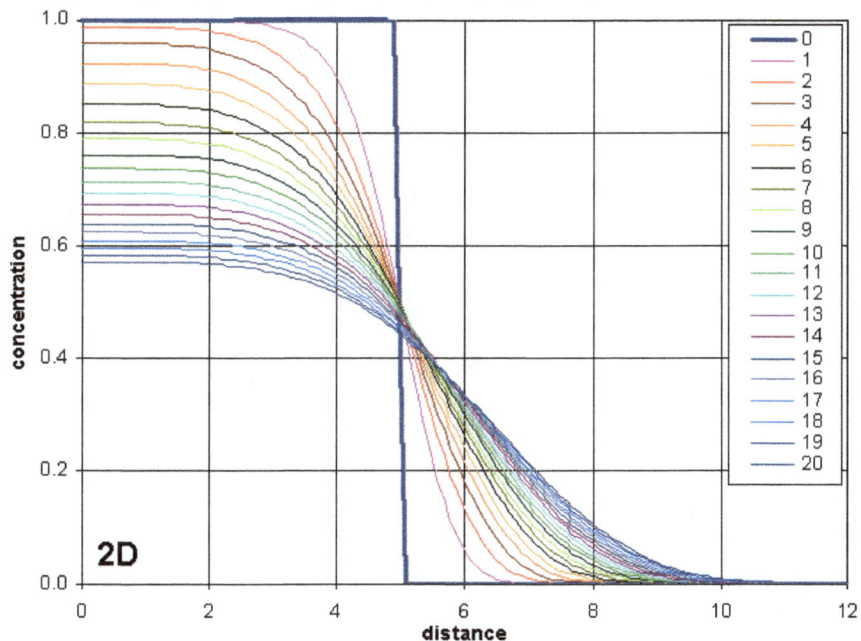

Figure 164. Diffusion in 2D for Semi-Infinite Source

The conservation of mass in this case is given by:

$$m_0 = \int_0^\infty 2\pi C(t,r) r\, dr = \pi a^2 C_0 \qquad (26.6)$$

With this approximation, spreading increases with time, that is, the radius increases and the area under the curve remains constant. The maximum concentration, which is at the center in the absence of transverse velocity, also decreases with time, though much more slowly. The extent of the plume (or the leading edge where the concentration has dropped by three or four orders of magnitude) can be calculated using a bisection search. The VBA code is in the spreadsheet and also listed below:

```
Function radius(t As Double, c As Double, D As Double, a
    As Double) As Double
    Dim iter As Integer, c0 As Double, r1 As Double, r2 As
    Double
    c0 = conc(radius, t, c, D, 0, a)
    r1 = a
    r2 = a
```

```
    While (conc(r2, t, c, D, 0, a) > c0 / 1000)
      r2 = r2 * 2
    Wend
    For iter = 1 To 32
      radius = (r1 + r2) / 2
      If (conc(radius, t, c, D, 0, a) > c0 / 1000) Then
        r1 = radius
      Else
        r2 = radius
      End If
    Next iter
End Function
```

The results are shown in this next figure:

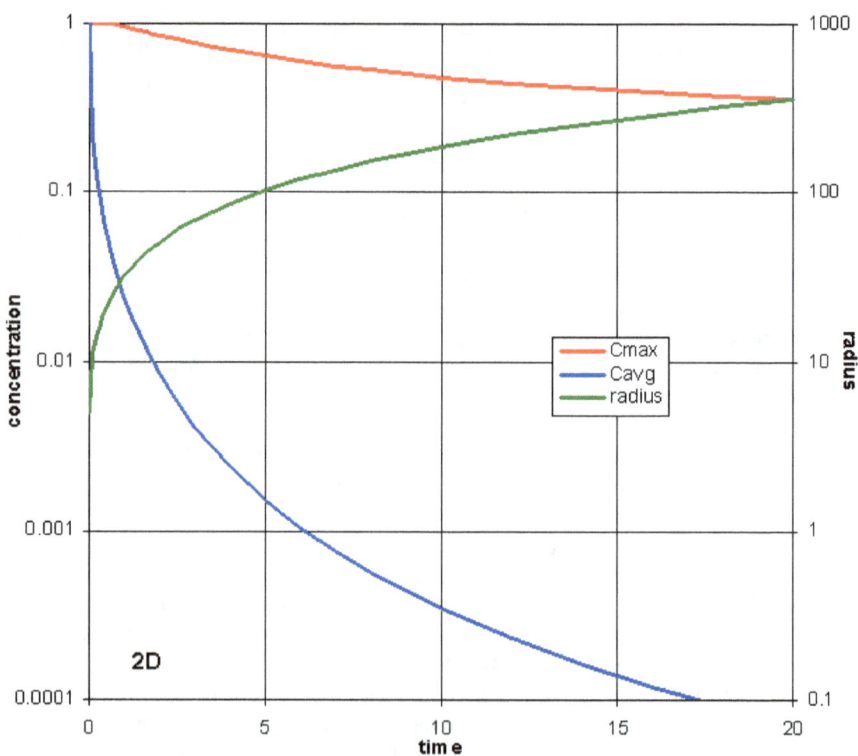

Figure 165. Concentration Parameters

In three-dimensional spherical coordinates, this becomes:

$$\frac{\partial C}{\partial t} = \frac{D}{r^2}\frac{\partial(r^2 C)}{\partial r} \tag{26.7}$$

The spherical solution is also an infinite series and impractical for our purposes. Fortunately, substituting x=r³/a² into Equations 26.2 or 26.4 works again. The conservation of mass in 3D is:

$$m_0 = \int_0^\infty 4\pi C(t,r) r^2 \, dr = \frac{4\pi a^3}{3} C_0 \qquad (26.8)$$

The resulting profiles are shown in this next figure. All of the calculations can be found in the online archive in folder examples\point in spreadsheet point_source.xls, including the conservation of mass, which is along the bottom (i.e., row 163).

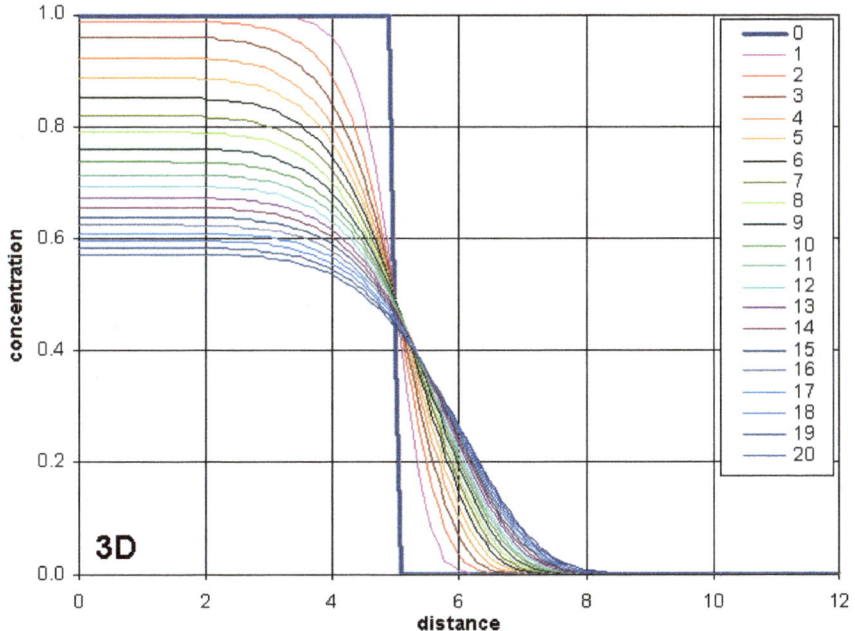

Figure 166. Diffusion in 3D for Finite Source

We can now use the modified Equation 26.4 to approximate an expanding point source 2D or 3D plume subject to a continuous cross wind, which is the first-order estimate of airborne contaminant transport.

<u>Example 1</u>

Consider an airborne plume having initial concentration of 1000 ppm and initial radius of 100 m. The diffusivity coefficient might be approximately 0.2 m²/s and the wind speed 0.8 m/s. This is a gentle breeze, not a howling wind. As a rough estimate, the plume spread might be mostly radial, that is, cylindrical instead of spherical. If this were the case, the average concentration would drop with the square of the radius. The peak concentration would also drop, but slower, as indicated in the preceding figure marked 2D.

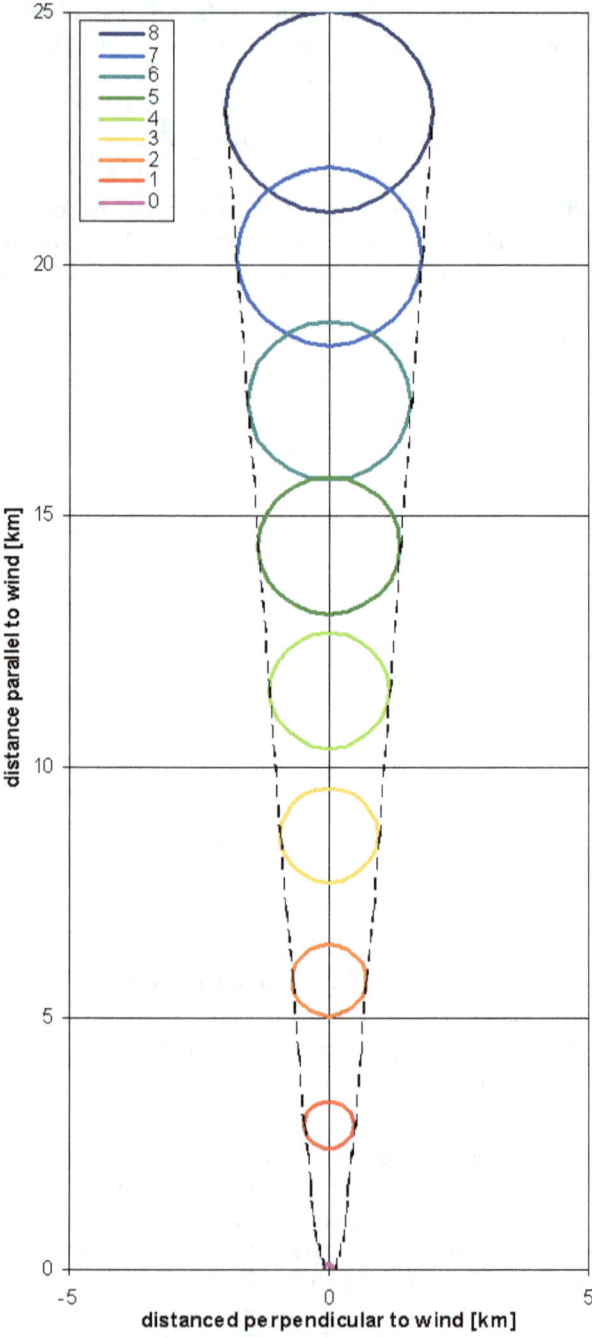
Figure 167. Plume Snapshots over Time

In this simplified calculation, the shift in center of the plume varies linearly with time (i.e., *x=vt*). The radial extent of the plume could be found in several ways, the easiest being a bisection search for the radius where the concentration has dropped of by a factor of 1000 or 0.1% of the maximum. It is easy enough to calculate the center and radius over time in a spreadsheet and produce a plot of the spreading, moving plume. Of course, this is an ideal representation and a more realistic approach would consider turbulent mixing, at least empirically.

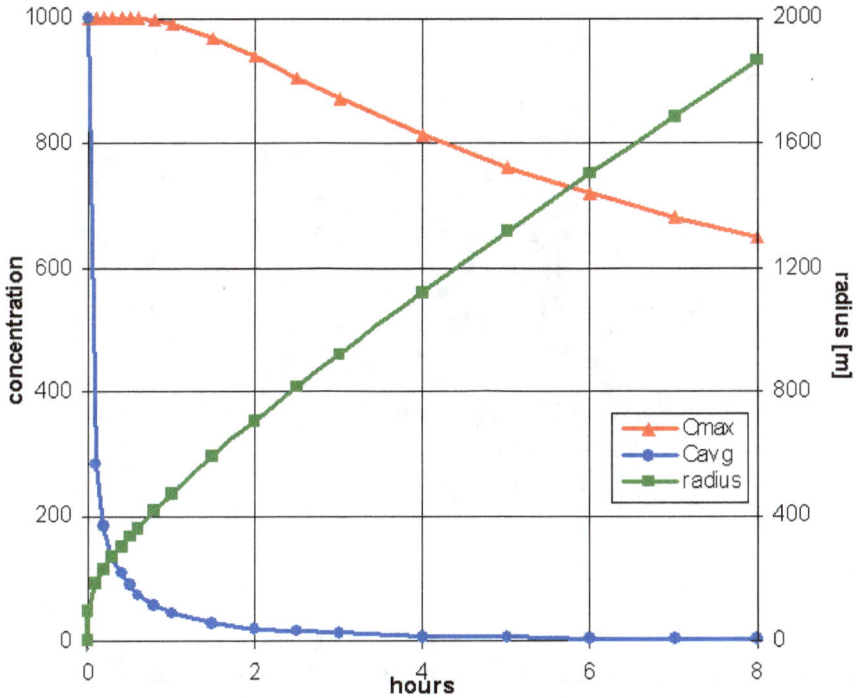

Figure 168. Plume Parameters over Time

Continuous Releases

If you search the Internet for atmospheric plume modeling, you will find many articles that seem to be related to the topic of finite point source releases, but are not. The vast majority of those articles are about continuous (i.e., semi-infinite) releases. You will not find time as a variable in any of the equations presented therein. You may find a statement that those efforts are related to steady-state conditions. We will cover steady-state plumes later in the text. For now, we will only consider transient plumes.

Chapter 27. Dispersion

Diffusion and diffusivity generally describe a slow and continuous process, while dispersion and dispersivity often describe a more chaotic or turbulent process. The length scales for these two processes may be quite different. While diffusion arises from microscopic, even molecular processes, dispersion arises from macroscopic processes. Diffusion might be illustrated by the following:

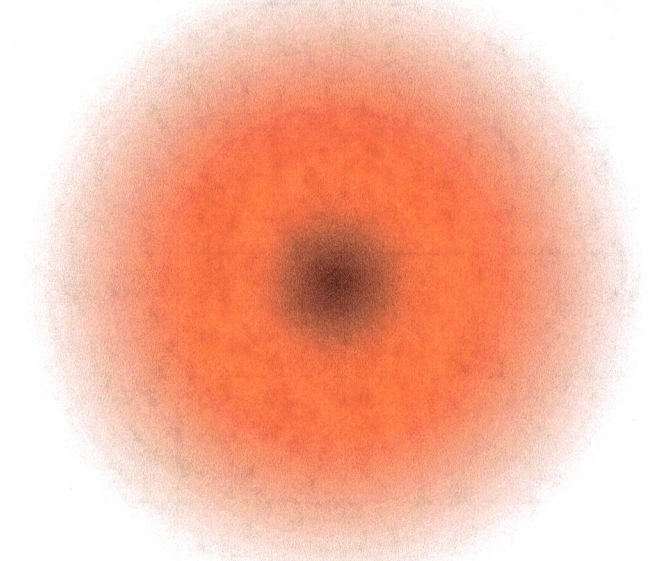

Figure 169. Typical Dispersive Field

which might be approximated by the familiar bell-shaped (Gaussian) curve:

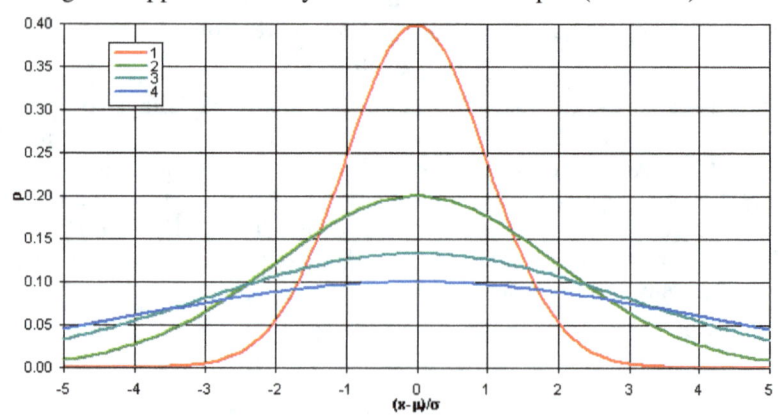

Figure 170. Gaussian Concentrations

Dispersion is better illustrated by this figure, which defies analytical description:

Figure 171. Typical Dye Plume

The only way to generate concentrations anything like the preceding figure is with computer modeling: finite differences, finite elements, finite volume, or particle tracking. By far, the last of these is the most efficient and effective. We will not cover the specifics of the particle tracking model here, as this is thoroughly described in my book, *Particle Tracking*. Results for a groundwater contaminant plume much like the ones presented in the preceding chapter are shown in the following figure. The upper region is after 15 years and the lower one is after 30 years. Agreement with approximate analytical solutions was quite good, which was an important part of the validation process.[15,16]

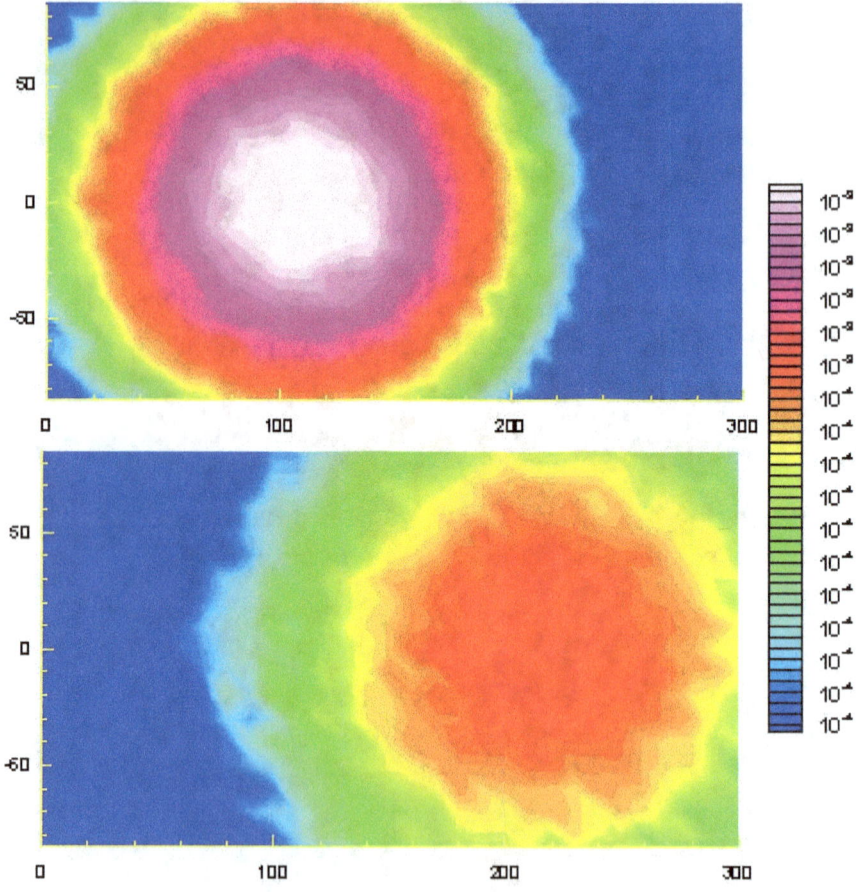

Figure 172. Analytical Model Results

Analytical models just use a larger diffusion coefficient to approximate this behavior, but will always produce a smooth result unlike the one shown above. We will see much more realistic representations of dispersion in Chapter 32 with particle tracking.

---

[15] http://dudleybenton.altervista.org/pub/PTRAX1.pdf
[16] http://dudleybenton.altervista.org/pub/PTRAX2.pdf

Chapter 28. Slot Jets

The next problem we consider (a thermal plume discharged into a flowing river) is more complex because there are several equations to solve in parallel, including: conservation of mass, momentum, and energy, as well as width and position. There is also the conservation of salt if this is a brine plume or the receiving water is salty. The plume can be positively, negatively, or neutrally buoyant with respect to the ambient, and composed of heated or cooled fresh or salty water. The ambient can be stratified or uniform in temperature, flowing, stagnant, fresh or salty. The model is described in detail in my TWRA paper.[17]

At large power plants, thermal and wastewater effluents are frequently discharged into a receiving ambient through a diffuser. The primary purpose of this is to dilute the effluent. The most common diffuser shapes are conceptually similar to a solitary round jet or a linear cluster of jets, which ideally approaches a slot. The discharge is most often hotter or equal to the ambient in temperature. The discharge and the ambient can contain varying amounts of salt. The discharge can either rise, fall, spread, or some combination of these as it mixes with the ambient. The trajectory as well as the final dilution is of interest to the engineer. Analytical and empirical models are often adequate, even for discharge into a flowing ambient, but these can't effectively account for differences in temperature and salinity or variations in the ambient, including velocity, temperature, and salinity. In such cases a computer model is the best approach. A realistic computer model must be based on first principles, analysis, and observations.

Analytical And Empirical Models

Analytical and empirical models are briefly presented before the numerical model for three reasons: 1) the analytical and empirical models provide a check on the numerical model, especially the asymptotic behavior, 2) laboratory and field data are most often presented in the literature in conjunction with analytical or empirical models and this introduces the data for eventual comparison with the numerical model, and 3) the simplifying assumptions for the analytical and numerical models are basically the same. It will be shown that the analytical models have compared favorably with laboratory and field data, suggesting that the simplifying assumptions are not unreasonable. The numerical model handles slot and round jets; therefore, the analytical models for both geometries are included. Most laboratory and field data are either for a slot or a round jet; so including both geometries increases the data available for comparison.

Analytical models are limited to simple geometries and boundary conditions for which analytical solutions are possible. Empirical models are based on analytical models, dimensionless numbers, and data correlation. As with all models, some simplifications are necessary, in order to make the solution

---

[17] Benton, D. J., "Development of a Two-Dimensional Plume Model for Positively and Negatively Buoyant Discharges into a Stratified Flowing Ambient," *Tennessee Water Resources Symposium*, 1989. http://dudleybenton.altervista.org/pub/Plume2D.pdf

tractable, including: flow regime, geometry, and dominant mechanisms. Several factors are used to classify plumes, including: laminar or turbulent, round or slot, and buoyant or neutral. Simplifications are most often related to these classifications. The two limiting cases presented in analytical plume models are momentum-dominated and buoyancy-dominated.

Laminar Jet

Perhaps the most simple analytical model for a momentum-dominated plume was presented by Schlichting[18], clarified by Bickley[19], and described by White.[20] Schlichting derived an analytical solution for a neutrally-buoyant laminar slot jet discharging into a semi-infinite static medium. The dilution, S, is given by Equation 28.1:

$$S = 1 + \left(36Re\frac{h}{b}\right)^{1/3} \qquad (28.1)$$

where h is the depth [m], b is the slot width [m], and Re is the slot Reynolds number, Equation 28.2:

$$Re = \frac{wb}{v} \qquad (28.2)$$

where w is the velocity at the slot [m/s] and v is the kinematic viscosity [m²/s]. Note that the dilution depends on the Reynolds number and increases with the one-third power of depth. Even this simple analytical model has been shown to be reasonably accurate for Reynolds numbers up to 30, Andrade.[21] Although laminar flow and such small Reynolds numbers are not of direct interest, Schlichting's solution is important; because it provided the basis for solving a turbulent jet.

Squire[22] used Schlichting's laminar slot jet solution to develop a laminar round jet solution, Equation 28.3:

$$S = 1 + \frac{32}{Re}\left(\frac{h}{d}\right) \qquad (28.3)$$

where d is the initial diameter of the jet [m] and Re is the jet Reynolds number, Equation 28.4:

---

[18] Schlichting, H., Zeitschrift für Angewandte Mathematik und Mechanik (Journal of Applied Mathematics and Mechanics), Vol. 13, pp. 260 263, 1933.
[19] Bickley, W. G., Philosophical Magazine, Vol. 23, pp. 727 731, 1937.
[20] White, F. M., Viscous Fluid Flow, McGraw-Hill, New York, pp. 290 292, 350 351, 1974.
[21] Andrade, E. N., Proceedings of the Physical Society of London, Vol. 51, pp. 784 793, 1939.
[22] Squire, H. B., Quarterly Journal of Mechanics, Vol. 4, pp. 321 329, 1951.

$$Re = \frac{wd}{v} \quad (28.4)$$

Note that the dilution depends on the Reynolds number and increases linearly with depth.

Turbulent Jet

Prandtl[23] used his mixing length concept to modify Schlichting's laminar theory to account for turbulence. Prandtl's assumptions were later substantiated by Reichardt[24] and later Görtler[25] showed that the resulting equations could be solved by Schlichting's method, to derive Equation 28.5:

$$S = 1 + 0.48 \left(\frac{h}{b}\right)^{\frac{1}{2}} \quad (28.5)$$

Note that the dilution does not depend on the Reynolds number and increases with the square-root of the depth. Görtler also developed a relationship for a round jet, Equation 28.6:

$$S = 1 + 0.424 \left(\frac{h}{d}\right) \quad (28.6)$$

Note that the dilution does not depend on the Reynolds number and increases linearly with the depth.

Turbulent Plume

The analytical solution for a buoyancy-dominated plume discharged vertically upward from a slot at negligible velocity into a semi-infinite medium was developed by Rouse et al.[26], applied by Cederwall[27], and can be expressed by Equation 28.7:

---

[23] Prandtl, L, Proceedings of the Second International Congress on Applied Mechanics, Zurich, pp. 62 75, 1926.

[24] Reichardt, H., "Gesetzmässigkeiten der freien Turbulenz (Regularities of Free Turbulence)," Verein Deutscher Ingenieure (Association of German Engineers), Forschungs p. 414 (Research Paper No. 414), 1942.

[25] Görtler, H., "Berechnung von Aufgaben der freien Turbulenz auf Grund eines neuen Näherungsansatzes (The Task of Computating Free Turbulence Due to the Proximity of a Beginning)," Zeitschrift für Angewandte Mathematik und Mechanik (Journal of Applied Mathematics and Mechanics), Vol. 22, No. 5, pp. 244-254, 1942.

[26] Rouse, H., C. S. Yih, and H. W. Humphreys, "Gravitational Convection from a Boundary Source," Tellus, Vol. 14, 1952.

[27] Cederwall, K., "Gross Parameter Solutions of Jets and Plumes," ASCE Journal of the Hydraulics Division, Vol. 101, No. HY5, 1975.

$$S = 1 + \frac{0.59}{F_D^{\frac{2}{3}}}\left(\frac{h}{b}\right) \qquad (28.7)$$

where F_D is the slot densimetric Froude number, Equation 28.8:

$$F_D = \frac{w}{\sqrt{gb\left(\frac{\rho_A - \rho_P}{\rho_A}\right)}} \qquad (28.8)$$

where g is the acceleration of gravity [m/s²], ρ_A is the ambient density [kg/m³], and ρ_P is the initial plume density [kg/m³]. Note that the dilution depends on the Froude number, not the Reynolds number, and increases linearly with depth. Fischer et al.[28] develop a similar relation for a round plume, Equation 28.9:

$$S = 1 + \frac{0.117}{F_D^{\frac{2}{3}}}\left(\frac{h}{d}\right)^{\frac{5}{3}} \qquad (28.9)$$

where F_D is the jet densimetric Froude number, Equation 28.10:

$$F_D = \frac{w}{\sqrt{gd\left(\frac{\rho_A - \rho_P}{\rho_A}\right)}} \qquad (28.10)$$

Note that the dilution depends on the Froude number, not the Reynolds number, and increases with the five-thirds power of the depth.

Buoyant Jet

Actual discharges are typically somewhere between a momentum-dominated jet and a buoyancy-dominated plume. Fischer et al. provide a comparison between measured data and the analytical expressions for a round turbulent jet (Equation 28.6) and a round turbulent plume (Equation 28.9). This transformation of variables allows jet and plume data to be plotted on the same curve as shown in the following figure:

---

[28] Fischer, H. B., E. J. List, J. Imberger, and N. H. Brooks, Mixing in Inland and Coastal Waters, Academic Press, San Diego, California, 1979.

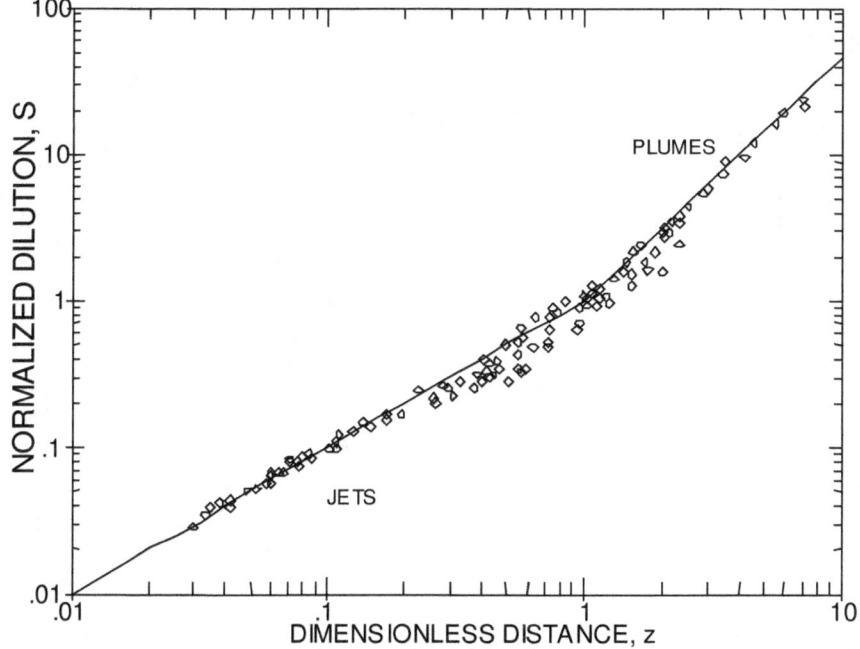

Figure 173. Plume Regimes

The rearranged analytical model is given in Equation 28.11.

$$\varsigma = \begin{cases} \delta & , \quad \delta \leq 1 \\ \delta^{\frac{5}{3}} & , \quad \delta > 1 \end{cases} \quad (28.11)$$

where δ is the dimensionless depth, Equation 28.12:

$$\delta = \frac{0.484}{F_D}\left(\frac{h}{d}\right) \quad (28.12)$$

and ς is the normalized dilution, Equation 28.13:

$$\varsigma = \frac{1.69}{F_D}(S-1) \quad (28.13)$$

<u>Inclined Jet</u>

The initial angle of inclination of a discharge has a pronounced impact on the trajectory of a plume and a lesser impact on the dilution. The impact of inclination angle has received much less attention in the literature than the other aspects presented here. One of the few publications containing laboratory data for a jet with variable inclination is that of Albertson et al.[29] and is reproduced in the next figure.

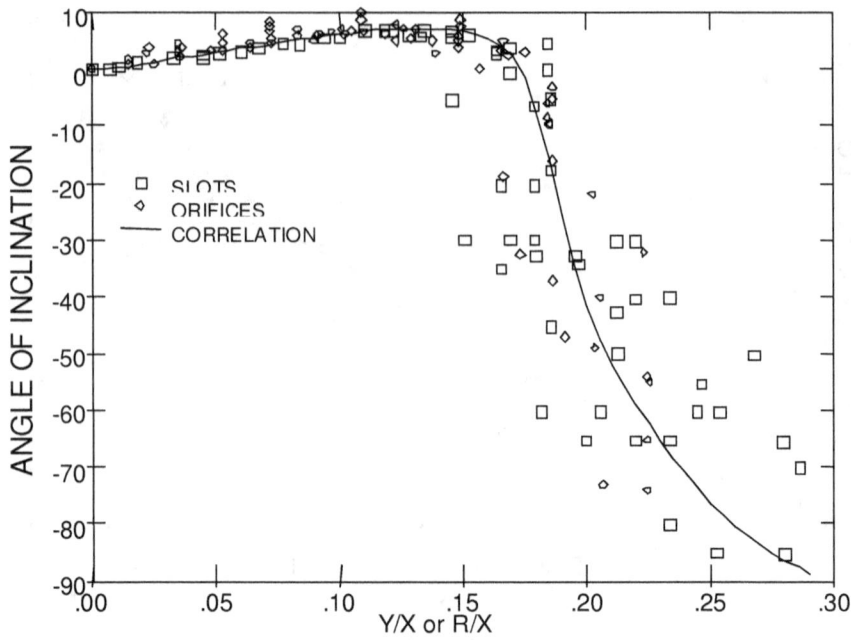

Figure 174. Impact of Inclination

As expected, the greatest impact on the trajectory occurs when directing a buoyant discharge downward (-90°). The above curve can be calculated by Equation 28.14:

$$\alpha = \frac{-15.3*\beta \; (5.78*\beta-1) \; (231*\beta+1) \; (117*\beta^2-13.1*\beta+1) \; (32.2*\beta^2-11.1*\beta+1) \; (8.94*\beta^2-5.9*\beta+1)}{(28.5*\beta^2-10.6*\beta+1)} \tag{28.14}$$

---

[29] Albertson, M. L., Y. B. Dai, R. A. Jensen, and H. Rouse, "Diffusion of Submerged Jets," Transactions of the American Society of Civil Engineers, Paper No. 2409, 1948.

where $\beta=y/x$ for slot jets and $\beta=r/x$ for round jets. Equation 28.14 is implicit in α and explicit in β as more than one angle of inclination can result in the same impact ratio. Even though Equation 28.14 is a high-order curve-fit it has been arranged so as to have only two real roots and stable behavior even with limited precision calculations.

Impact of Crossflow

Discharge into a flowing ambient is an important aspect of most jets and plumes. The simplest and most analyzed type of ambient is a uniform crossflow. Purely analytical solutions to this problem have remained illusive; but empirical solutions have been developed. Discharges may behave like a jet in the near field and like a plume in the far field, reacting differently to a crossflow throughout its trajectory. In some cases, for example Wright[30], List[31], and Wood et al.[32], four or more separate empirical relationship are developed, one for each regime: near field momentum-dominated, near field buoyancy-dominated, far field momentum-dominated, far field buoyancy-dominated, etc.

Perhaps the simplest general approach for a momentum-dominated *slot* discharge into a uniform crossflow is given by Adams[33], which can be expressed as Equation 28.15:

$$S = \frac{1}{2}\left(\frac{Q_R}{Q_D} + \left[\left(\frac{Q_R}{Q_D}\right)^2 + 2\frac{h}{b}\right]^{\frac{1}{2}}\right) \quad (28.15)$$

where h is the ambient depth [m], Q_D is the discharge flow [m³/s], and Q_R is the ambient (or river) flow [m³/s] over the diffuser. A comparison between Equation 28.15 and data can be seen in this next figure, which shows laboratory data from McIntosh et al.[34] and Seo et al.[35] as well as field data from Harleman et al.[36], Almquist et al.[37], and McIntosh et al.[38].

---

[30] Wright, S., "Mean Behavior of Buoyant Jets in a Crossflow," ASCE Journal of Hydraulics, Vol. 103, pp. 499-513, 1977.

[31] List, E. J., "Turbulent Jets and Plumes," Annual Review of Fluid Mechanics, Vol. 14, pp. 189-212, 1982.

[32] Wood, I., R. Bell, and D. Wilkinson, "Ocean Disposal of Wastewater," Advanced Series on Ocean Engineering, Vol. 8, World Scientific Publishers, Singapore, 1993.

[33] Adams, E. E., "Submerged Multiport Diffusers in Shallow Water with Current," Master's Thesis, Ralph M. Parsons Laboratory for Water Resources and Hydrodynamics, Department of Civil Engineering, MIT, Cambridge, Massachusetts, 1972.

[34] McIntosh, D. A., B. E. Johnson, and E. B. Speaks, "Validation of Computerized Thermal Compliance and Plume Development at Sequoyah Nuclear Plant," TVA Report WR28 1 45 115, 1983.

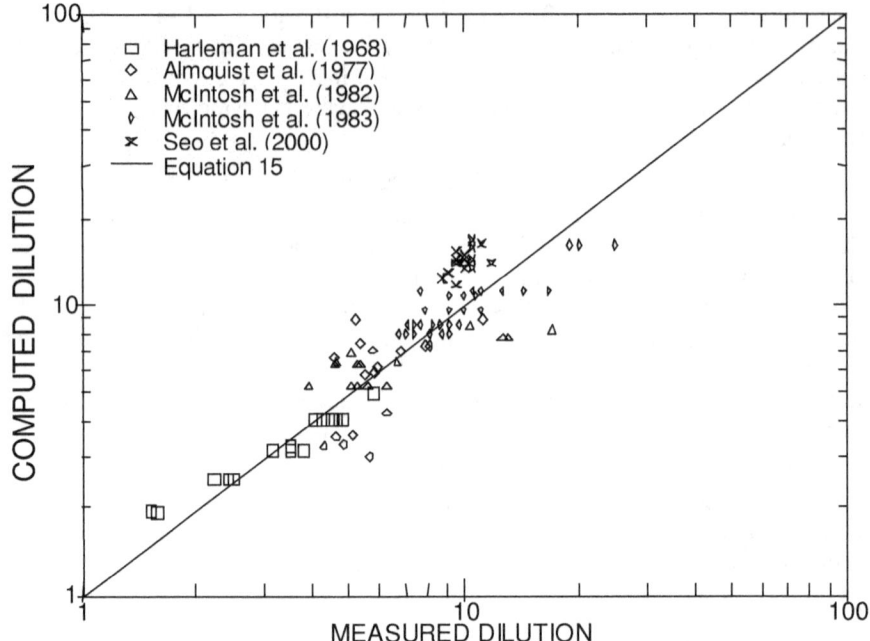

Harleman et al. suggest a simple formula for the dilution of buoyant *slot* plumes in a crossflow:

$$Q_F = 0.56 \left[g D^3 L^2 Q_D \left(\frac{\rho_A - \rho_D}{\rho_A} \right) \right]^{\frac{1}{3}} \qquad (28.16a)$$

---

[35] Seo, I. W., D. S. Kim, and H. S. Kim, "The Mechanics of Tee Diffuser for Thermal Discharges in Crossflow," 4th International Conference on Hydro-Science and - Engineering, 2000.

[36] Harleman, D. R. F., L. C. Hall, and T. G. Curtis, "Thermal Diffusion of Condenser Water in a River During Steady and Unsteady Flows," Hydrodynamics Laboratory Report No. 111, Massachusetts Institute of Technology, Cambridge, Massachusetts, 1968.

[37] Almquist, C. W., C. D. Ungate, and W. R. Waldrop, "Field Model Results for Multiport Diffuser Plume," Proceedings of the ASCE Conference on Verification of Mathematical and Physical Models in Hydraulic Engineering, College Park, Maryland, 1978.

[38] McIntosh, D. A., B. E. Johnson, and E. B. Speaks, "A Field Verification of Sequoyah Nuclear Plant Diffuser Performance Model: One-Unit Operation," TVA Report WR28 1 45 110, 1982.

$$Q_M = \frac{1}{2}\left(Q_R + \sqrt{Q_R^2 + \frac{5D\,Q_D}{3L}}\right) \qquad (28.16b)$$

$$if\ (Q_R \geq Q_F)\ S = \frac{Q_F}{Q_D} \qquad (28.16c)$$

$$if\ (Q_R < Q_F)\ S = S = \frac{Q_M}{Q_D} \qquad (28.16d)$$

Equation 28.16 is introduced here because it will subsequently be compared to laboratory and field data. Stolzenbach[39] suggests a slightly more complex formula for the dilution of buoyant *slot* plumes in a crossflow:

$$Q_H = \frac{1}{2}\left(Q_R + \sqrt{Q_R^2 + \frac{10L}{D}Q_D^2}\right) \qquad (28.17a)$$

$$if\ (Q_R \geq Q_H)\ S = \frac{Q_H}{Q_D} \qquad (28.17b)$$

$$Q_F = D\left[L^2 Q_D (T_D - T_A)\frac{d\rho}{dT}\right]^{\frac{1}{3}} \qquad (28.17c)$$

$$if\ (Q_R \geq Q_F)\ S = \frac{Q_R}{Q_D} \qquad (28.17d)$$

$$if\ (Q_R < Q_F)\ S = \frac{Q_F}{Q_D} \qquad (28.17d)$$

Equation 28.17 is introduced here because it will subsequently be compared to laboratory and field data. Almquist suggests a set of formulae for the dilution of buoyant *slot* plumes in a crossflow that considers twelve different regimes based on the relative importance of buoyancy vs. momentum and positive, negative, and zero ambient flow:

$$S_{FBDZ} = 1 + \frac{C_B b_0^{\frac{1}{3}} h}{Q_Z} \qquad (28.18a)$$

---

[39] Stolzenbach, K. D., "Analytical and Experimental Studies of Discharge Designs for the Cayuga Station at the Somerset Alternate Site," Ralph M. Parsons Laboratory for Water Resources and Hydrodynamics Report No. 211, 1976.

$$\sigma = 1 + C_M \left(\frac{vh}{Q_Z} + \sqrt{\frac{v_2 h^2}{Q_Z^2} + \frac{2M_0 h}{Q_Z^2}} \right) \qquad (28.18b)$$

$$Q_B = (\sigma - 1)Q_Z L \qquad (28.18c)$$

$$Q_X = \frac{Q_R h_L}{h_1} \qquad (28.18d)$$

$$S_{FMDP} = 1 + \frac{Q_X}{Q_Z L} \qquad (28.18e)$$

$$S_{FMDE} = 1 + C_M \sqrt{\frac{2M_0 h}{Q_Z}} \qquad (28.18f)$$

$$S_{TRNZ} = 1 + \frac{C_T b_0^{\frac{1}{6}} M_0^{\frac{1}{4}} h^{\frac{3}{4}}}{Q_Z} \qquad (28.18g)$$

$$S_{TRNP} = S_{TRNZ} + (S_{FMDP} - S_{TRNZ})v_B \qquad (28.18g)$$

$$Q_Z = \frac{Q_D}{L} \qquad (28.18h)$$

$$M_O = \frac{Q_Z^2}{0.3709} \qquad (28.18i)$$

$$b_0 = \frac{g(\rho_A - \rho_D)}{\rho_A} Q_Z \qquad (28.18j)$$

$$h_L = \frac{D b_0^{\frac{2}{3}}}{M_0} \qquad (28.18k)$$

$$v = \frac{Q_R}{LD} \qquad (28.18l)$$

$$v_B = \frac{3v}{2} b_0^{\frac{2}{3}} \qquad (28.18m)$$

$$m = \frac{v^2 h}{M_0} \qquad (28.18n)$$

$$b = g\rho Q_Z \qquad (28.18p)$$

where subscript B indicates buoyancy dominated, h is the depth, h_L is the effective mixing length, subscript M indicates momentum dominated plus dimensionless momentum ratio, L is the diffuser length, Q indicates discharge or river flow, depending on the subscript, subscript R indicates reverse flow, S is dilution the, subscript T indicates transition, and subscript Z indicates zero flow.

Equation 28.18 is introduced here because it will subsequently be compared to laboratory and field data and this set of formulae have special historical significance in the development of the plume model. Huang et al.[40] developed an expression for a *round* buoyant plume discharging into a uniform crossflow, which can be rearranged to form Equation 28.19:

$$S = \frac{4}{\pi}\left(\frac{h}{d}\right)^2 \left(\frac{u}{w}\right)\left[0.1\zeta^{\frac{1}{3}} + \frac{0.51}{\left(1 + 0.1\zeta^2\right)}\right] \quad (28.19)$$

where ζ is the dimensionless depth, Equation 28.20:

$$\zeta = \left(\frac{\pi}{4 F_D^2}\right)\left(\frac{d}{h}\right)\left(\frac{w}{u}\right)^3 \quad (28.20)$$

Agreement between Equation 28.19 and the data of Lee and Cheung[41] is shown in this next figure.

---

[40] Huang, H., R. Fergen, J. Proni, and J. Tsai, "Initial Dilution Equations for Buoyancy-Dominated Jets in Current," Journal of Hydraulic Engineering, Vol. 124, No. 1, pp. 105 108, 1998.

[41] Lee, J. H. W. and V. Cheung, "Generalized Lagrangian Model for Buoyant Jets in Current, ASCE Journal of Environmental Engineering, Vol. 116, No. 6, pp. 1085 1105, 1991.

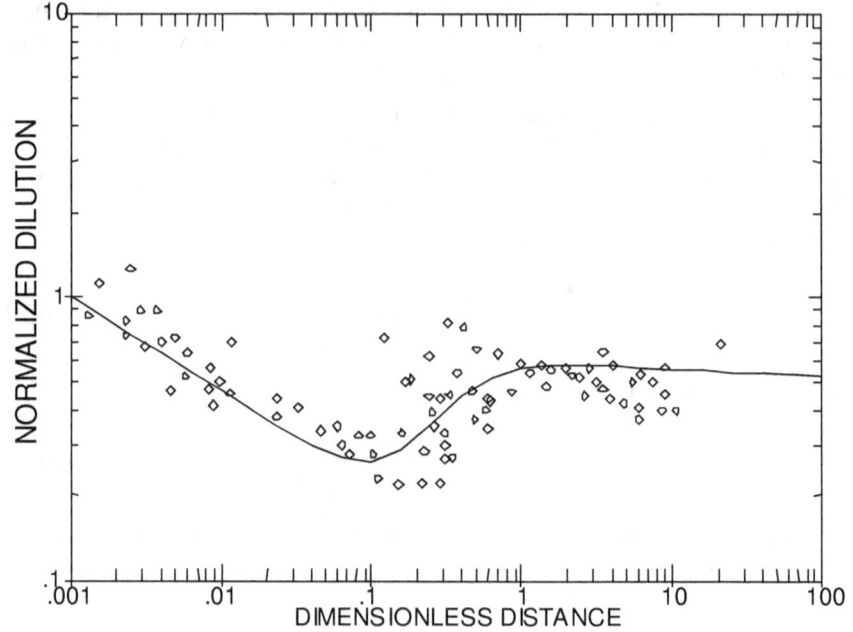

Impact of Stratification

Stratification is present in the ambient, most often due to a temperature gradient. Several authors already cited have also dealt with this complication, for instance, Fischer, Wood, and their associates. Most of the papers on discharge into a stratified ambient have focused on buoyant plumes rising in an ambient with linearly varying density. The data and correlations are most often presented in the form of maximum height reached before stagnation. A straightforward approach

$$\frac{z}{d} = \frac{1}{F_D}\left[\frac{1.5}{J^{\frac{1}{4}}} + \frac{3.7}{J^{\frac{3}{8}}}\right] \qquad (28.21)$$

yielding dilution is given by Wood et al.. Their correlations for maximum height reached by round and slot jets are given by Equations 28.21 and 28.22, respectively.

$$\frac{z}{b} = \frac{1}{F_B^{\frac{4}{3}}}\left[\frac{4.6}{J^{\frac{1}{3}}} + \frac{0.67}{J^{\frac{1}{2}}}\right] \qquad (28.22)$$

where J is the normalized stratification (similar to the densimetric Froude number) and is given by Equation 28.23:

$$J = -\frac{w^2 \dfrac{d\rho_A}{dz}}{g(\rho_A - \rho_P)} \tag{28.23}$$

The correlations of Wood et al. for dilution of round and slot jets are given by Equations 28.24 and 28.25, respectively.

$$S = \frac{1}{F_{\not{D}}^{\frac{4}{}}}\left[\frac{0.55}{J^{\frac{1}{4}}} + \frac{0.46}{J^{\frac{5}{8}}}\right] \tag{28.24}$$

$$S = \frac{1}{F_{\not{D}}^{\frac{2}{}}}\left[\frac{0.26}{J^{\frac{1}{6}}} + \frac{0.76}{J^{\frac{1}{2}}}\right] \tag{28.25}$$

Agreement between Equations 28.21 and 28.22 and the experimental data for maximum height reached of Wong[42] and Abraham and Eysink[43] are shown in this next.

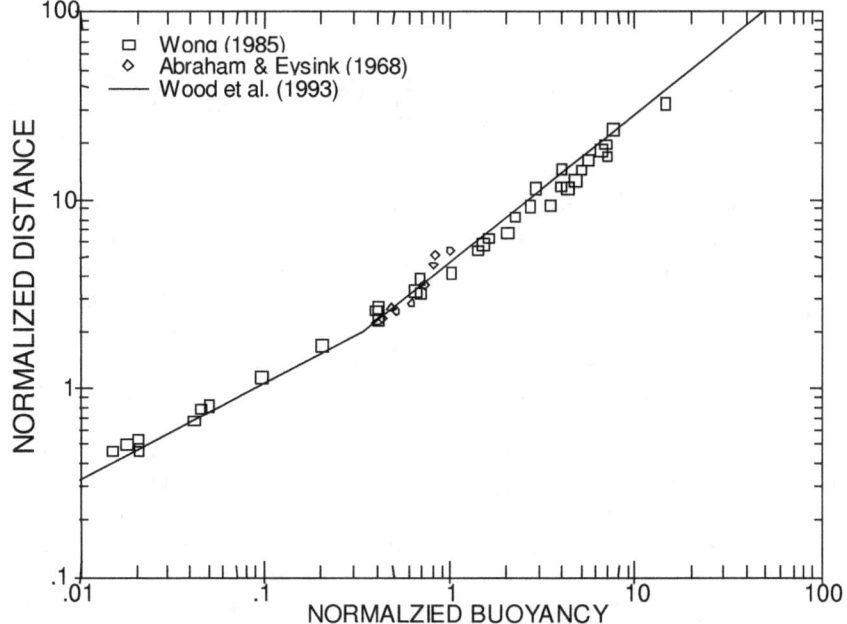

---

[42] Wong, D.,, "Buoyant Jet Entrainment in Stratified Flows," Ph.D. Thesis, University of Michigan, Ann Arbor, Michigan 1986.
[43] Abraham, G., and W. Eysink, "Jets Issuing into a Fluid with a Density Gradient," Journal of Hydraulic Research, Vol. 7, pp. 145 147, 1968.

Agreement between Equations 28.24 and 28.25 and the experimental data for dilution of Wong are shown in this next figure.

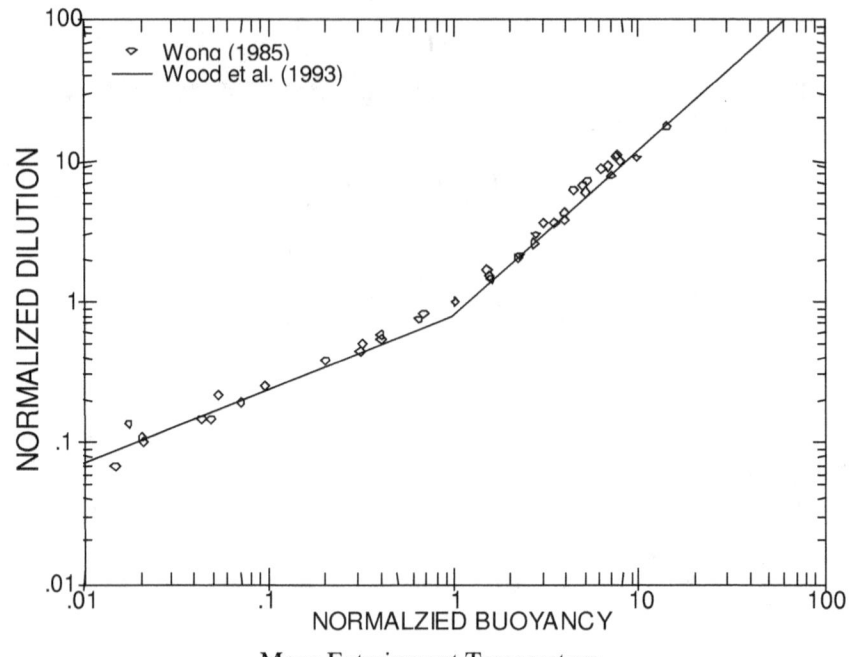

Mean Entrainment Temperature

Perhaps the simplest way to account for a thermally stratified ambient is to use some mean entrainment temperature. The rate of entrainment varies over the extent of the plume such that using a linear average of the ambient temperature does not yield sufficiently accurate results. The appropriate weighting with which to average the ambient temperature can be deduced from the analytical relationships already presented. Prandtl's relationship for a momentum-dominated or turbulent slot discharge, Equation 28.5, shows the dilution (and, thus, entrainment) increasing with the square root of the depth. Cederwall's relationship for a buoyancy-dominated or turbulent slot plume discharge, Equation 28.7, shows the entrainment increasing linearly with the depth. The relationship of Fischer et al., as shown in the first figure in this chapter, is consistent with these two, showing the entrainment for buoyancy-dominated plumes varying more steeply with depth than momentum-dominated jets. McIntosh suggests that for a slot discharge in general the dilution and entrainment increase with the three-fifths power of the depth. The weighting which, upon integration, results in the positive three-fifths power of depth is the negative two-fifths power, or Equation 28.26:

$$T_e = \frac{3}{5} \int_0^1 T \left(\frac{z}{h}\right)^{-\frac{2}{5}} d\left(\frac{z}{h}\right) \qquad (28.26)$$

As this integral has a singularity at $z=0$, simplistic methods of numerical integration will not work; however, there are ways to work around this problem. A trapezoidal rule of integration using the average of each pair of $z's$ is adequate. Gauss quadrature is a common and accurate method of numerical integration which does not require the end point at $z=0$. If the temperature varies in some simple algebraic manner, such as linearly, an analytical solution can be obtained. A linearly varying temperature can be expressed by Equation 28.27:

$$T = T_o + (T_h - T_o)\left(\frac{z}{h}\right) \quad (28.27)$$

Substituting Equation 28.27 into Equation 28.26 and solving yields Equation 28.28:

$$T_e = T_o + \frac{3}{8}(T_h - T_o) \quad (28.28)$$

A simple arithmetic average would yield one-half instead of three-eights in Equation 28.28. For a linearly varying temperature this is equivalent to using the temperature at three-fifths of the total depth.

Jet Interference

Although the preceding analytical and empirical relationships suggest that the entrainment is greater for a round discharge than a slot discharge there are practicalities of scale and fabrication, which may not be reflected in the equations and must be considered. It is often impractical to design, construct, or operate a single round jet for a large discharge. One of the implicit assumptions in these analytical and empirical relationships is that the diffuser is small compared to the ambient and the laboratory models on which they are based are so constructed. If a prototype diffuser violates this assumption the actual performance may be significantly less than suggested by the scale model.

Multi-port diffusers are much more common for these reasons. A multi-port diffuser is easier to construct than a slot; but it is not a true slot nor a solitary jet, the geometric bases for developing the analytical and empirical relationships. Considerable attention has been given to the actual behavior of jets in close proximity, such as would be the case in a multi-port diffuser, which is typically a pipe with rows of round holes. Holley and Jirka[44] include a section entitled "Jet Interference" in which they discuss various aspects of the phenomenon and ways of accounting for it in a model. They basically suggest a distance at which the jets can be considered merged, which is about twice the distance between them, and an equivalent slot width, which may be based on a simple equivalent area or up to twice this width to account for diminished entrainment due to interference. Several

---

[44] Holley, E. R., and G. H. Jirka, "Mixing in Rivers," Environmental and Water Quality Operational Studies Technical Report E 86 11, U. S. Army Corps of Engineers Waterways Experiment Station, Vicksburg, Mississippi, 1986.

three-dimensional computational fluid dynamics computer models of this jet interference and merging zone have been presented in the literature but these are beyond the scope of the present discussion.

Analytical and Empirical Model Assumptions

The following assumptions are implicit in developing all of these analytical and empirical models:

1) The discharge plume/jet is a coherent object, distinct from the ambient, which can be characterized by a profile or distribution and mean parameters.
2) The discharge plume/jet is small relative to the ambient such that it entrains the ambient, a process which changes its character but does not change the character of the ambient so much that the profile or distribution and mean parameters characterizing the ambient are no longer meaningful.
3) Interaction between the plume/jet and the ambient occurs at the boundary between the two and is limited to entrainment.
4) Small-scale turbulent phenomena can be characterized by bulk parameters such as mixing length and turbulent viscosity.
5) Large-scale turbulent phenomena can be averaged over time and accounted for by empirical factors such as entrainment coefficients.

Considering these assumptions and the associated simplifications involved in developing these models, the agreement shown in Figures 1 through 6 is remarkable. Besides providing a basis for testing the asymptotic behavior of the numerical plume model, these analytical solutions are offered as evidence that the assumptions yield acceptable results.

Numerical Model

The present numerical model was developed to address a particular need: dilution in a thermally stratified ambient, and subsequently expanded for other uses, for instance, discharge of brine. The formulation of the numerical model was influenced by this historical context. The numerical model is based on a concept, which can be illustrated by the following figure.

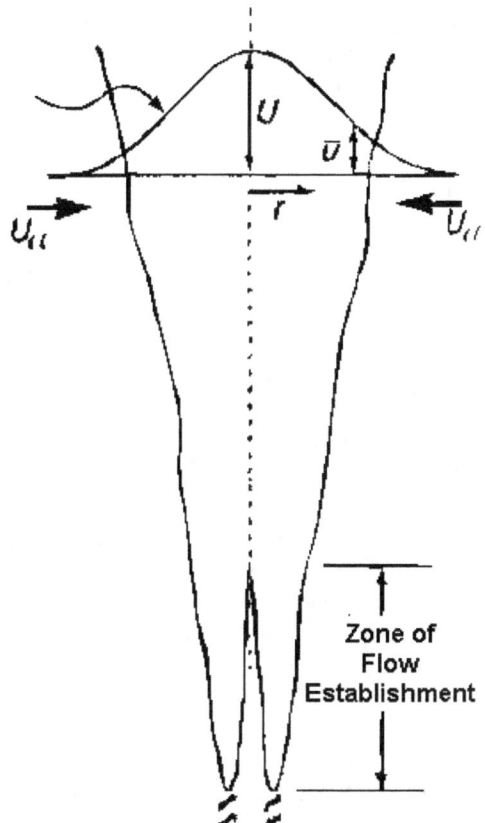

Figure 175. Entraining Plume Conceptual Model

The first three assumptions/simplifications listed previously are inherent in this conceptual model. The discharge is distinct from the ambient. The ambient is assumed large in comparison to the plume. The extent of the plume is identified by the boundary, where the entrainment is assumed to occur. This figure shows the discharge from adjacent jets merging to form a single plume.

Large-Scale Turbulence and Time-Averaging

Large-scale turbulence can be clearly seen in the figure below, which is a photograph taken by Lee et al.[45] of a thermal plume revealed with blue dye. Spatial variability in plumes like the one shown are part of common experience. Large plumes caused by discharges into water bodies such as rivers and lakes are often described as *boils*, even if there is no heat involved. The discharges from some power plants such as TVA's Browns Ferry and Sequoyah Nuclear appear on the surface of the river as irregular areas with different rippling than the

---

[45] Lee, J. H. W., W. Rodi, and C. F. Wong, "Turbulent Line Momentum Puffs," ASCE Journal of Engineering Mechanics, Vol. 122, pp. 19 29, 1996.

surrounding water, making them visually distinct. These *boils* slowly drift back and forth in the vicinity of the diffuser. The rate and extent of drift depends on the local conditions, but the movement is certainly noticeable over a period of fifteen minutes to an hour.

Figure 176. Entrainment Visualized with Dye

Considering this commonly observed spatial and temporal variability in plumes, it is remarkable that any agreement can be achieved between a model and field data; but two field data sets have already been introduced which have acceptable agreement. The key lies in time averaging. Combining the technologies of laser-Doppler anemometry, laser-induced fluorescence, and digital image processing, Lee et al. have been able to produce pictures, which graphically reveal this phenomenon. These next two figures show the instantaneous and time averaged concentration, respectively, resulting from the discharge of a single jet as viewed from the end.

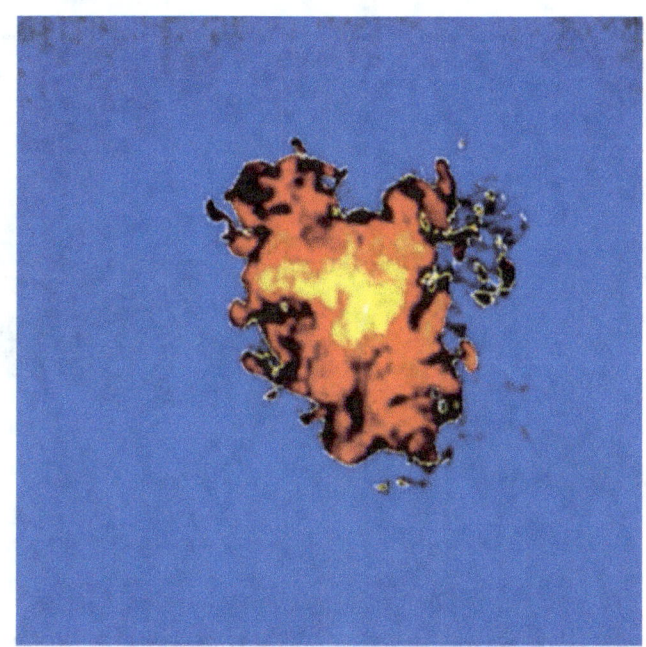

Figure 177. Turbulent Plume End View

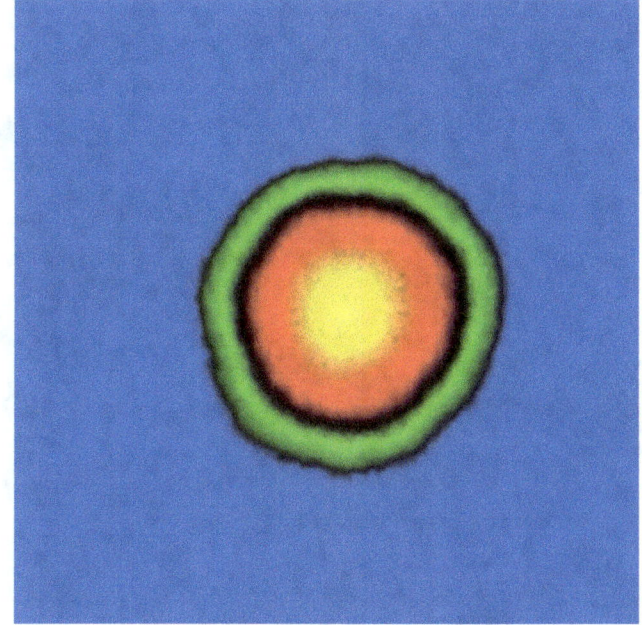

Figure 178. Laminar Plume End View

The figure below shows the instantaneous concentration of the same jet as viewed from the side.

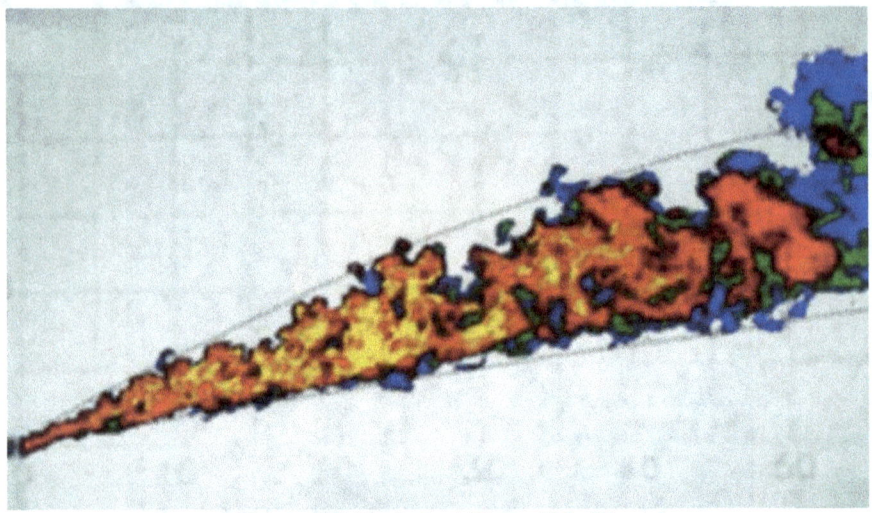

Figure 179. Turbulent Plume Side View

The figure below shows the time-averaged side view. The contrast between these two temporal perspectives is quite striking.

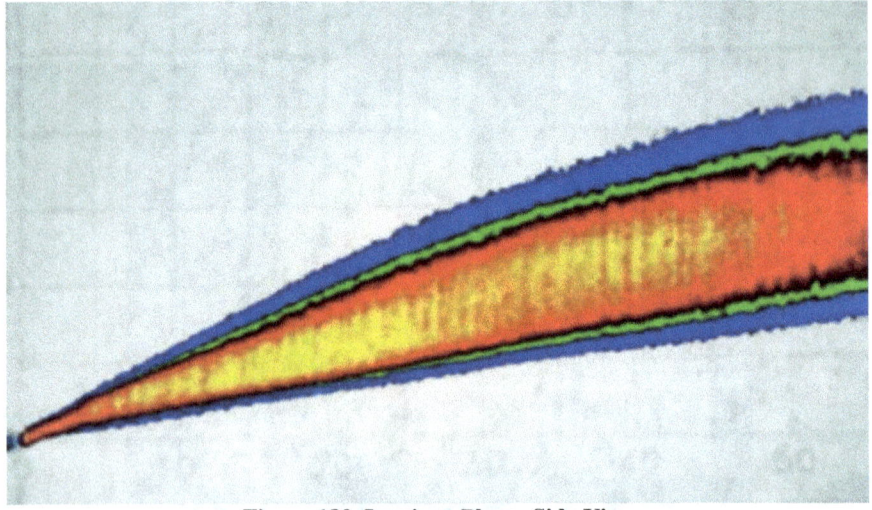

Figure 180. Laminar Plume Side View

In addition to the qualitative comparison of these two figures, the next two provide a quantitative comparison. First, the instantaneous then the time averaged:

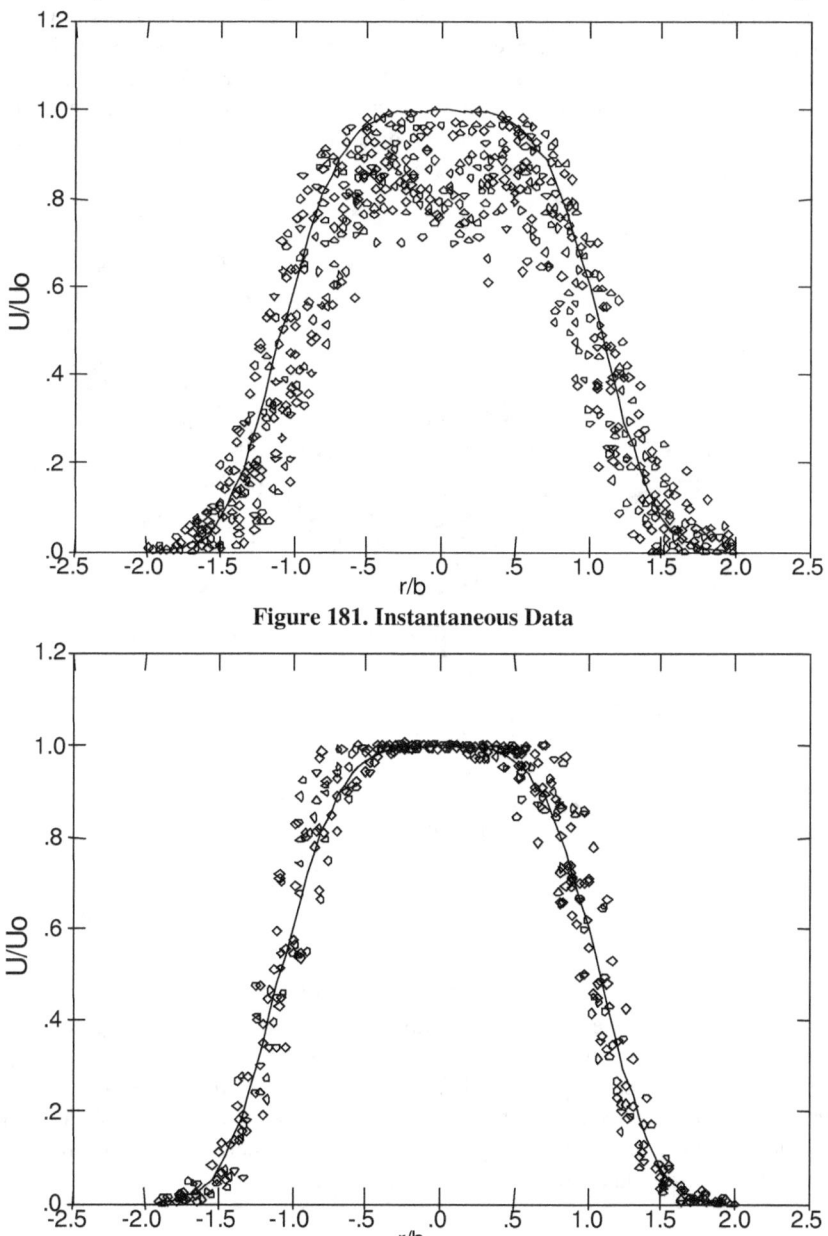

Figure 181. Instantaneous Data

Figure 182. Time-Averaged Data

These data cover approximately the first fourth (left side) of the jet. These data are presented as normalized radius (local/total) vs. normalized concentration (local/centerline). The difference between the second pair of figures is significant, though not as dramatic as the previous pair, as they represent only the initial part of the jet. Also shown in the figures is the theoretical (Gaussian) distribution. The variance between data and theory for the instantaneous measurements is more than four times that for the time-averaged measurements. Lateral entrainment is illustrated in this next figure.

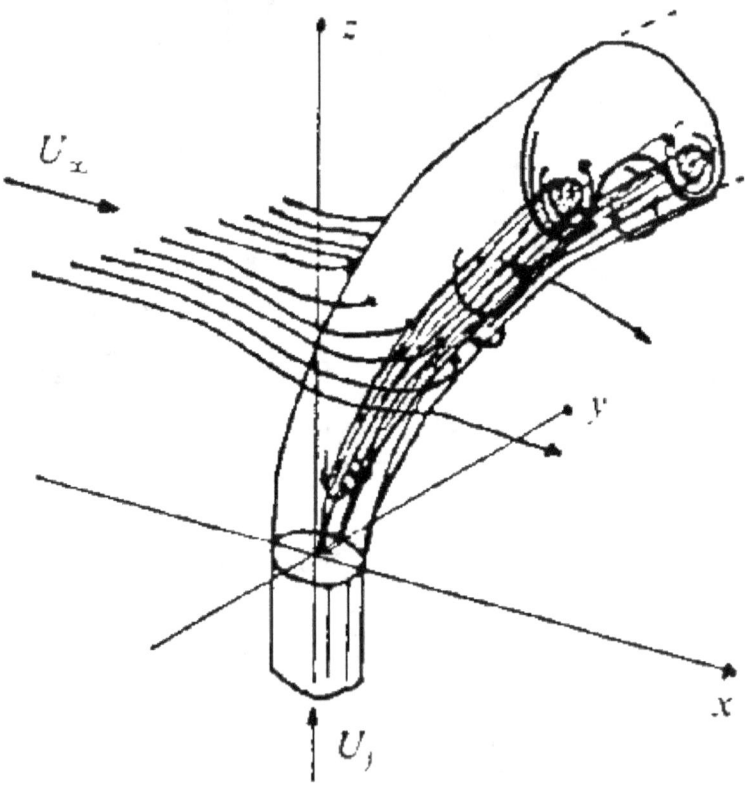

Figure 183. Entrainment Illustrated

Measurements taken on an instantaneous basis are unlikely to agree locally with any model; whereas, measurements taken on the averaged time scale might be adequately predicted by a model. It appears likely from the preceding figures that instantaneous measurements taken at the edges of the plume would differ significantly from the time-averaged measurements taken in the same locations and that the discrepancy would be significantly greater near the edges than near the centerline.

Observation of plumes such as the ones at TVA's Browns Ferry and Sequoyah Nuclear Plants indicates that the time scale over which point measurements must be averaged in order to obtain values sufficiently representative of the large-scale turbulent variability is at least fifteen minutes, and may be as long as an hour. The field data already introduced are hourly averages. Consideration of the preceding figures also indicates that field measurements should concentrate on the core of the plume. Comparison of the near-centerline velocities in latter two indicates that, even in the core of the plume, field measurements averaged over a time interval less than the time-scale of the turbulence may not be sufficiently characteristic.

Quasi-Steady-State Modeling

The numerical model, as well as all the analytical and empirical models already presented, assume that the modeling time scale will be such that the turbulent processes, though certainly not steady-state, physically approach some coherent average which can be mathematically described at a level of detail illustrated by the preceding figures. For large plumes such as the ones at TVA's Browns Ferry and Sequoyah Nuclear Plants this time scale is on the order of an hour. It is assumed that any significant changes in discharge temperature or flow and ambient temperature or flow occur slowly enough that the turbulent processes have enough time to sufficiently approach this coherent average in order to be adequately described by simply averaging these values. A longer modeling period, say on the order of a day, would be treated as a sequence of quasi-steady states, thus the term, *quasi-steady-state* modeling is applicable here.

Although it is beyond the scope of this modeling to consider transients and the specifics of turbulence, it is worth noting that turbulence appears to behave randomly and is treated as a random process in three-dimensional transient computational fluid dynamics models that consider the local details of turbulence. It is highly unlikely that any three-dimensional transient model simulating random turbulence could ever quantitatively match the *swirls*, as shown previously in this chapter. Certainly, something qualitatively like the swirling dye pattern could be generated, but a point-by-point comparison of computed and measured parameters would likely be as random as the turbulence itself.

Internal Distributions

It is often assumed that a plume will have some internal distribution of velocity, temperature, or concentration relative to the centerline. Laboratory and field measurements indicate this is the case. The most common distribution is Gaussian, or an exponential decay, as this is an analytical solution to similar problems and can easily be integrated. A Gaussian distribution also fits reasonably well with data.

Consider any scalar value represented by, C. The average value along the centerline of a jet can be computed by Equation 28.29:

$$\overline{C} = \frac{\int_0^\infty C(r) e^{\left(-\frac{r^2}{b^2}\right)} 2\pi \left(\frac{r}{b}\right) \frac{dr}{b}}{\int_0^\infty e^{\left(-\frac{r^2}{b^2}\right)} 2\pi \left(\frac{r}{b}\right) \frac{dr}{b}} \qquad (28.29)$$

where r is the distance from the centerline and b is the local width of the jet. If the distributions along the centerline are similar, that is, when appropriately normalized they have the same shape, then Equation 28.26 results in a constant. For this reason it is often assumed that the ratio of the mean and maximum values of any scalar quantity is a constant over the entire plume, and so it is in this model. There is, of course, a difference between the arithmetic and integrated average values of nonlinear quantities, such as velocity squared, for which some empirical correction could be applied.

Problems arise with internal distributions when considering entrainment. If the entrainment is assumed to be proportional to the difference between the ambient velocity and the maximum velocity within the plume, then the average velocity within the plume will never equal the ambient velocity, regardless of the duration of interaction. If the entrainment is assumed to be proportional to the difference between the ambient velocity and the average velocity within the plume, then the eventual average velocity of the plume will equal the ambient. However, the maximum velocity within the plume will exceed the ambient velocity. No internal distributions are assumed in this model.

Numerical Model Assumptions

Based on the preceding arguments the following assumptions are made in developing the numerical model:

1) The discharge plume/jet is a coherent object, distinct from the ambient, which can be characterized by a similar profile or distribution and mean parameters.
2) The discharge plume/jet is small relative to the ambient such that it entrains the ambient, a process which changes its character but does not change the character of the ambient so much that the profile or distribution and mean parameters characterizing the ambient are no longer meaningful.
3) Interaction between the plume/jet and the ambient occurs at the boundary between the two and is limited to entrainment.
4) Small-scale turbulent phenomena can be characterized by bulk parameters such as mixing length and turbulent viscosity.
5) Large-scale turbulent phenomena can be averaged over time and accounted for by empirical factors such as entrainment coefficients.
6) The temporal behavior of the plume/jet is assumed to be quasi-steady-state.

7) The internal structure and properties of the plume/jet (density, velocity, temperature, salinity, etc.) can be characterized by average parameters that vary along the centerline with the trajectory.

Plume Differential Element

The computational element on which the numerical model is developed is based on these seven assumptions. The plume variables and flows are illustrated in this next figure, where ***m*** is the mass flow [kg/s], ***E*** is the energy flow [kJ/s], and ***C*** is the flow of any extensive scalar property [units/s]. The corresponding intensive properties are also shown in the figure, ***e*** is the specific energy [kJ/kg] and ***c*** is the specific scalar property [units/kg]. The subscripts ***A***, ***P***, and ***E*** indicate the ambient, plume, and entrainment, respectively. A Taylor series expansion is used to represent the change of variables through the element and forms the basis for generating the ordinary differential equations. Ordinary differential equations, rather than partial differential equations, arise from Assumption 7, which implies that there is only one independent variable, the distance along the centerline, ***s***.

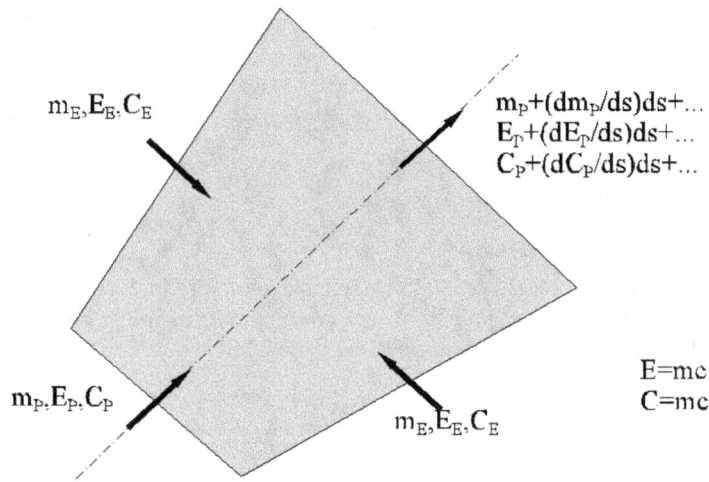

Figure 184. Differential Element

Governing Differential Equations

The applicable governing principles are: the conservation of mass, energy, and linear momentum. When there is salt in either the discharge or ambient, it is necessary to have a separate conservation equation for the salt. The conservation of mass for a slot and round jet is given in differential form by Equations 28.30 and 28.31, respectively:

$$\frac{d\left(b\rho_P\sqrt{u_P^2+w_P^2}\right)}{ds} = m_E \qquad (28.30)$$

$$\frac{d\left(\pi r^2 \rho_P \sqrt{u_P^2 + w_P^2}\right)}{ds} = 2\pi r \, m_E \qquad (28.31)$$

where u_P and w_P are the x and z components of the velocity along the centerline [m/s] and m_E is the entrainment mass flux [kg/m/s]. The conservation of energy for a slot and round jet is given in differential form by Equations 28.32 and 28.33, respectively:

$$\frac{d\left(b \rho_P C T_P \sqrt{u_P^2 + w_P^2}\right)}{ds} = m_E C T_E \qquad (28.32)$$

$$\frac{d\left(\pi r^2 \rho_P C T_P \sqrt{u_P^2 + w_P^2}\right)}{ds} = 2\pi r \, m_E C T_E \qquad (28.33)$$

where C is the specific heat [kJ/kg/°C] and T is the temperature [°C]. The conservation of linear momentum has two orthogonal vector components, x and z [m], and for a slot jet these are given in differential form by Equations 28.34 and 28.35:

$$\frac{d\left(b \rho_P u_P \sqrt{u_P^2 + w_P^2}\right)}{ds} = m_E u_E \qquad (28.34)$$

$$\frac{d\left(b \rho_P w_P \sqrt{u_P^2 + w_P^2}\right)}{ds} = m_E w_E + b\left(\rho_E - \rho_P\right) g \qquad (28.35)$$

where the last term in Equation 28.35 is the contribution of buoyancy. For a round jet the two orthogonal components of the conservation of linear momentum are given in differential form by Equations 28.36 and 28.37:

$$\frac{d\left(\pi r^2 \rho_P u_P \sqrt{u_P^2 + w_P^2}\right)}{ds} = 2\pi r \, m_E u_E \qquad (28.36)$$

$$\frac{d\left(\pi r^2 \rho_P w_P \sqrt{u_P^2 + w_P^2}\right)}{ds} = 2\pi r \, m_E w_E + \pi r^2 \left(\rho_E - \rho_P\right) g \qquad (28.37)$$

The trajectory of the plume (x_P, z_P) is given in differential form by Equations 28.38 and 28.39:

$$\frac{d x_P}{ds} = u_P \qquad (28.38)$$

$$\frac{d z_P}{ds} = w_P \qquad (28.39)$$

The only term not already defined is the entrainment flux, m_E, which will be described after the method of solution, as any number of relationships for entrainment could be used in this formulation.

Solution of the Governing Equations

These ten ordinary differential equations express the behavior of the plume. In addition to the differential equations, it is necessary to prescribe the initial conditions and the conditions of the ambient. Once these are defined, the trajectory and dilution of the plume can be determined by solving the equations numerically. A satisfactory method is fourth order Runge-Kutta. A limited geometric progression of the step size can be used to speed the process without loss of accuracy.

Entrainment

The entrainment flux, m_E, has been computed in diverse ways by various investigators so that several relationships exist in the literature. Typically, an entrainment coefficient, α, is defined as in Equation 28.40 for slot or round jets:

$$m_E = \alpha \rho_E |\Delta v| \quad (28.40)$$

where Δv is the magnitude of the difference between the velocity of the plume and ambient, considered as vectors. A simple and satisfactory relationship for the entrainment coefficient based on the densimetric Froude number, F_D, is given by Fischer et al. for slot and round jets and can be expressed by Equations 28.41 and 28.42, respectively:

$$\alpha = 0.0520 \, e^{\left(\frac{1.62}{Fr^{1.5}}\right)} \quad (28.41)$$

$$\alpha = 0.0535 \, e^{\left(\frac{1.43}{Fr^{2}}\right)} \quad (28.42)$$

While the entrainment at the upper and lower or upstream and downstream boundaries of the plume are no doubt different, there is insufficient experimental information to separate the two. McIntosh inferred a relationship for the entrainment coefficient from the data reported in 1983, which can be expressed by Equation 28.43 and would apply to a slot jet:

$$\alpha = \begin{cases} 0.27, & Fr < 0.75 \\ \dfrac{0.27}{Fr^{2.5}}, & 0.75 \leq Fr \leq 1 \\ 0.55, & Fr > 1 \end{cases} \quad (28.43)$$

These relationships are shown in the figure at the top of the next page.

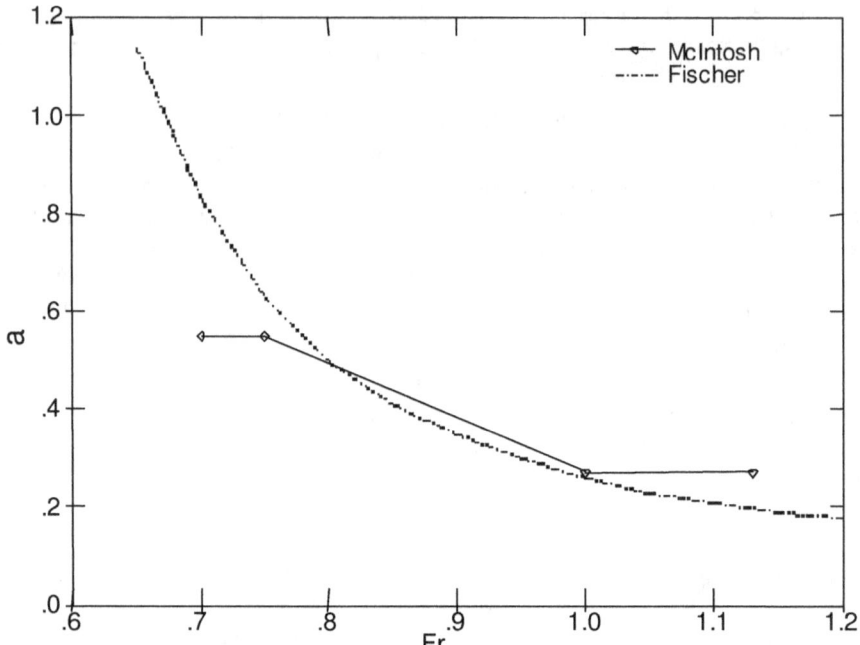

Figure 185. Entrainment Factor vs. Froude Number

Plume Termination

Just as merging jets and real slots are handled in many different ways at the beginning of the plume, so the end of the plume is handled in various ways by different developers. The current plume termination scheme is neither the simplest nor the most complex. When the upper edge of the plume reaches the surface the entrainment on that side ceases and the vertical momentum associated with it stops. When the lower edge of the plume reaches the bottom the entrainment on that side ceases and the momentum associated with it stops. If both sides of the plume cease entraining the calculations cease. If the centerline of the plume reaches the surface or the bottom the calculations cease. After more than a fixed number of steps the calculations cease.

Plume Shape and Boundary

The primary purpose of this model is to accurately predict the entrainment or dilution. The secondary purpose is to estimate the extent of the plume, especially if it stagnates, as when a buoyant plume doesn't reach the surface or a dense plume doesn't reach the bottom. It is not the purpose of this model to accurately predict the shape or boundary of the plume. If prediction of the shape or boundary of the plume is required a three-dimensional model would be more appropriate.

Low Flow Conditions

If the flow in the river persists at a rate less than that required to completely remove the heat (or salt) an unsteady build-up will occur. Such a situation would violate the assumption of quasi-steady modeling and invalidate the results. The plume model could be used to analyze a system with feedback, such as a closed loop, but there still must be an ultimate heat (or salt) sink in order to obtain valid results.

Reverse river flow episodes can cause the effluent to travel upstream. During the subsequent forward flow episode the effluent is recycled through the plume. Recycling of the effluent through the plume is not taken into account in the development. Modeling extended periods of reverse flow is beyond the scope of this model as this would be quite speculative. Field data collection is usually avoided during reverse flow and it is not clear how a laboratory model could be constructed in order to accurately represent the prototype under such conditions.

Results

The results which led to the development of the plume model are shown in the following figure, where the red line indicates monitored values and the green line indicates computed values.

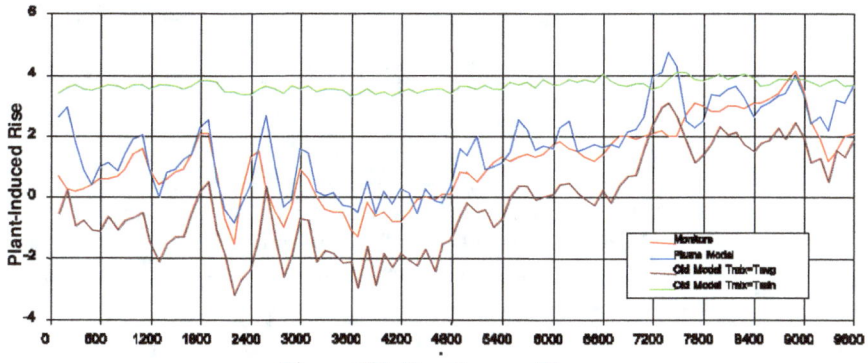

Figure 186. Results over Time

On May 5, 1981, the Sequoyah Nuclear Plant Operational Scheduling Model was predicting a steady plant-induced temperature rise in the river of 2.5°C; but the continuous monitors located in the river were reporting no more than a 1°C difference in the upstream and downstream temperatures in the compliance zone at an average depth of 1.5 meters below the surface. During the evening of May 5 and the morning of May 6 the monitored plant-induced rise fluctuated between positive and negative values. Throughout the rest of May 6 and the morning of May 7 the plant-induced rise remained fairly steady between ±0.5°C. For the rest of May 7 and all of May 8 the plant-induced rise slowly climbed to about 2°C. Throughout this entire period the plant was discharging almost 2300 MWt (megawatts-thermal) of waste heat into the river. Everything was checked, nothing was malfunctioning, but something was missing from the analysis: stratification.

Impact of Stratification

Not only are the temperatures in river upstream and downstream of the compliance zone monitored, but so is the temperature of the water being withdrawn from the river and the water being discharged back into the river as it enters the diffuser. During this period a steady flow of 35.5m³/s of water was being heated 15.5°C. By the time this heated discharge reached the surface it was often colder than the surrounding water. The vertical temperature profiles upstream of the plant are shown at six-hour intervals in the next two figures.

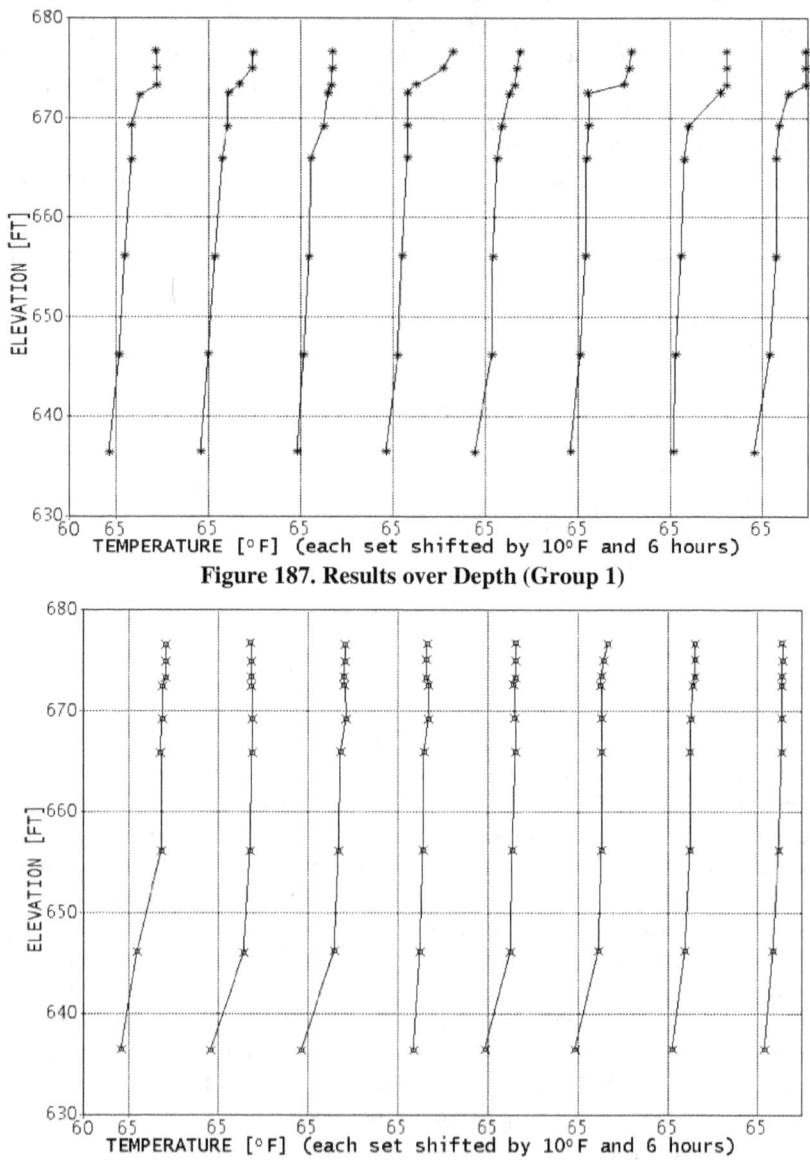

Figure 187. Results over Depth (Group 1)

Figure 188. Results over Depth (Group 2)

The compliance temperature is the average of the top three points, which are maintained by a float at 1.0, 1.5, and 2.0 meters below the surface. When the red line in the preceding figure passes through zero the average temperature at a depth of 1.5m below the surface upstream and downstream of the plant were the same. In which case the stratification was significant enough to absorb the entire waste heat output of the plant. When the red line in the preceding figure is below zero the stratification was more than enough. The flows during this period are shown in the figure below:

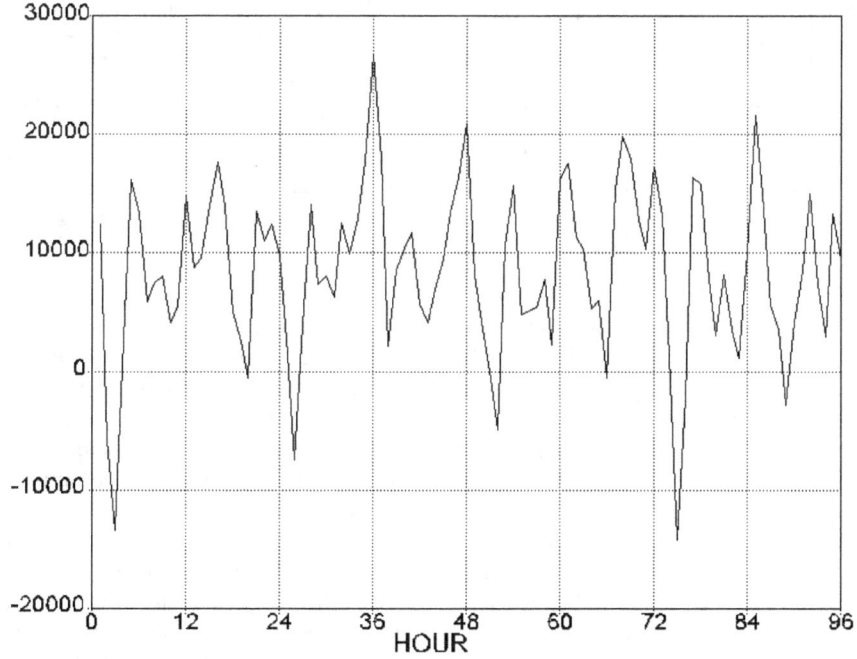

Figure 189. Variation of Flow over Time

As mentioned previously several empirical models were also tested as were various temperature averaging schemes, but satisfactory agreement between model and monitors was not obtained. The best results of this effort using the empirical model of Almquist are shown by the brown line in the preceding figure.

The unadjusted results for all the empirical models tested are shown in this next figure.

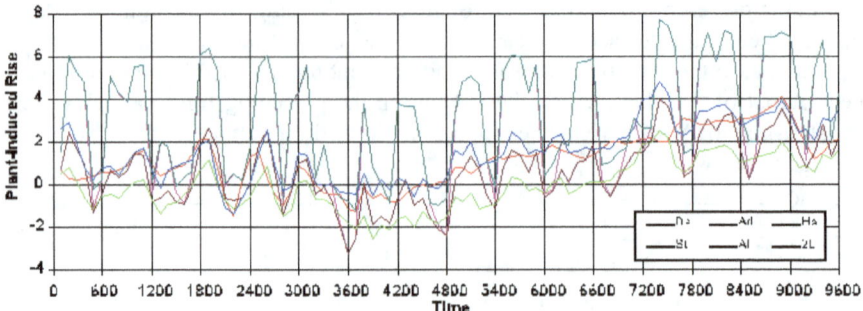

Figure 190. Model Results & Data

The adjusted results for the same empirical models using the best upstream temperature averaging are shown in the figure above. None of these models were developed for use in a stratified ambient.

Plume Stagnation

The analytical and empirical efforts to predict plume behavior in a stratified ambient already presented focused on how high a plume will rise before stagnating. For the purposes of computing plant-induced rise this is the most important quantity to predict accurately. If the plume does not reach the compliance zone it does not matter what the dilution is. The second most important quantity is predicting at what temperature (or salinity) is the ambient being entrained. The actual total volumetric entrainment is the least important variable in a strongly stratified ambient. This is why no upstream temperature averaging scheme will close the gap between the brown (computed) and red (measured) lines in the previous figure. Even with the exact volumetric dilution and mean entrainment temperature a plume may behave as the one at Sequoyah did from 6:00 p.m. (1800 hours in the figures) on May 5 to 6:00 a.m. (3000 hours in the figures) on May 6 when the stagnation zone drifted in and out of the monitors during essentially steady operation causing the apparent plant-induced rise to cycle from positive to negative. A plume model might predict such a response; but the empirical models tested would not.

Comparison with Monitor Data

The preceding figure shows the plume model closely tracking monitor data over a 96-hour period during which time the ambient temperature varied considerably as seen in the two previous figures. The shifting phase difference between the monitor data and plume model is a result of the distance between the upstream and downstream monitors, the varying river flow (as shown in the last figure), and the resulting travel time. The upstream ambient temperature stratification changed from gradual over the depth at 6:00 a.m. on May 5 (the left-most vertical line in the previous figure) to gradual over most of the depth with a strong gradient near the surface at 6:00 p.m. on May 6 (the second right-most vertical line in the previous figure). During the next 48 hours the upstream temperature was nearly uniform over the upper half of the river with a gradient over the lower half (the vertical lines in the same figure).

This 96-hour period provides a diverse test of the plume model with variable stratification and flows ranging from -500 to +750 m^3/s. The plant-induced rise varied from positive to negative. The plume reached the surface during most of this period but stagnated several times, never reaching the top three sensors and registering its presence. The correlation coefficients, R^2, for the models are as follows: Adams 0.14, Harleman 0.14, Stolzenbach 0.51, Almquist 0.45, the one-dimensional plume model 0.32, and the two-dimensional plume model 0.69.

Comparison with Laboratory Data

This next figure shows agreement between the laboratory data of Harleman et al. and the empirical models and the numerical plume model. The correlation coefficients, R^2, for the models are as follows: Adams 0.03, Harleman 0.03, Stolzenbach 0.28, Almquist 0.29, the one-dimensional plume model 0.27, and the two-dimensional plume model 0.48.

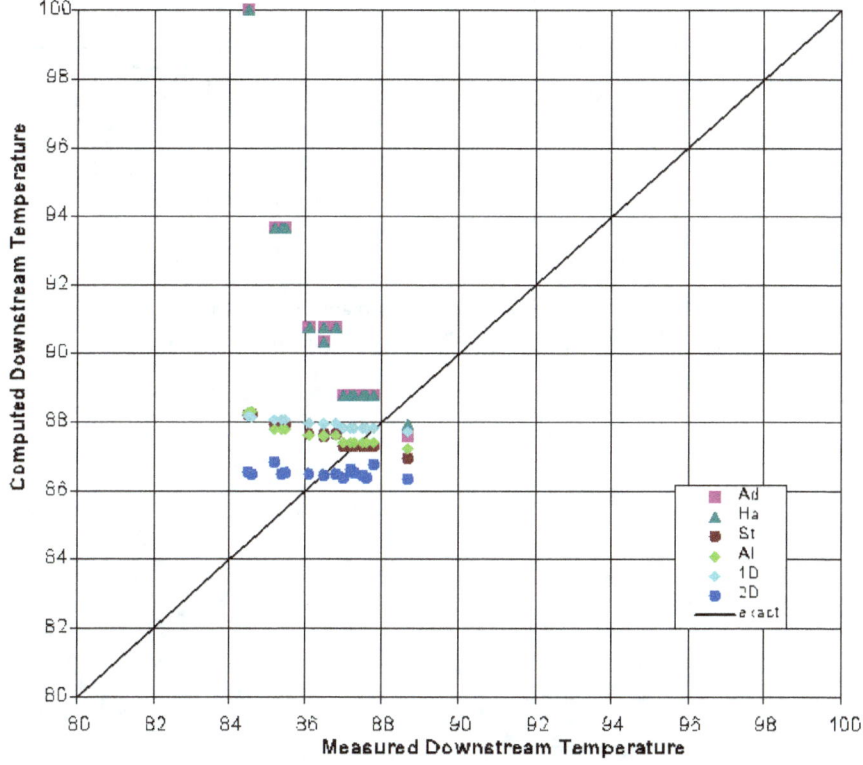

Figure 191. Comparison with Laboratory Data

This next figure shows agreement between the laboratory data of Almquist[46] and the empirical models and the numerical plume model. The correlation coefficients, R^2, for the models are as follows: Adams 0.03, Harleman <0.01, Stolzenbach 0.55, Almquist 0.42, the one-dimensional plume model 0.12, and the two-dimensional plume model 0.82.

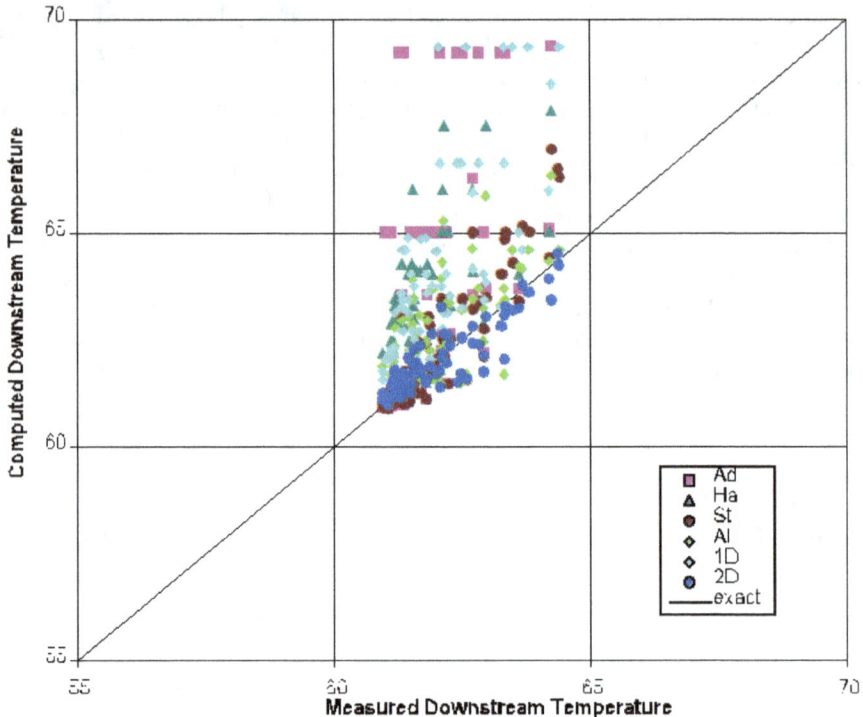

Figure 192. Comparison with Field Data (Analytical Model)

---

[46] Almquist, C. W., "Submerged Multiport Diffuser Analysis and Design for Hartsville Nuclear Plant," TVA Report WR28 1 89 100, 1978.

This next figure agreement between the laboratory data of McIntosh and the empirical models and the numerical plume model. The correlation coefficients, R^2, for the models are as follows: Adams 0.38, Harleman 0.54, Stolzenbach 0.82, Almquist 0.62, the one-dimensional plume model 0.74, and the two-dimensional plume model 0.81.

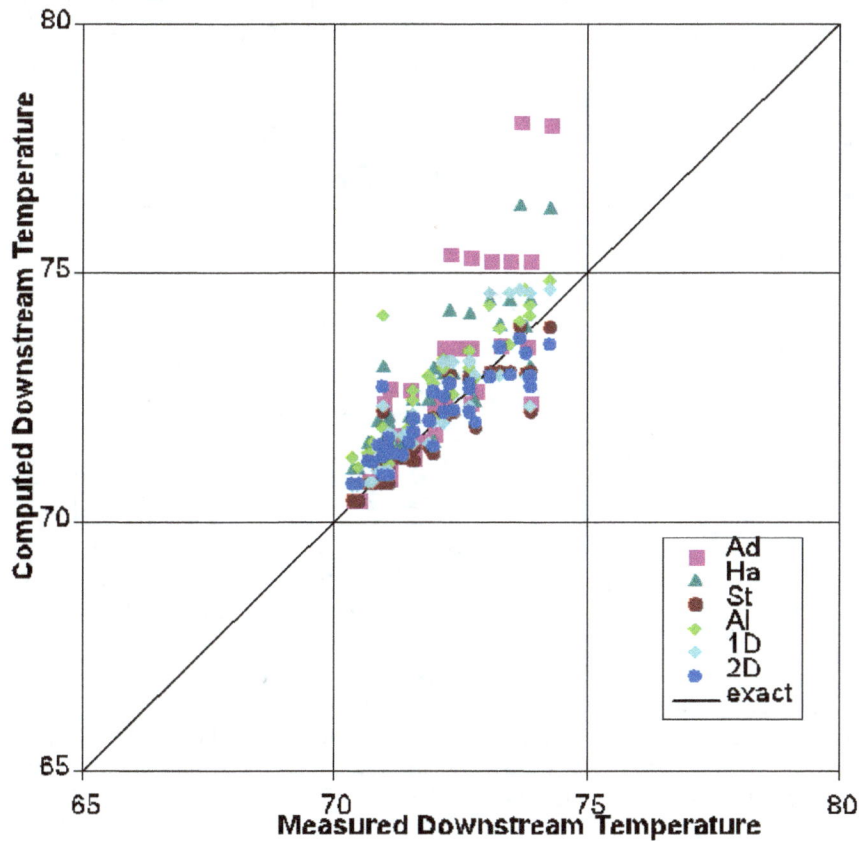

Figure 193. Comparison with Field Data (Empirical Model)

Comparison with Field Data

The figure on the top of the next page shows agreement between the field data of Almquist et al. and the empirical models and the numerical plume model. The correlation coefficients, R^2, for the models are as follows: Adams 0.04, Harleman 0.01, Stolzenbach 0.46, Almquist 0.54, the one-dimensional plume model 0.49, the two-dimensional plume model 0.60, and the three-dimensional finite-difference fluid dynamics model, EFDC, 0.51. The three-dimensional model was applied to this field data set as part of another project.

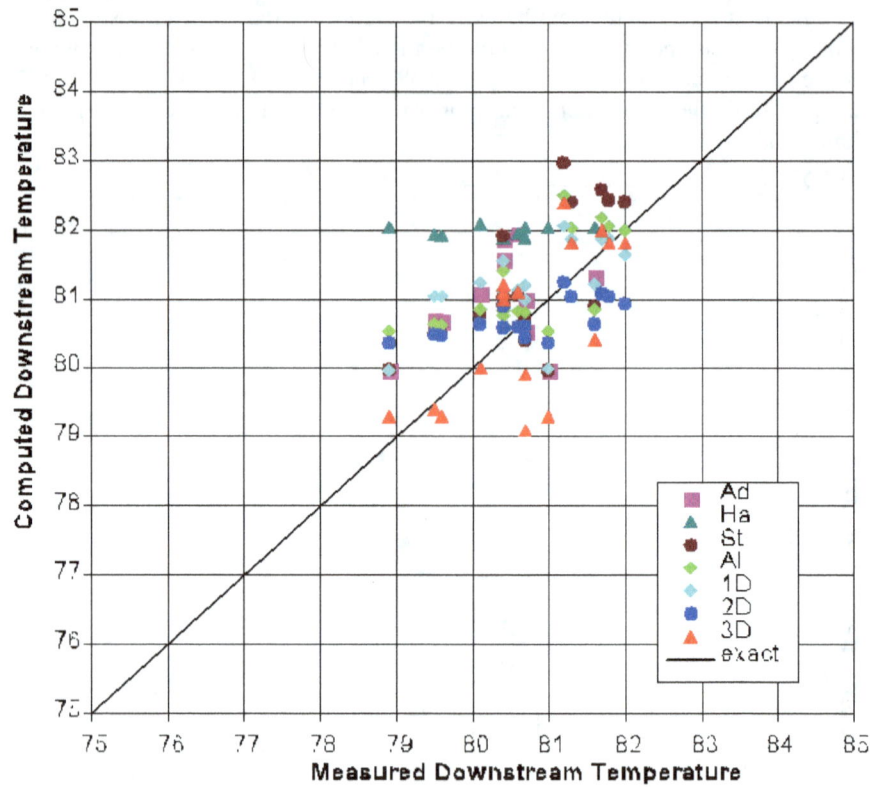

Figure 194. Comparison with Field Data (Numerical Model)

The figure on the next page shows agreement between the field data of McIntosh et al. and the empirical models and the numerical plume model. The correlation coefficients, R^2, for the models are as follows: Adams 0.91, Harleman 0.88, Stolzenbach 0.99, Almquist 0.98, the one-dimensional plume model 0.91, and the two-dimensional plume model 0.93.

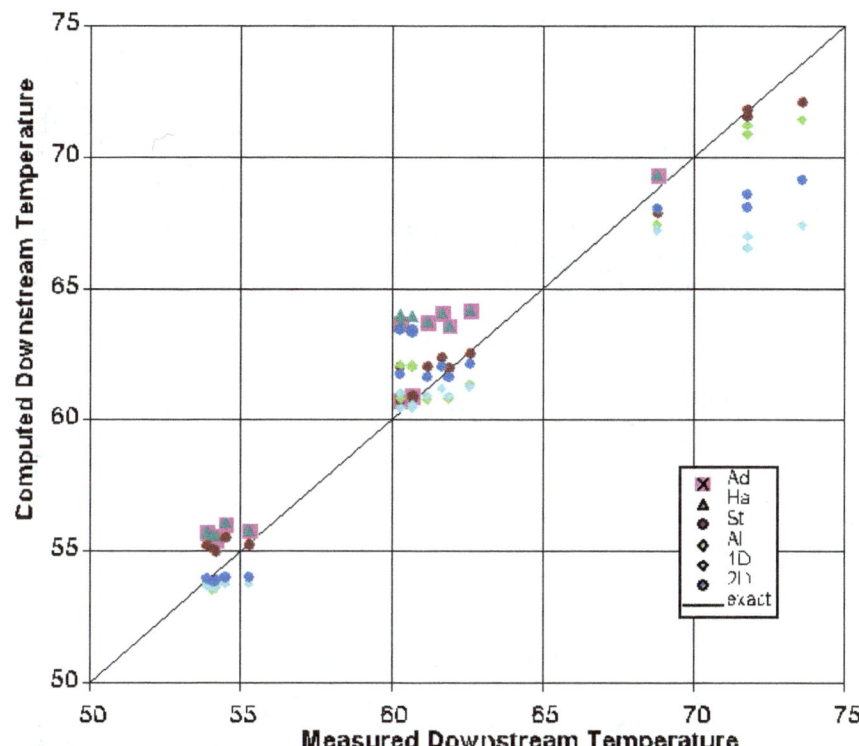

Figure 195. Comparison with All Data Sets

Asymptotic Behavior

The next three figures show the asymptotic behavior of the computed dilution of the plume for Froude numbers of 1, 10, and 100 respectively. Here a thermal plume is discharged upward into an infinite, uniform, stagnant ambient. Also illustrated in these figures are the results of three analytical methods already presented (Almquist, Cederwall, and Roberts) and the numerical model, SLOTJET1, by Alavian et al.[47]. The plume results agree well with Cederwall and lie between the results of Almquist and Roberts.[48]

---

[47] Alavian, V., P. Ostrowski, and J. A. Parsly, "Two Computer Models for Diffuser Performance Evaluation," TVA Report WR28 1 900 162, 1988.

[48] Roberts, P. J. W., "Line Plume and Ocean Outfall Dispersion," ASCE Journal of Hydraulics, 1979.

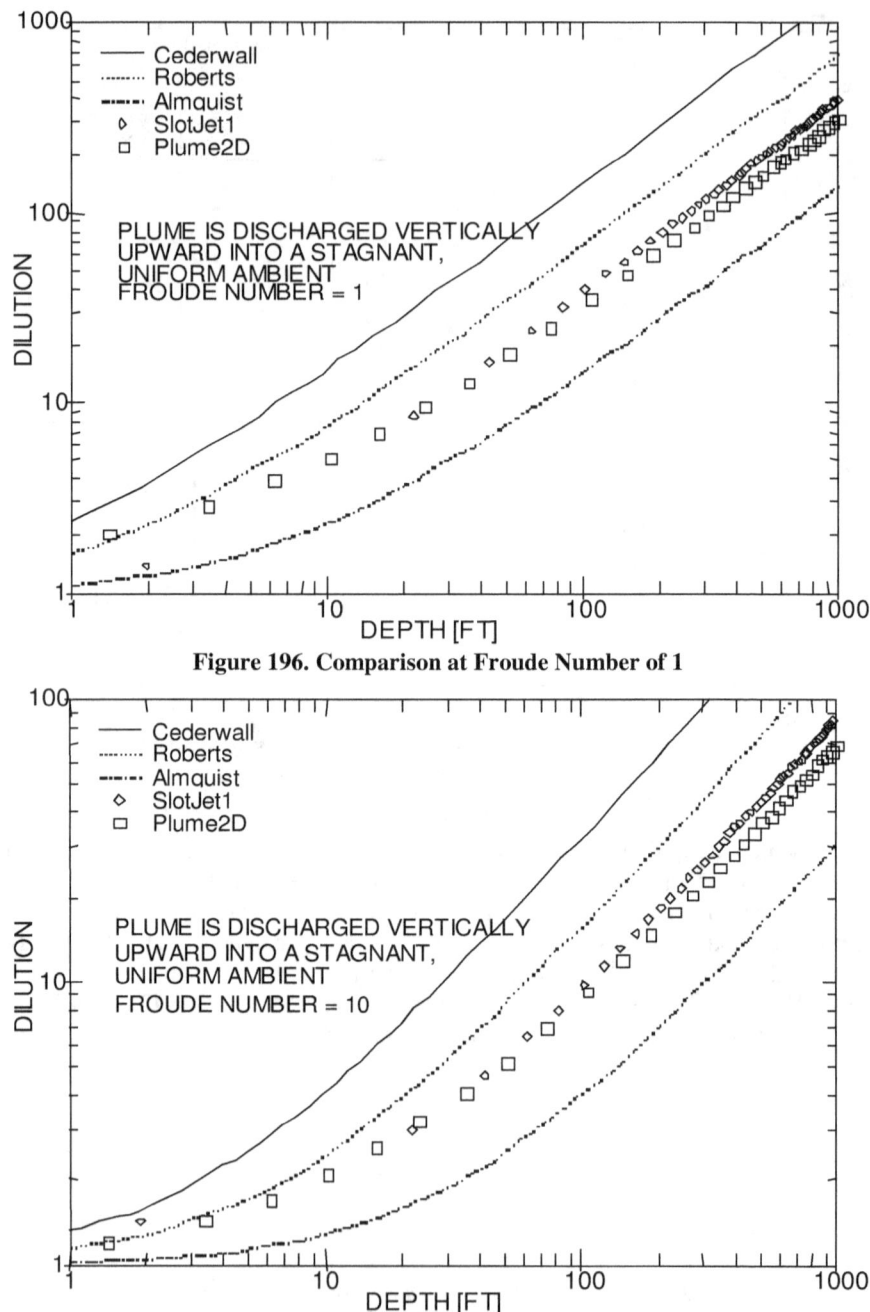

Figure 196. Comparison at Froude Number of 1

Figure 197. Comparison with Froude Number of 10

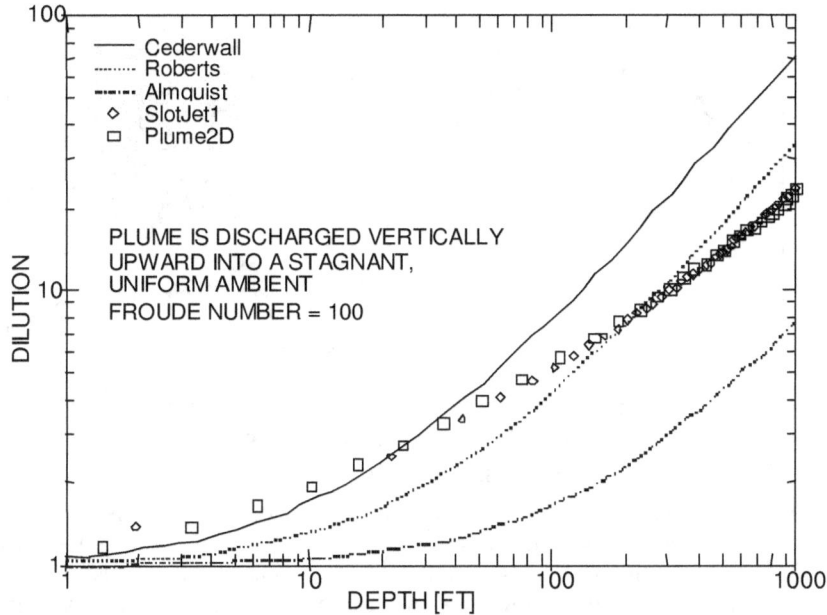

Figure 198. Comparison with Froude Number of 100

<u>Sample Trajectories</u>

The next two figures show the computed trajectory of the plume centerline for a range of ambient river flows. The first is for a positively buoyant thermal plume and the second is for a negatively buoyant salty plume.

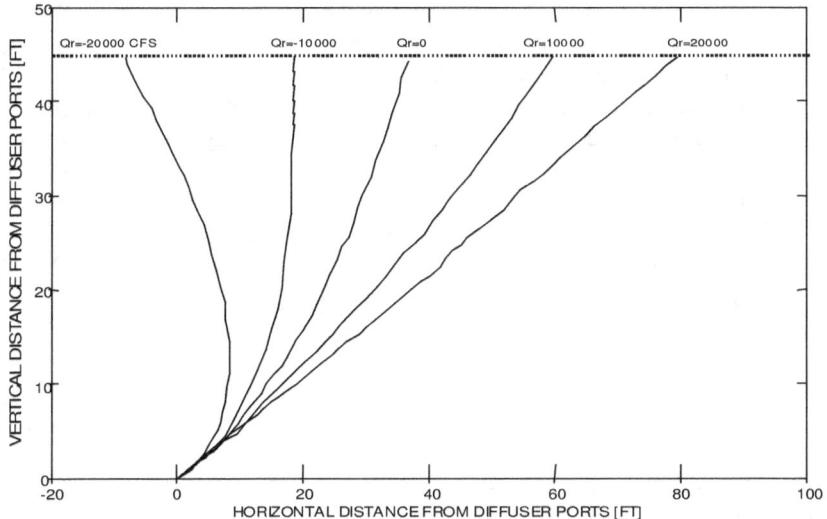

Figure 199. Impact of Velocity on Plume Trajectory

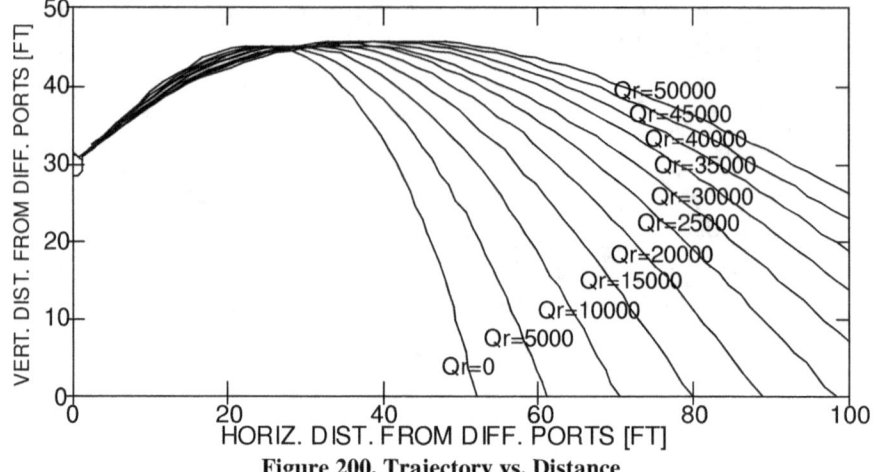

Figure 200. Trajectory vs. Distance

The development of a two-dimensional plume model for a round or slot jet has been presented. The formulation of this model is based on the conservative form of the governing equations. The model can handle combinations of heated and salty discharges and flowing ambients. The model results compare favorably with laboratory and field data. The model is consistent with several analytical models for the asymptotic behavior a thermal plume in an infinite medium. The model closely tracks monitor data for variable flow, ambient temperature, and upstream thermal stratification. Additional details can be found in the TVA and TWRA reports.[49,3]

<u>Computer Code</u>

The two-dimensional numerical plume model has been implemented in the form of a computer code. You can find the source (plume2D.c) and all of the associated input and output files in the online archive in folder examples\plume2d. Some input data are necessary to drive the model, including: the composition of the plume and the ambient temperature, salinity, and velocity. The plume may be discharged from a slot or a round jet and the equations must handle both cases. Output files are created to facilitate displaying the results graphically. Sample input files include: plume1.inp, plume2.inp, plume3.inp, and plume4.inp.

---

[49] Benton, D. J., "Development of a Two-Dimensional Plume Model for Positively and Negatively Buoyant Discharges into a Stratified Flowing Ambient," TVA Report WR28-1-45-105, 1986.

The variables are listed in the following table:

variable	units	description
α	-	entrainment coefficient
b	ft	width/diameter of the plume
depth	ft	depth
difd	ft	diffuser diameter
difl	ft	diffuser length
dify	ft	diffuser elevation (up from bottom)
dil	-	mixing ratio (dilution)
Froude	-	densimetric Froude number
Qr	ft³/sec	river flow
Qdis	ft³/sec	discharge flow
ρ	lbm/ft³	density
S	-	salinity
T	°F	temperature
ang	°	angle of inclination from the horizontal
u	ft/s	horizontal velocity of the plume
v	ft/s	vertical velocity of the plume
w	ft/s	velocity along centerline w=sqrt(u^2+v^2)
x	ft	horizontal distance from diffuser ports
y	ft	vertical distance from diffuser ports
z	ft	distance along the centerline

The governing equations in symbolic form are:

p(1)=x	horizontal coordinate of centerline
p(2)=y	vertical coordinate of centerline
p(3)=r*w*b²*π/4	round jet mass flux
p(3)=r*w*b	slot jet mass flux/unit diffuser length
p(4)=p(3)*u	horizontal momentum flux
p(5)=p(3)*v	vertical momentum flux
p(6)=p(3)*t	thermal energy flux
p(7)=p(3)*s	salt flux

Results for the first example (a rising plume1.inp) are shown in this figure:

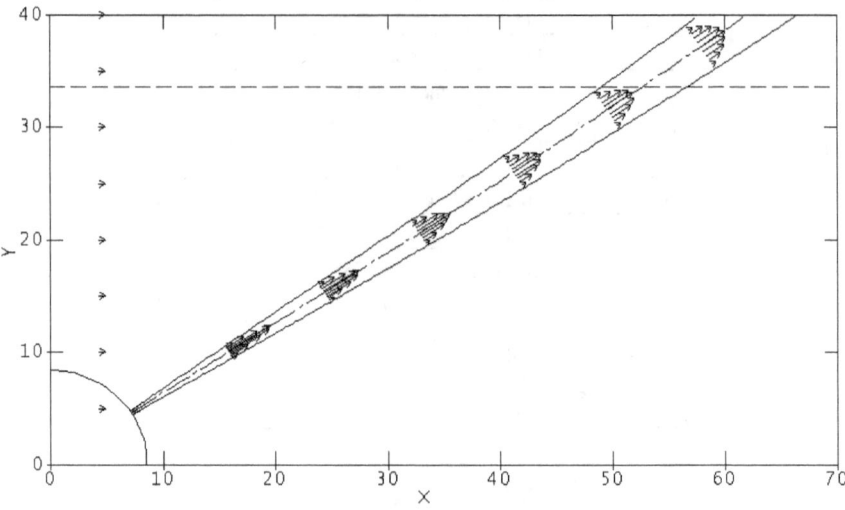

Figure 201. Rising Plume

Results for the fourth example (a sinking plume4.inp) are shown in this next figure:

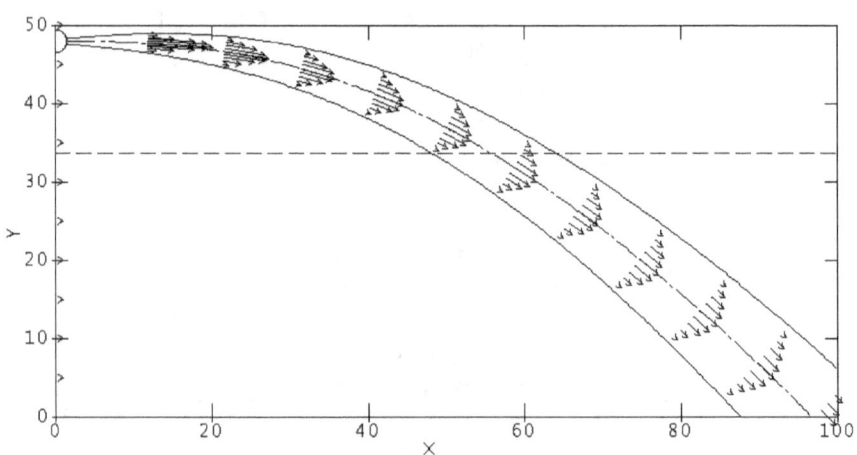

Figure 202. Sinking Plume

Chapter 29. 3D Thermal Plume

In the preceding chapter we only considered the receiving water on a gross basis. We will now take a much more detailed, three-dimensional approach. We first consider the geometry: three long pipes big enough to drive a truck through if they weren't under water along the bottom of a river. The bathymetry is shown in this first figure:

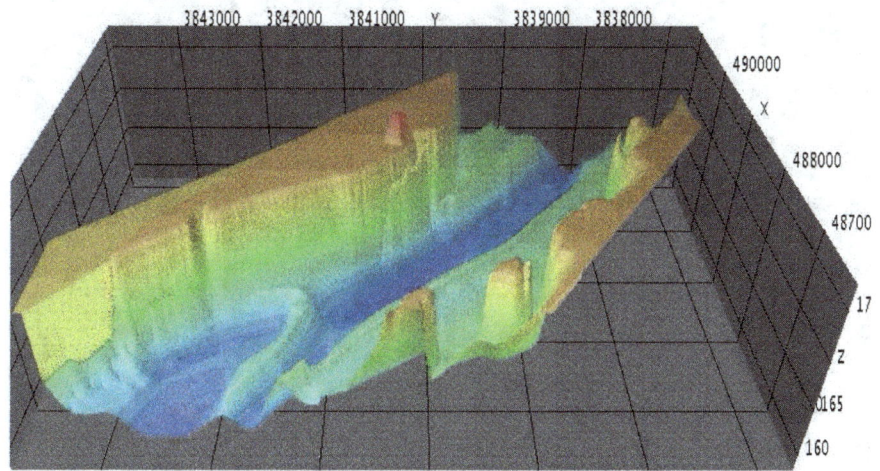

Figure 203. River Bathymetry

The diffusers are 20.5 ft (6.25 m), 19.0 ft (5.79 m), and 17.0 ft (5.18 m) in diameter. The lengths are 1010 ft (307.8 m), 1610 ft (490.7 m), and 2210 ft (673.6 m), respectively. Shown to scale, the pipes appear quite long and thin:

Figure 204. Diffuser Pipe in Profile

Without vertical exaggeration, it's difficult to represent the diffusers—even as large as they are—when compared to the river. The modeling is done at 1:1:1 scale for X:Y:Z.

The discharge flows are approximately 1450 ft³/s (41 m³/s) per diffuser. Water is discharged through 2-inch (0.05 m) ports spaced 6 inches (0.15 m) apart over approximately 600 feet (183 m), or about 7,000 ports per diffuser. The ports are drilled so as to span approximately 24° to 45° angle with respect to the horizontal. The navigation channel where the diffusers are located is about 850 ft (260 m) wide at this point, although the river banks are significantly farther apart than this. The diffusers are embedded in gravel and held in place by several concrete pours.

221

The vertical and horizontal aspect ratios are shown in the lower left corner of this next figure:

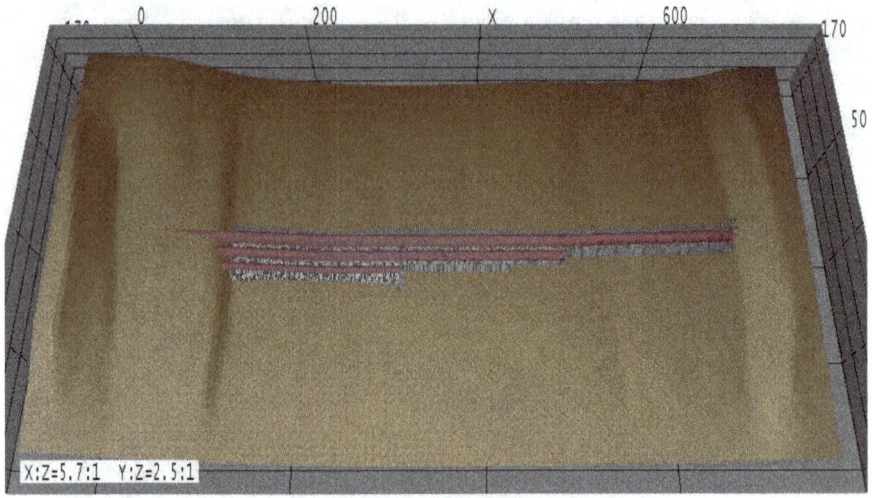

Figure 205. Embedded Diffusers

The area is shown in the following figure, including diffuser zone (in green), navigation channel, and overbank regions.

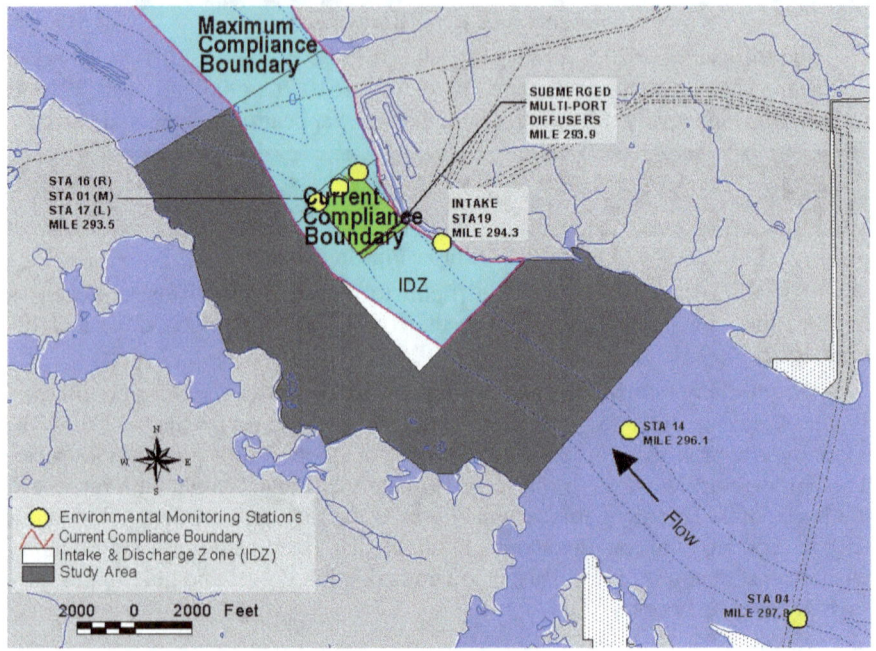

Figure 206. Aerial Map of Vicinity

The three-dimensional flow field, including discharge from the diffusers is calculated using the Environmental Fluid Dynamics Code (EFDC) is a public domain, an open source, surface water modeling system, which includes hydrodynamic, sediment and contaminant, and water quality modules fully integrated in a single source code implementation. EFDC has been applied to over 100 water bodies including rivers, lakes, reservoirs, wetlands, estuaries, and coastal ocean regions in support of environmental assessment and management and regulatory requirements.

EFDC was originally developed at the Virginia Institute of Marine Science (VIMS) and School of Marine Science of The College of William and Mary, by Dr. John M. Hamrick beginning in 1988. This activity was supported by the Commonwealth of Virginia through a special legislative research initiative. Further developments were made by Dr. Robert Byrne, the late Dr. Bruce Neilson, and Dr. Albert Kuo, of VIMS. Subsequent support for EFDC development at VIMS was provided by the U.S. Environmental Protection Agency (EPA) and the National Oceanic and the Atmospheric Administration (NOAA) Sea Grant Program. EFDC can be obtained from the EPA at:

https://www.epa.gov/ceam/environment-fluid-dynamics-code-efdc-download-page

The figure below shows the thermal plume in cross-section. Upstream is to the right and downstream is to the left. The magenta grid indicates the thermal compliance or designated *mixing* zone. This is at low flow, which is why the thermal plume can be seen extending upstream approximately as far as it does downstream. The picture for high flow is not that much different, as shown in the next figure. The vertical exaggeration in both figures is 50:1 (Z:X/Y). This figure also shows the primary grid in black. There is also a secondary (finer) grid surrounding the diffusers in order to capture the near field effects of the jets.

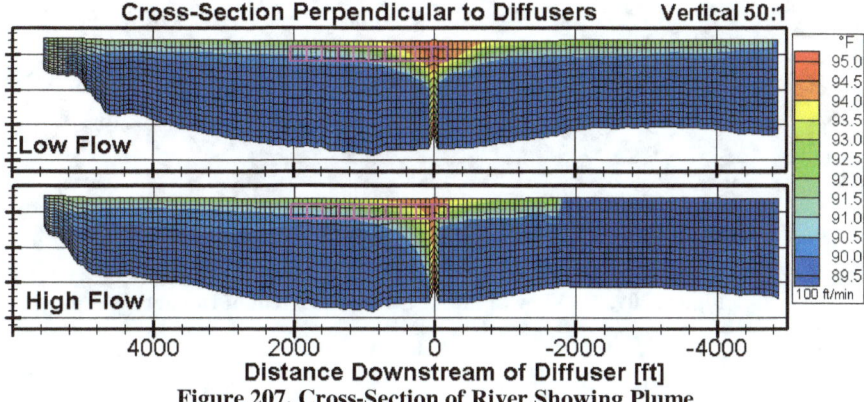

Figure 207. Cross-Section of River Showing Plume

While the discharge is approximately 120°F, the zone where this temperature persists is so small that it can't be seen at this resolution. The extent of the center thermal plume is indicated by the red mesh in the following figure, which shows velocity vectors as tiny blue arrows and the channel bottom as an undulating brown surface:

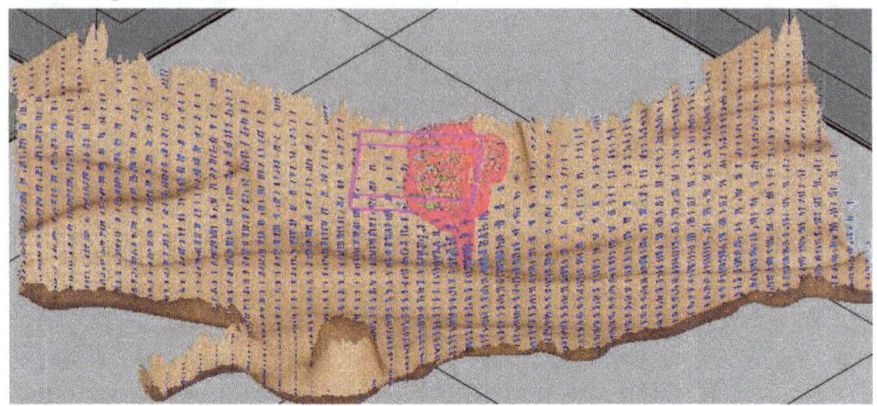

Figure 208. 3D View of Plume Showing Fish Lethal Zone

The combined plume from all three diffusers along with the compliance boundary is shown in this next figure in plan view:

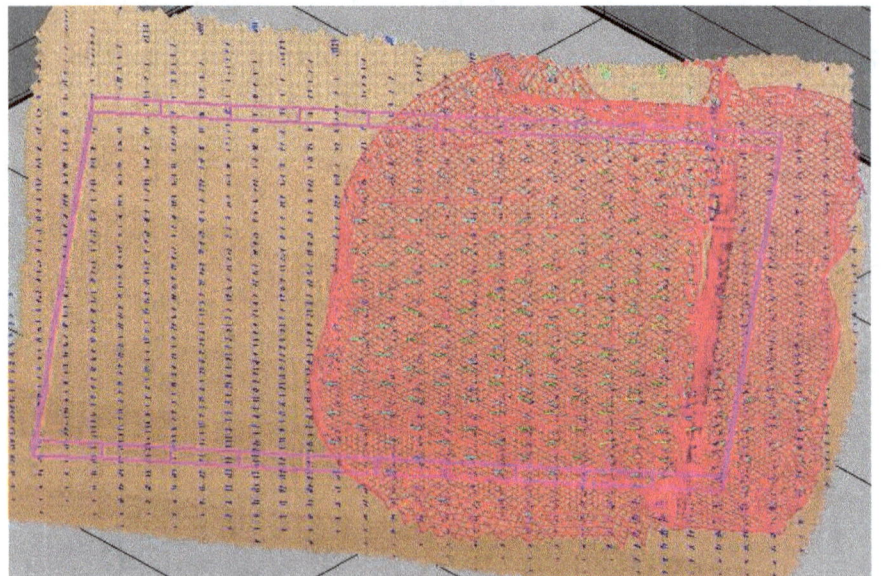

Figure 209. 3D View of Plume Showing Extent of Impact

The figure below is the same model results, only at a different angle showing depth:

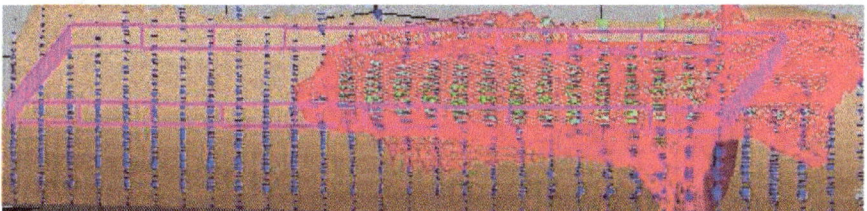

Figure 210. Side View Showing Depth of Plume

Another plan view with X:Z exaggeration of 5.7:1 and a Y:Z exaggeration of 2.5:1 is shown below. This view is looking upstream along the center of the navigation channel. The gravel and backset is also shown.

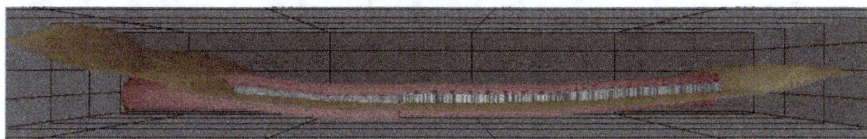

Figure 211. Side View of Embedded Diffusers

The extent of plume is shown as a blue shell and orientation of the diffusers is shown in this next figure:

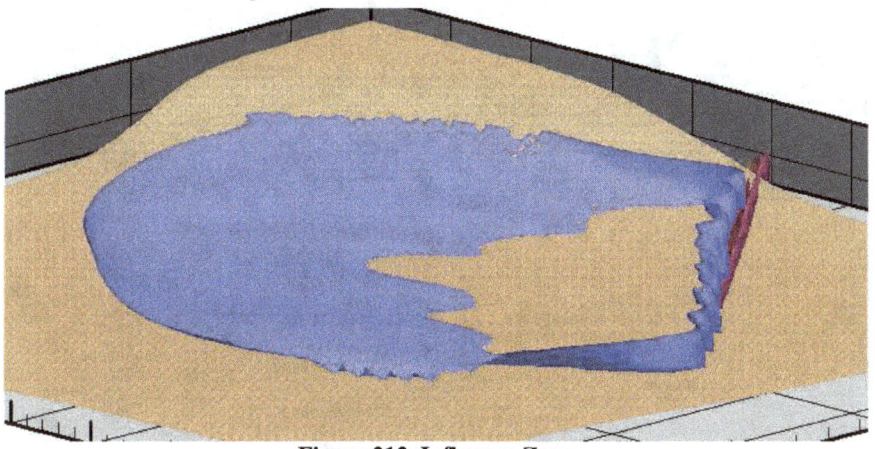

Figure 212. Influence Zone

Field Data

One of the extraordinary features of this project is that the modeling effort was supported by considerable field data, which was very costly to obtain, requiring many days effort by an experienced team. The low flow data are shown in this next figure:

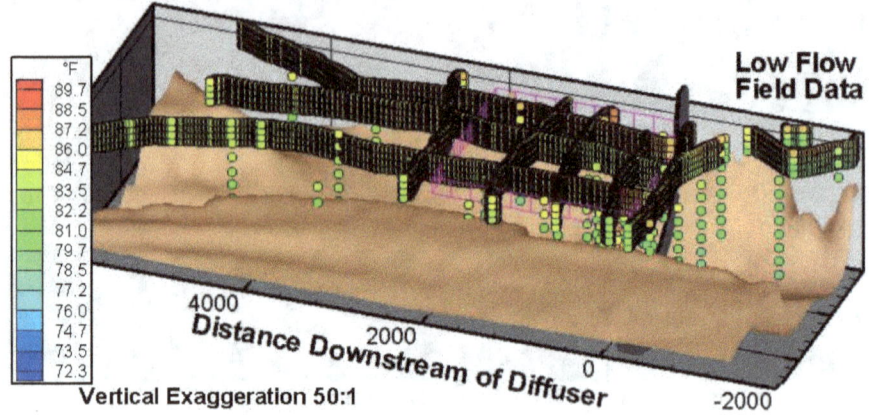

Figure 213. Field Data at Low Flow

and the high flow data are shown in this figure:

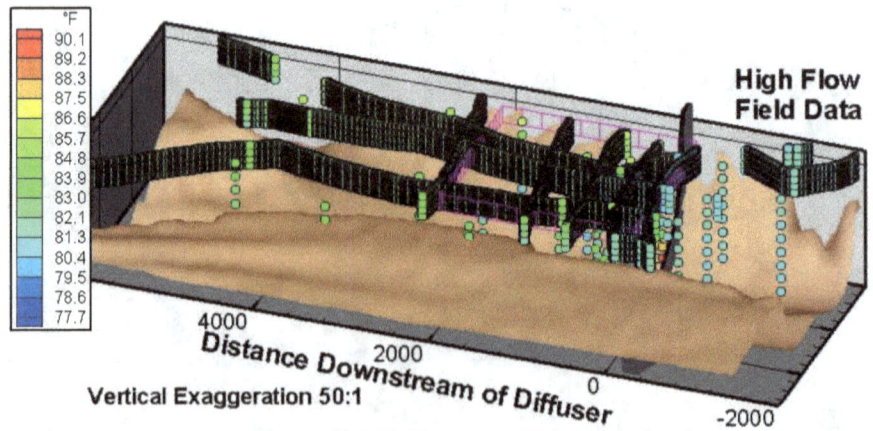

Figure 214. Field Data at High Flow

Agreement between field data for several flows, analytical models from the previous chapter, and the current 3D model are shown in this next figure:

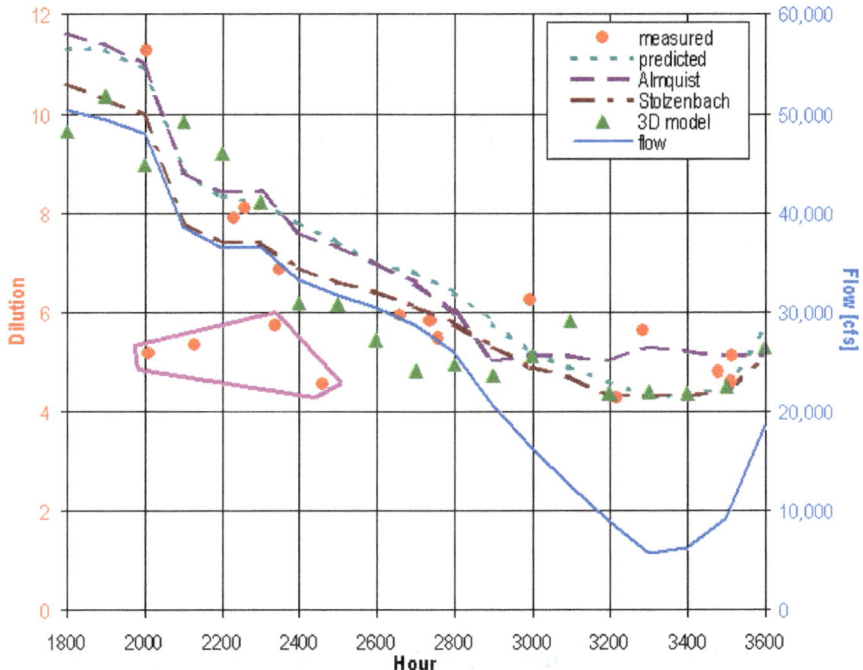

Figure 215. Dilution Data and Model Results

Agreement is acceptable (the green triangles are the near red circles) for high flow (left side of figure) and low flow (right side of figure) flows, except for the four data points inside the magenta box. The flow varied continuously over the 18-hour period, which is typical operation for this reservoir. While the previous analytical models (Almquist and Stolzenbach) and empirical calculation (dotted cyan curve, based on regression by Harper) and current 3D model all yield similar results for dilution, only the 3D model provided spatial details, which were of particular interest. The models of Almquist and Stolzenbach were based on laboratory (i.e., scale) models and not field data. The 3D model results were in general agreement with the field data. Graphical comparison of the field data and 3D model results were challenging, as these spanned several length scales and the field data were necessarily collected over several hours. The following graph shows the low flow data and model results:

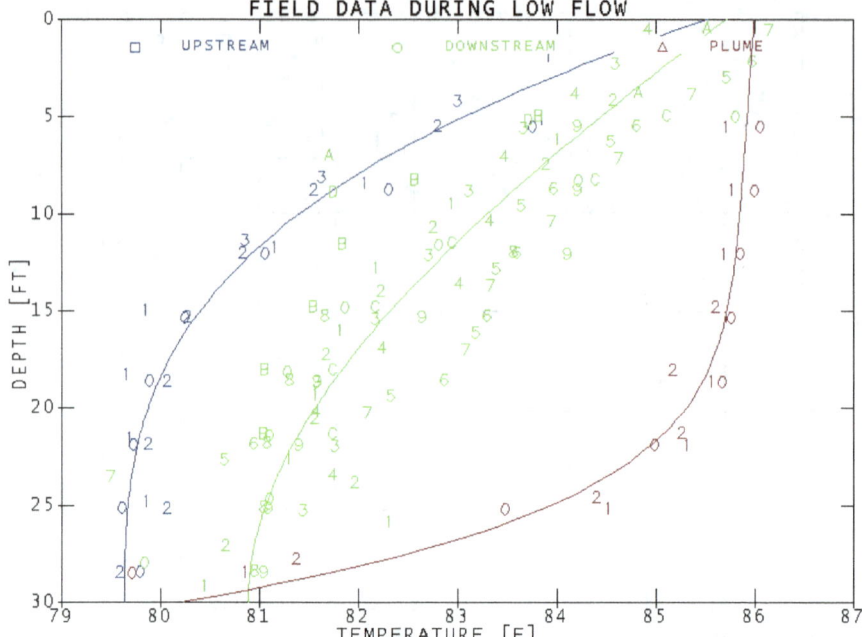

Figure 216. Spatial Data and Model Results (Low Flow)

Measurements near the plume (red numbers) agree quite well with the 3D model (red curve). Upstream measurements (blue numbers) agree quite well with the 3D model boundary conditions (blue curve). Downstream measurements (green numbers) are scattered about the 3D model results (green curve), due to large-scale eddies and other transient behavior. This is not surprising, considering the complexity of this system and the duration of the data collection (several hours). Recall the instantaneous and time-averaged plumes on page 31.

Field data and 3D model results for the high flow period are shown in this next figure:

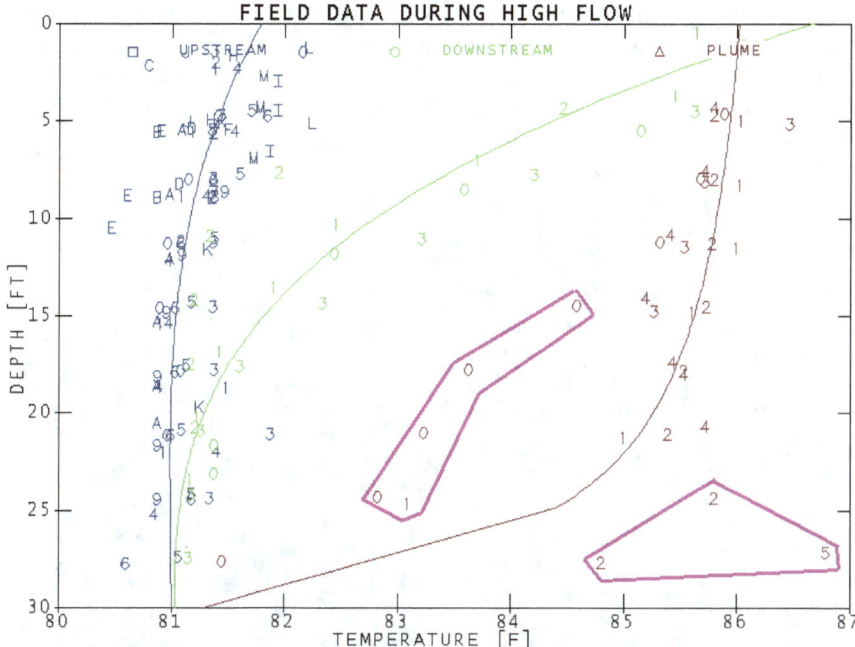
Figure 217. Spatial Data and Model Results (High Flow)

Upstream (blue letters and curve) and downstream (green numbers and curve) comparisons are similar to the low flow case, but we see more scatter in the near-plume results (red numbers and curve), as indicated by the data points in the two magenta boxes. Other than these 8 data points, the agreement is quite acceptable. Again, this is attributed to eddies and the data were collected over a span of time (18+ hours), while the 3D model runs are instantaneous snapshots.

Chapter 30. Three-Dimensional Geological Data

Most of the plumes I have modeled over the past thirty years have been contaminants in groundwater. I'm an engineer, not a geohydrologist. I first became involved with groundwater plumes by serving in the role of applied mathematician to assist a team of geohydrologists on the project known as MADE (MAcro Dispersion Experiment), which was conducted at the Columbus Air Force Base in Mississippi. Quite a few publications are available on this site, many of which are available on Research Gate.[50,51,52,53]

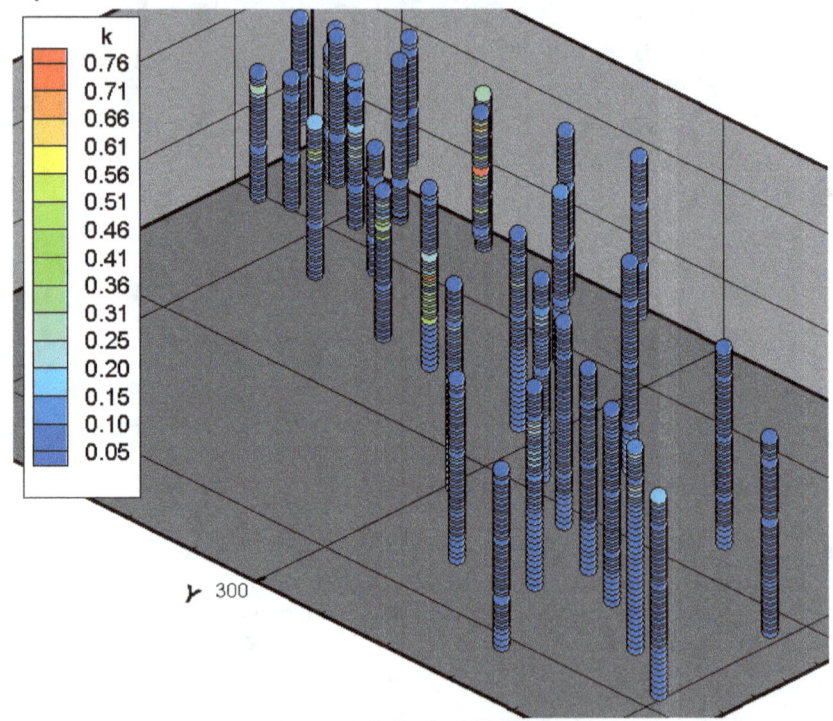

Figure 218. Typical Well Data

---

[50] Gelhar, L. W. and Axness, C. L, "Three-Dimensional Stochastic Analysis of Macro-Dispersion in Aquifers," Water Resources Research, January 1983.

[51] Stauffer, T. B, Boggs, J. M., and MacIntyre, W. G., "Ten Years of Research in Groundwater Transport Studies at Columbus Air Force Base, Mississippi," Biotechnology in the Sustainable Environment, December 1996.

[52] Li, L., Zhou, H., and Gomez-Hernandez, J., "A Comparative Study of Three-Dimensional Hydraulic Conductivity Upscaling at the MAcro-Dispersion Experiment (MADE) site, Columbus Air Force Base, Mississippi (USA), Journal of Hydrology, Vol. 404, No. 3, pp. 278-293, June 2011.

[53] Elfeki, A. M., and Rajabiani, "Simulation of Plume Behaviour at the Macro-Dispersion Experiment (MADE1) Site by Applying the Coupled Markov Chain Model," March 2013.

This important experiment is one of the very few large-scale groundwater experiments where tracers were intentionally injected into the ground. It is also important that work has continued at the site for so many years. Data collected at the site are uncharacteristically *good* and detailed compared to the many more unintentional contaminant sites we will cover subsequently. First, what do the data look like? For one thing, measurements come from wells. These first three figures are the first data set ever extracted from the MADE site. On the previous page we see a 3D view. This next view is in the horizontal (XY) plane.

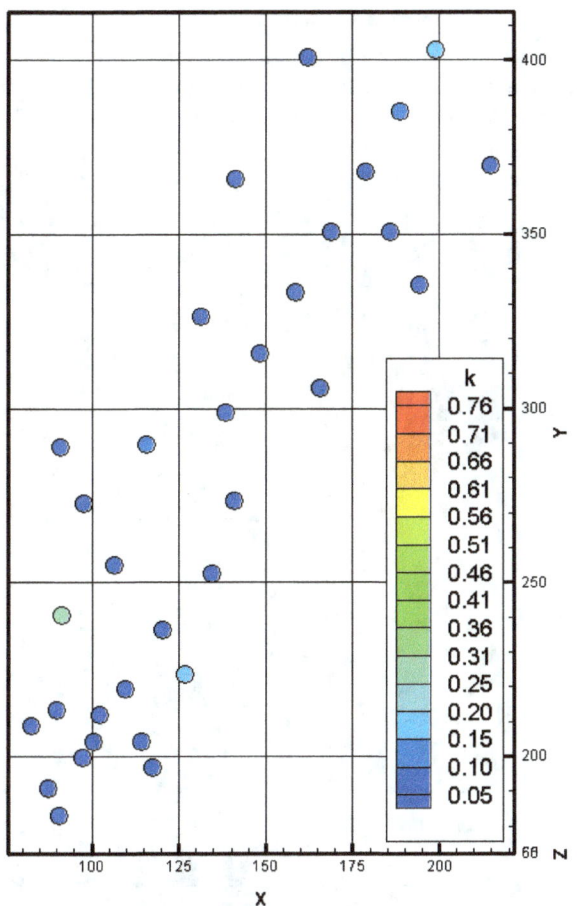

Figure 219. Well Data Top View

As you can clearly see, the data are not evenly spaced, which means that simplistic interpolation methods are not likely to be effective. We have already discussed the inverse distance method (Appendix D). This is just one of the methods tested on this data set.

231

This next view is in the vertical (YZ) plane:

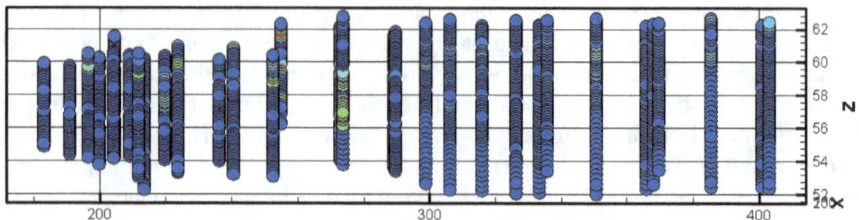

Figure 220. Well Data Side View

Note that this data is hydraulic conductivity and non-zero, in spite of the reported values. Reasonable bounds were placed on the data in order to obtain this representation. A bounding polygon was also applied. A 3D slicing of the interpolated field is shown below:

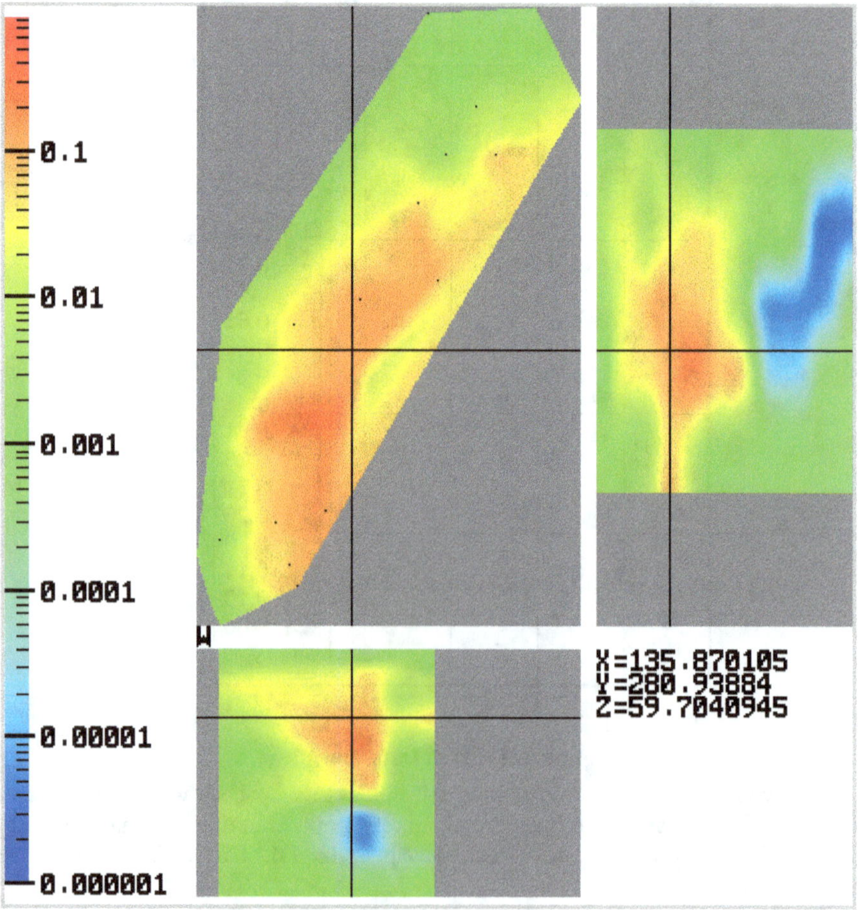

Figure 221. Planar Slices of Plume in 3D

The preceding result was obtained by first applying the inverse distance method with several control points, then smoothing the results repeatedly until the spatial artifacts of well sampling were not discernable (see Appendix E for more details). This is, of course, a subjective process, which was guided by the lead hydrologist and implemented by this author.

The second data set to come from the MADE site was also hydraulic conductivity but in a different area. This came as a surprise because the values were so different, which led to a revaluation of the methods and instruments. This data was also bounded in extent. The bounding polygon is shown in the figure below as white dots:

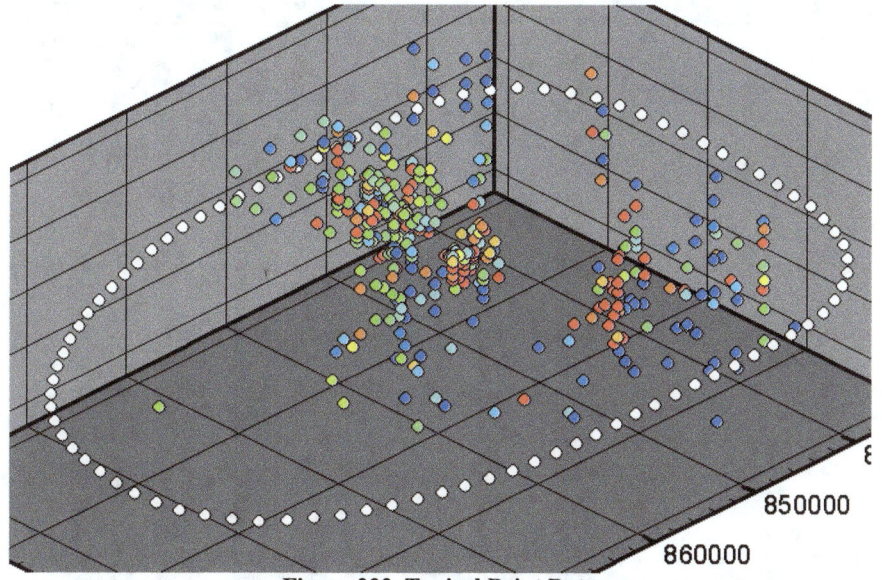

Figure 222. Typical Point Data

This data set proved more challenging than the first, which let the team to try other methods of analysis, including relaxation, finally settling on a combination of inverse distance and relaxation. When we get to the point of analyzing tracer concentration levels, we will use the same techniques developed for analyzing the hydraulic conductivity. This step also paved the way for numerically modeling the site, which was one of the primary goals of the study.

The second data set was so *blotchy* that it defied smooth approximation. The team eventually adopted a different approach. First, we will consider inverse distance interpolation. The results are shown in the figure at the top of the next page. This wasn't what we expected or wanted, but it does generally follow the trend that is suggested by the data. Still, it doesn't look like the sort of conductivity field you want to feed into a computer model. This was produced using Tecplot™. TP2 doesn't produce exactly the same results for nominally the same method (viz., inverse distance with octants and 2.5 power) because we modified the approach.

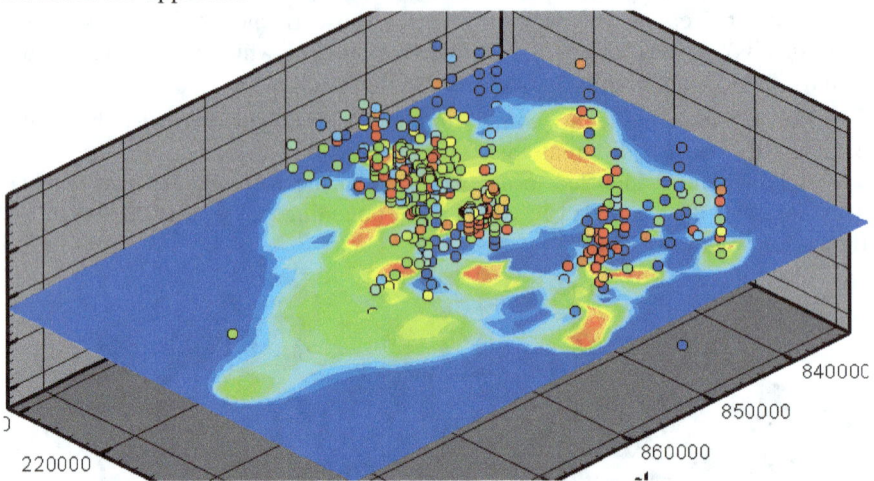

Figure 223. Horizontal Slice with Point Data

This next figure is what Tecplot™ produces using kriging:

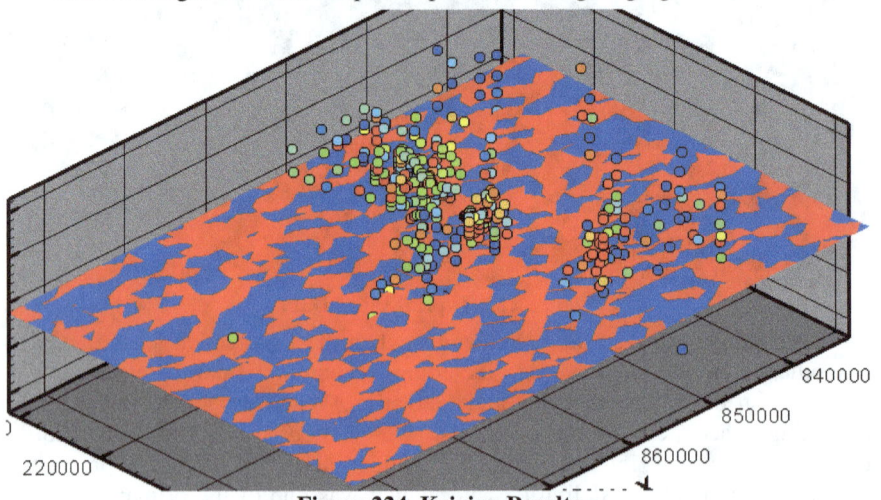

Figure 224. Kriging Results

This second figure is completely unreasonable—again, no criticism of Tecplot™, which faithfully implements the conventional formulas. Up until this point, the conventional wisdom was, "No problem. Just use kriging." Clearly, the conventional wisdom is not adequate, nor is traditional kriging. Inverse distance coupled with relaxation, was built into TP2 and the project moved forward toward success. See Appendices A, B, and C for more details.

Modified kriging (see Appendix F) yields the following results, which are adequate:

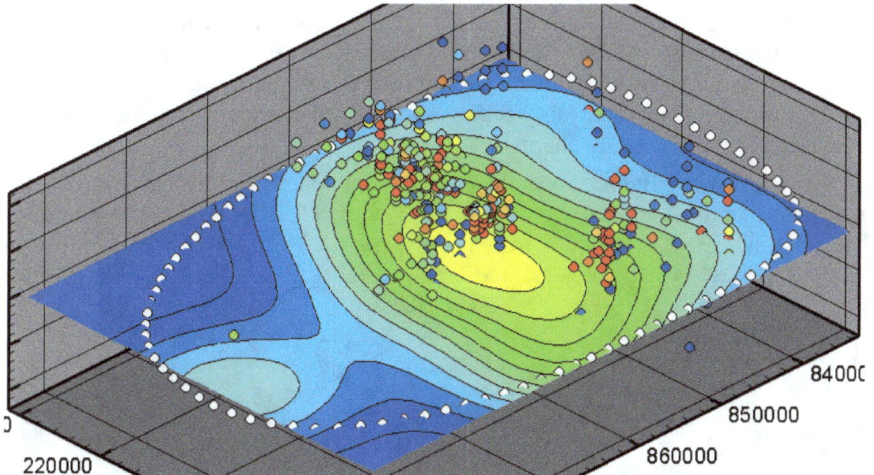

Figure 225. Smooth Results (Not Kriging)

Chapter 31. Contaminant Plumes in Groundwater

This topic is where most of my efforts have been devoted and for which I have the most data. Data comes in the form of point concentrations obtained from extraction wells. This represents a snapshot in time over the duration of sampling, which is typically much shorter than the residence time or the lifetime of the plume. Data analysis techniques (and challenges) are the same as discussed in the preceding chapter.

The following figure shows the contaminant concentrations what was by far the most plentiful data set we ever worked with (15,723 points). The extent of plume is indicated by the thick magenta polygon.

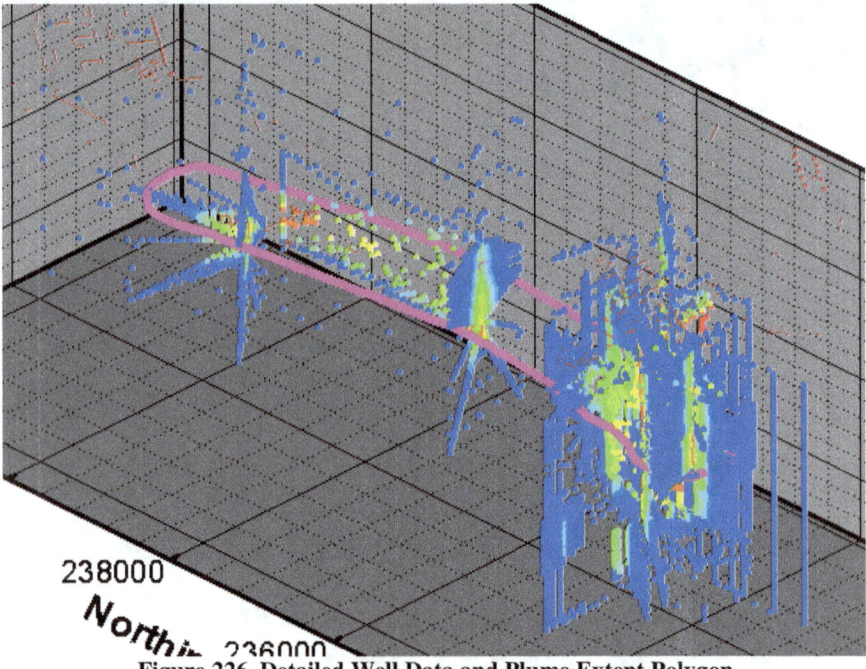

Figure 226. Detailed Well Data and Plume Extent Polygon

Both Tecplot™ and TP2 will extract iso-surfaces from a volume. Iso-surfaces are shells of constant value, in this case, concentration. These are quite useful for visualization. TP2 will also extract iso-lines (i.e., contours) from a surface. Iso-surfaces show the contaminant as a blob in space, which can be rotated, sliced, and viewed. Iso-surfaces are an important part of devising effective remediation strategies, especially when there are attendant structures, such as a stream, lake, or road.

Concentration iso-surfaces for the preceding data (log(C)=-5, -4, -3, -2, -1, 0, and 1) are shown in the figure below:

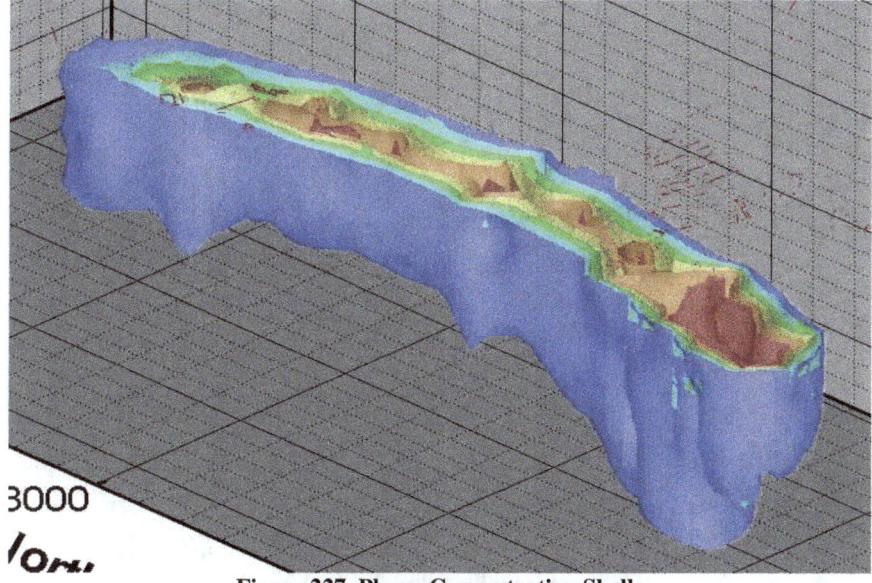

Figure 227. Plume Concentration Shells

In this view, you can see how the magenta bounding polygon in the previous figure controls the overall shape of the blob. We called this plume *larva*. Data for the next plume is shown in the figure below:

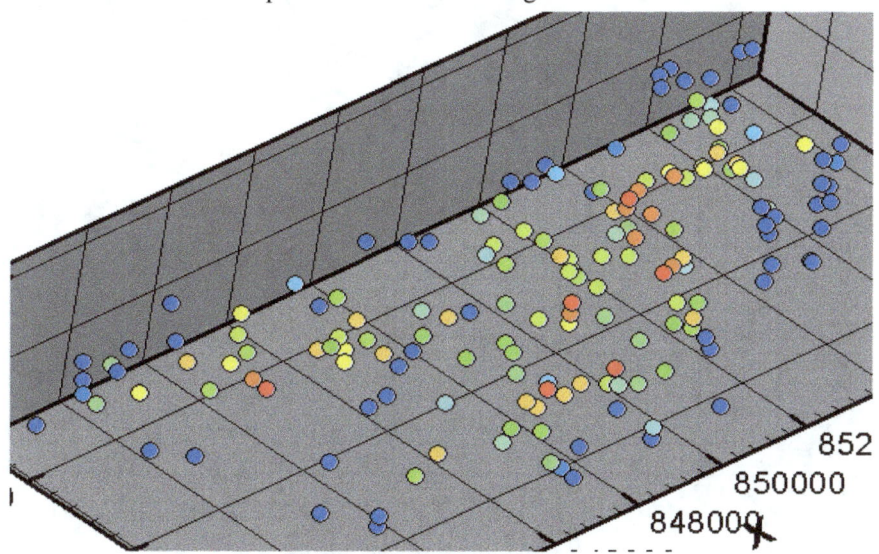

Figure 228. Sparse Point Data

Concentration iso-surfaces for what we named the *sandal* plume shown on the cover are shown in this next figure.

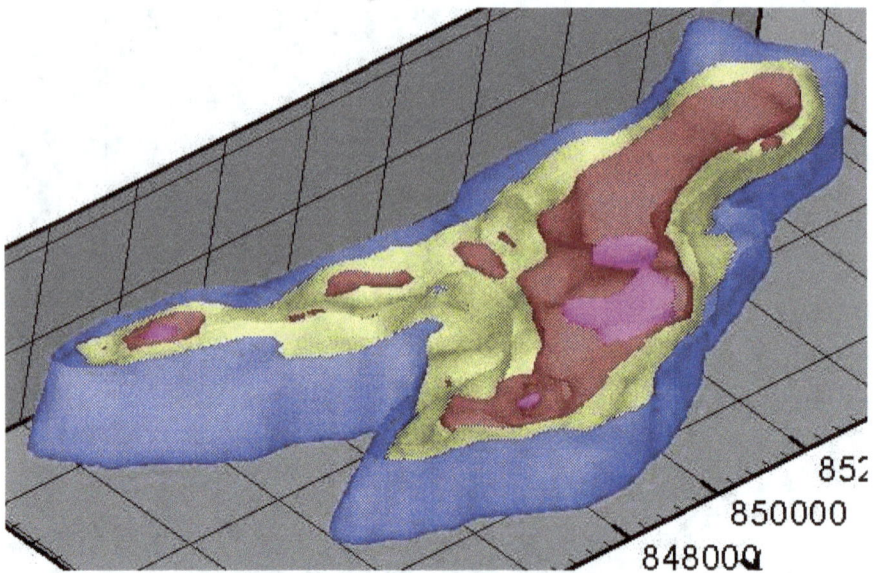

Figure 229. Plume Concentration Shells

Contours for the same 3D field are:

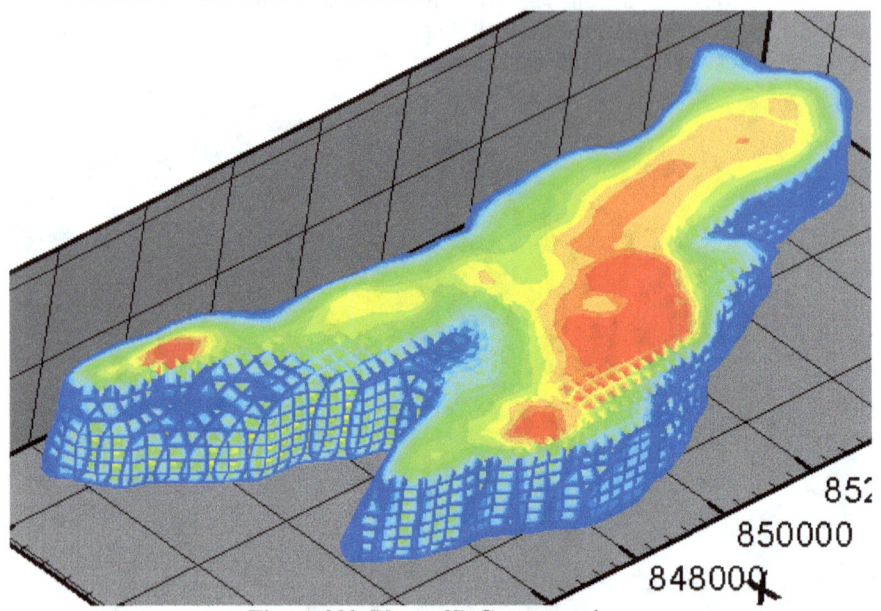

Figure 230. Plume 3D Concentrations

Data for the *drumstick* plume shown on page ii can be seen in this next figure:

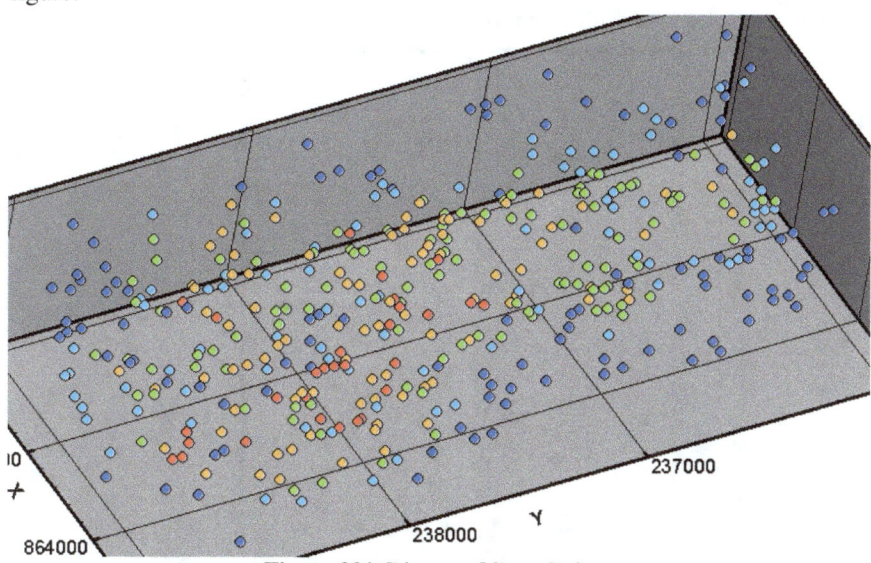
Figure 231. Dispersed Data Points

One slice of contours is shown below:

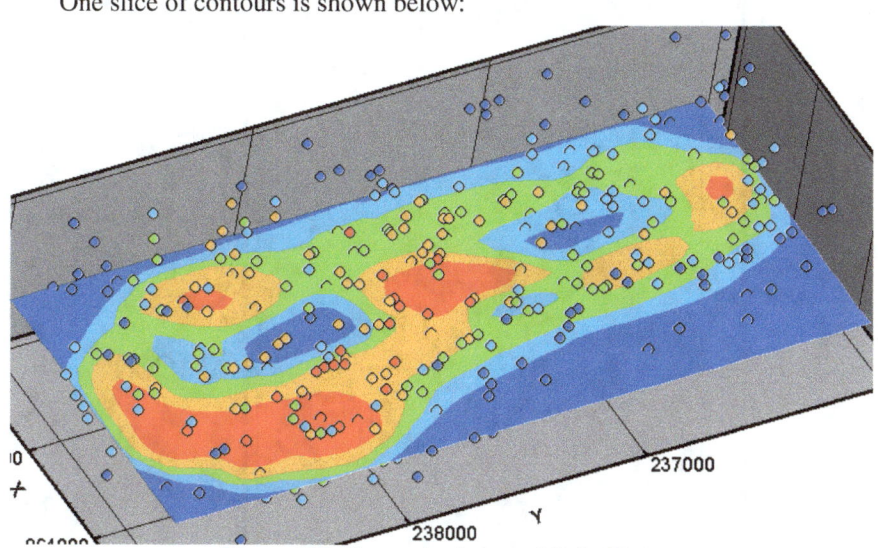
Figure 232. Concentrations and Point Data

Contours in a horizontal slice through the middle of the plume and boundary for the *banana* plume are shown in this next figure along with surface details, which are quite helpful in formulating a remediation strategy:

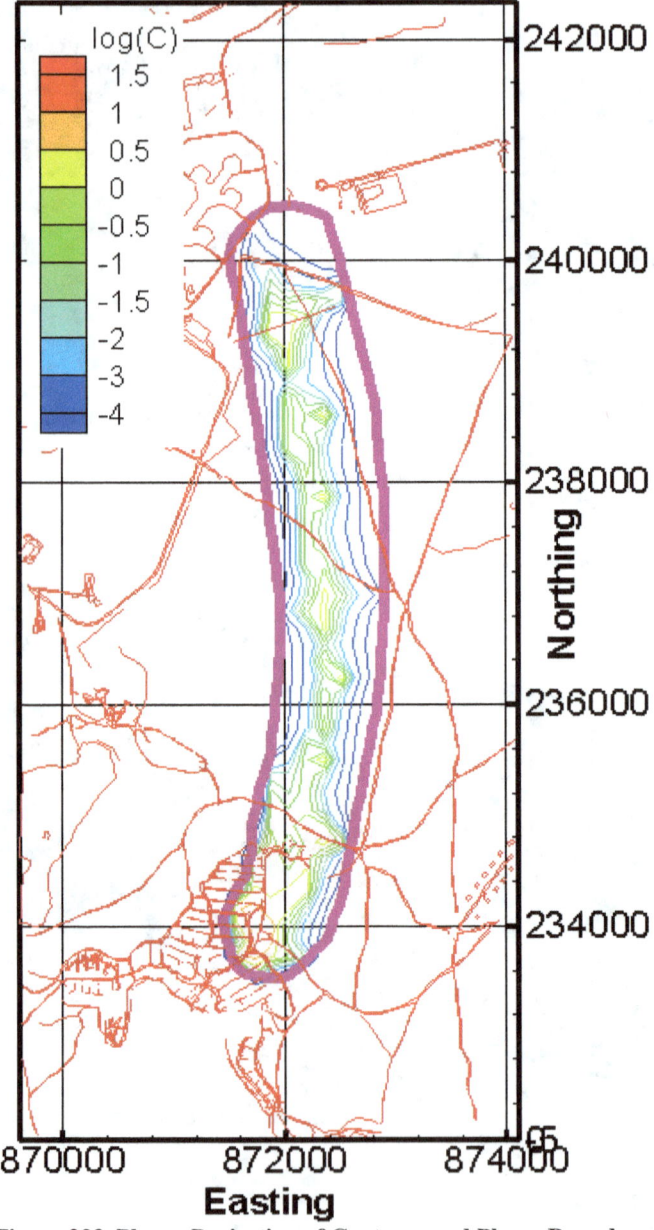

Figure 233. Planar Projection of Contours and Plume Boundary

Data for the *banana* plume are shown in the figure below:

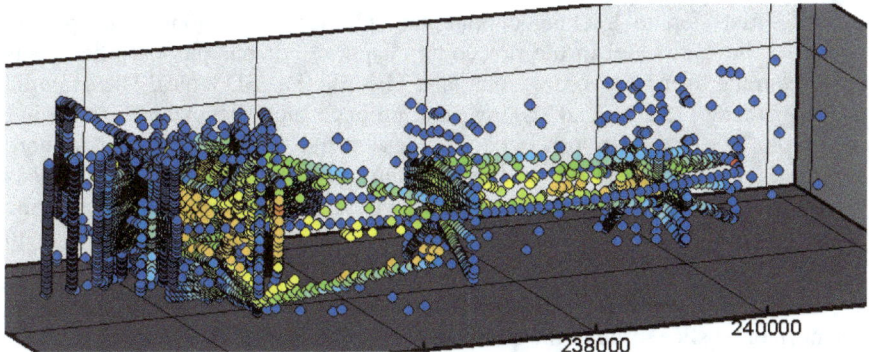

Figure 234. Plume Data for "Banana" Plot

The data were concentrated near what was initially thought to be the highest concentration center of the plume.

Chapter 32. Particle Tracking of Plumes

The first step in tracking contaminant plumes is developing the plume, which we have covered in the preceding chapters. For our purposes here, this must be in the form of a volume, that is, a TB3 file (i.e., 3D table). The example codes (invdist.c, relax.c, and kriging.c) all produce such files, which are similar to Tecplot™ files (also produced by these same codes), only much more compact. The second step in tracking contaminant plumes is generating seeds (initial particles: position and mass), which accurately represent the plume defined by the specified volume. Add a flow model and the particle tracker to complete the process.

The first example we will consider here is the crab *claw* plume. Data plus one horizontal slice through the contours are shown in this first figure:

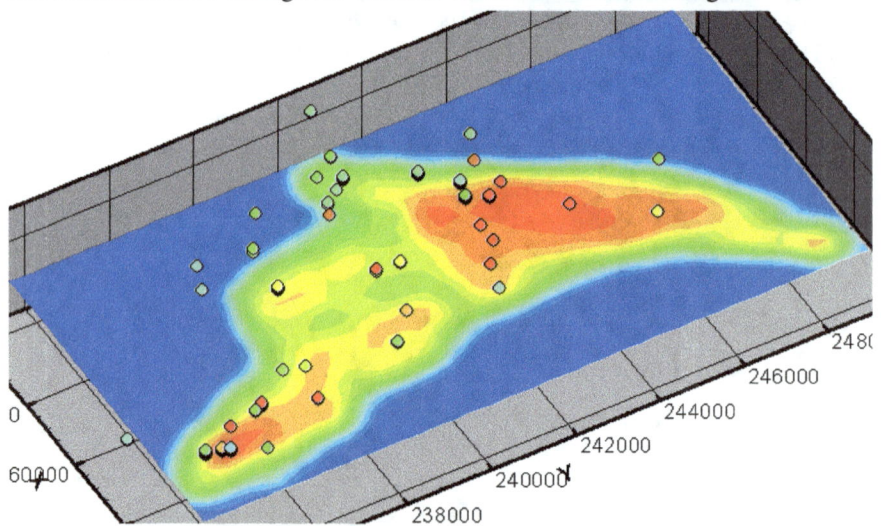

Figure 235. Plume Concentrations and Data

In order to create the seeds we need the volume plus a boundary polygon plus an algorithm. There are at least two ways of creating seeds to represent the concentration field: 1) uniformly distributed seeds of varying mass proportional to local concentration and 2) seeds of uniform mass distributed proportional to local concentration. Some combination of the two would be a third possibility. Generating random numbers in one dimension that have a specified probability distribution other than the normal (i.e., Gaussian) can be done, but in three dimensions and with multiple highs and lows? That is no trivial algorithm. The code (seed3d.c) can be found in the online archive in folder examples\seed3d.

Iso-surfaces of constant log(concentration) for this plume are shown in the figure below:

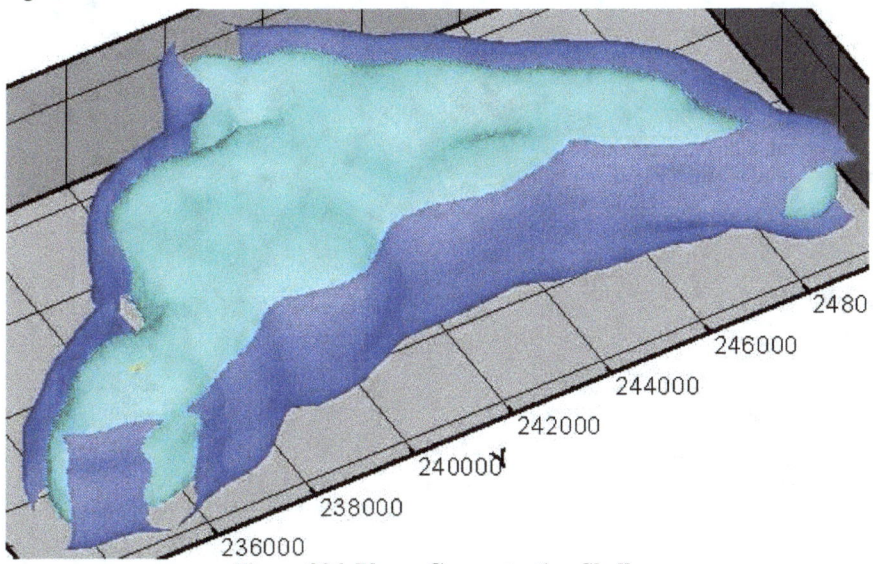

Figure 236. Plume Concentration Shells

Before we describe how to create the seeds, consider the following figure showing 1775 seeds of equal mass (our first attempt):

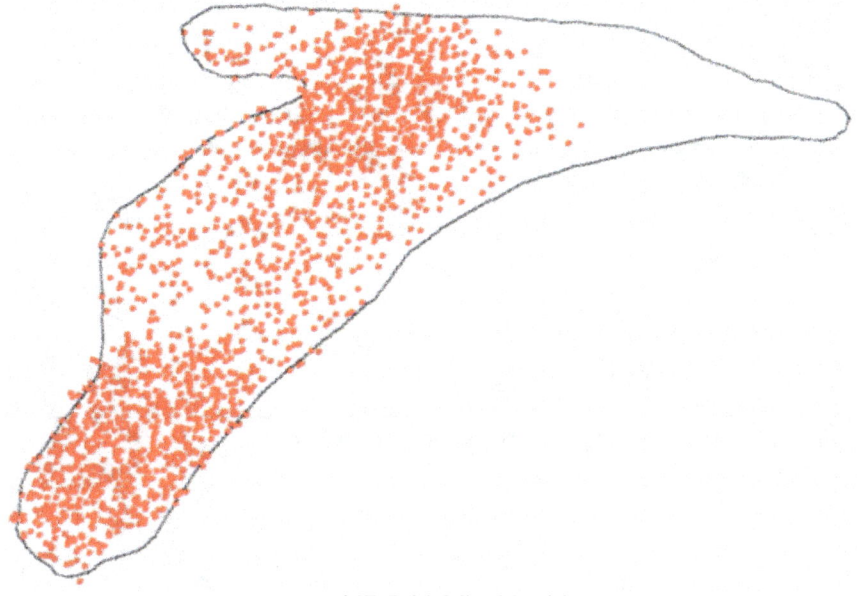

Figure 237. Initial Seed Positions

This next attempt with 50,004 seeds is much more representative of the actual plume as described by the concentration volume:

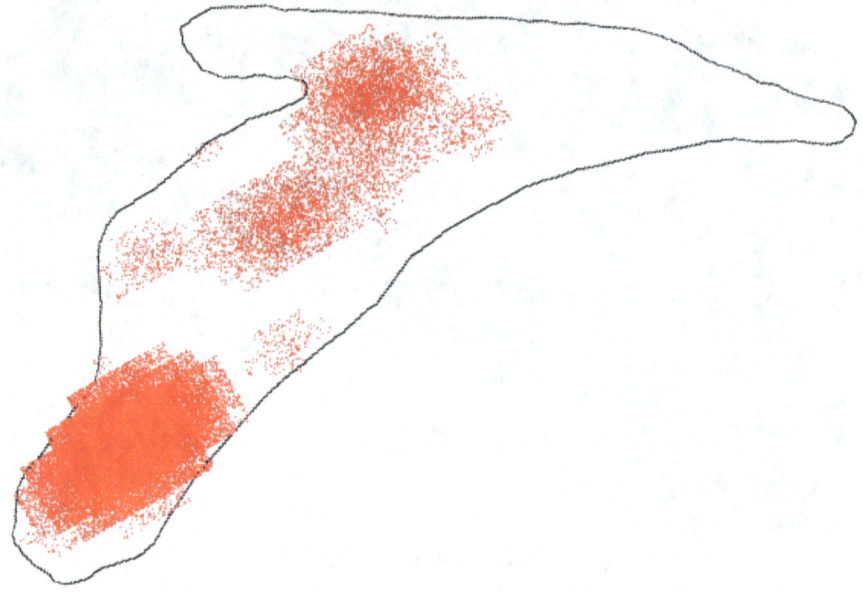

Figure 238. Seed Density Based on Initial Concentration

This *sprinkling* of particles is accomplished by first splitting the hexahedra (six-sided rectangular brick elements) of the volume up into tetrahedra, then filling these randomly based on the local concentration. The number of particles per tetrahedron is proportional to the concentration. The particles are scattered about inside each tetrahedra that meets the threshold concentration; otherwise, there are none. The concentration window is adjusted until the resulting number of particles roughly matches the target quantity. A side view of the seeds in the YZ plane reveals the 3D nature of this problem:

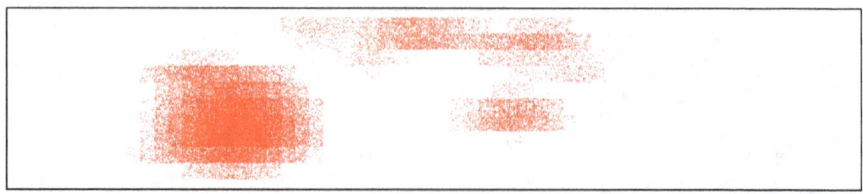

Figure 239. Seed Density Side View

Even after the seeds are created, there's still the matter of calculating the flow field and feeding all of this information into the particle tracker. For this plume, that looks like this:

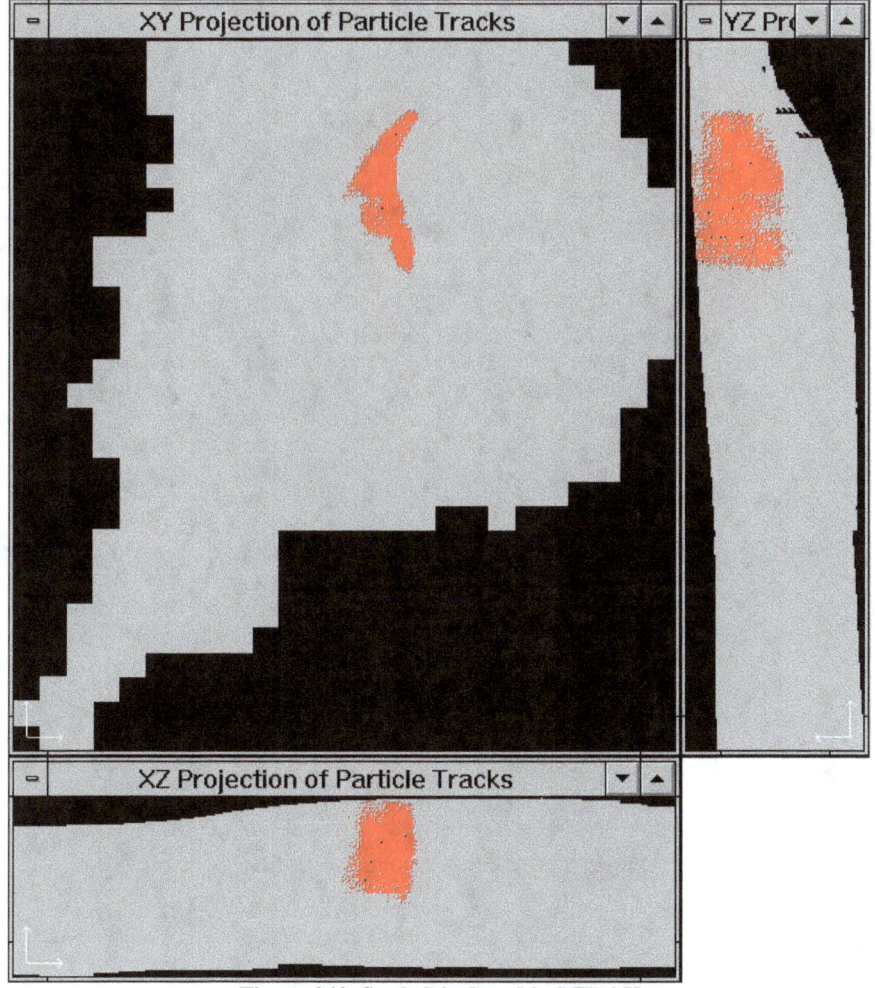

Figure 240. Seeds Displayed in PTRAX

This plume was part of a larger study involving 7 plumes. This next figure shows the expected position of the plume (shown in green instead of red) after 3 years with no remediation.

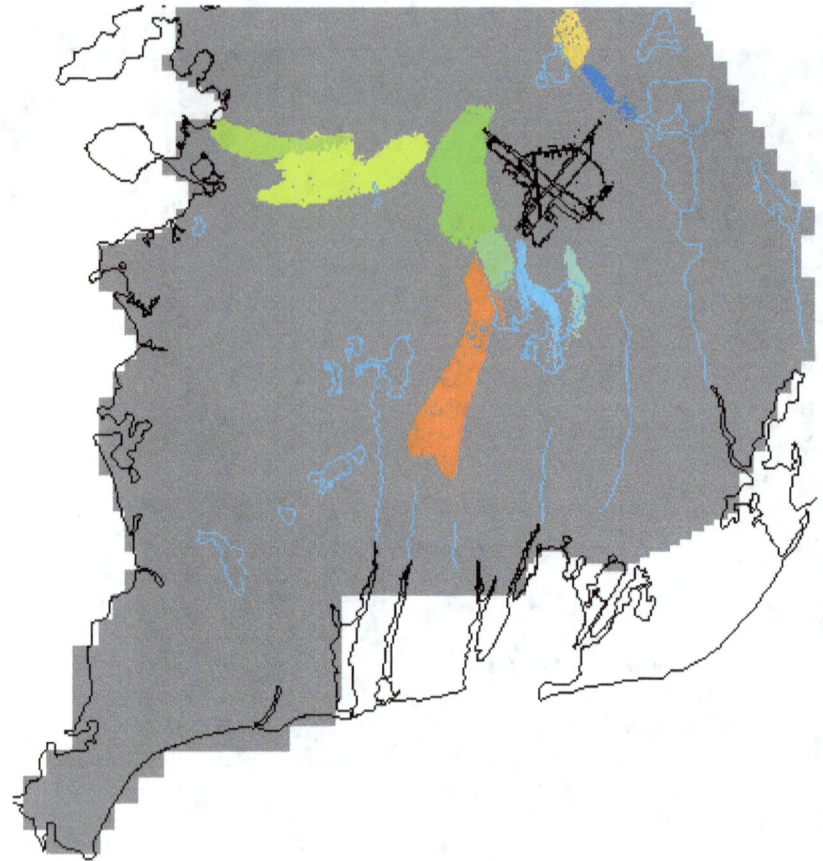

Figure 241. Multiple Plumes Initial State

This next figure shows the expected position of the plume (in green) after 30 years without remediation.

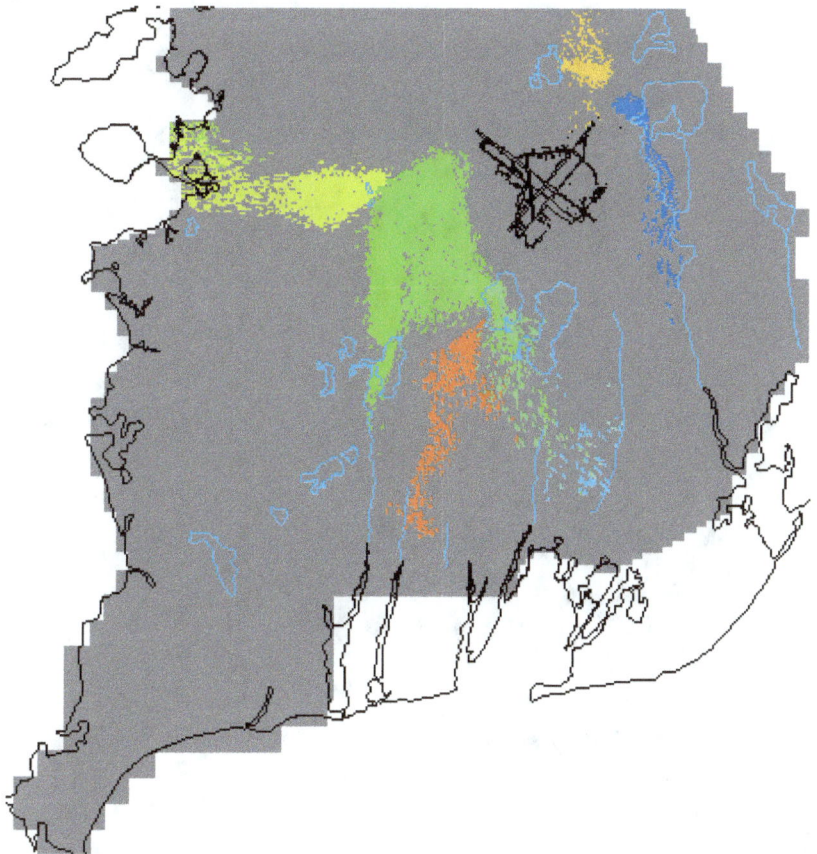

Figure 242. Multiple Plumes without Remediation after 30 Years

These seven plumes were remediated by installing a series of extraction wells attached to removal systems as indicated by the black +'s and o's in the following figure:

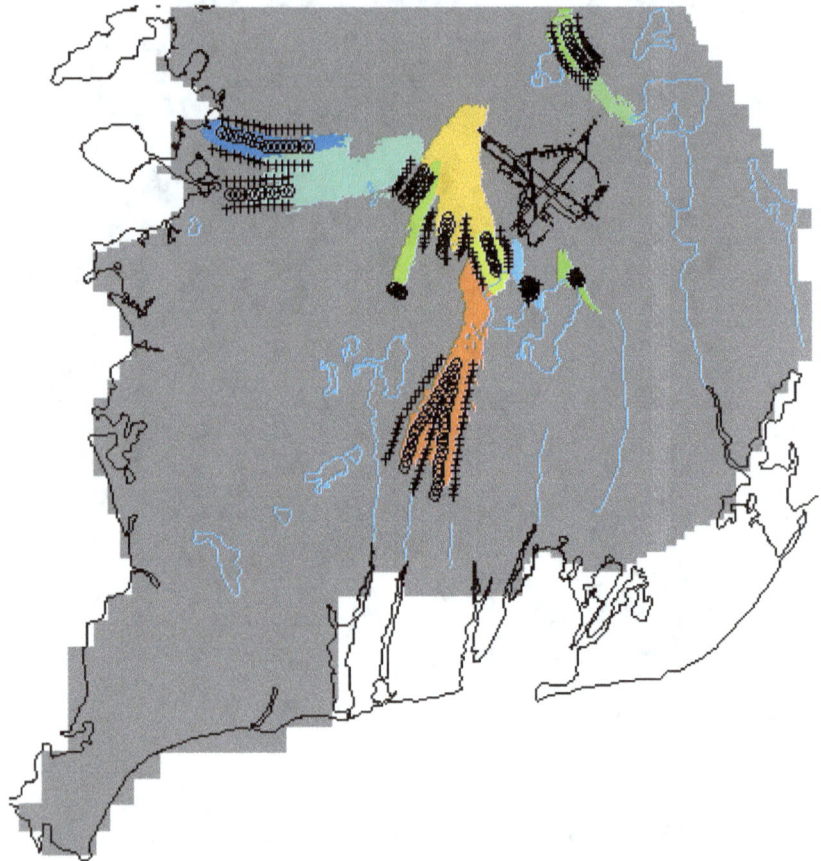

Figure 243. Remediation Wells

The expected results after 30 years are shown in this next figure:

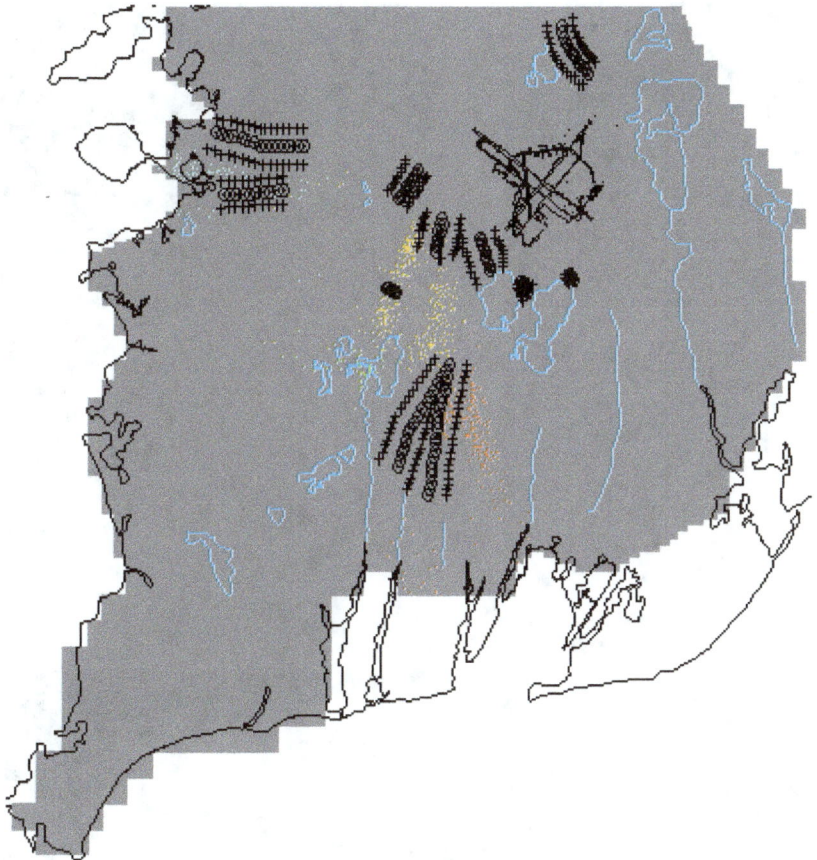

Figure 244. After 30 Years with Remediation

Twenty years have elapsed since this plan was implemented and the results have been quite satisfying, confirming the modeling efforts as well as the physical processes involved in capturing the contaminants.

We referred to this next plume as the *starship*:

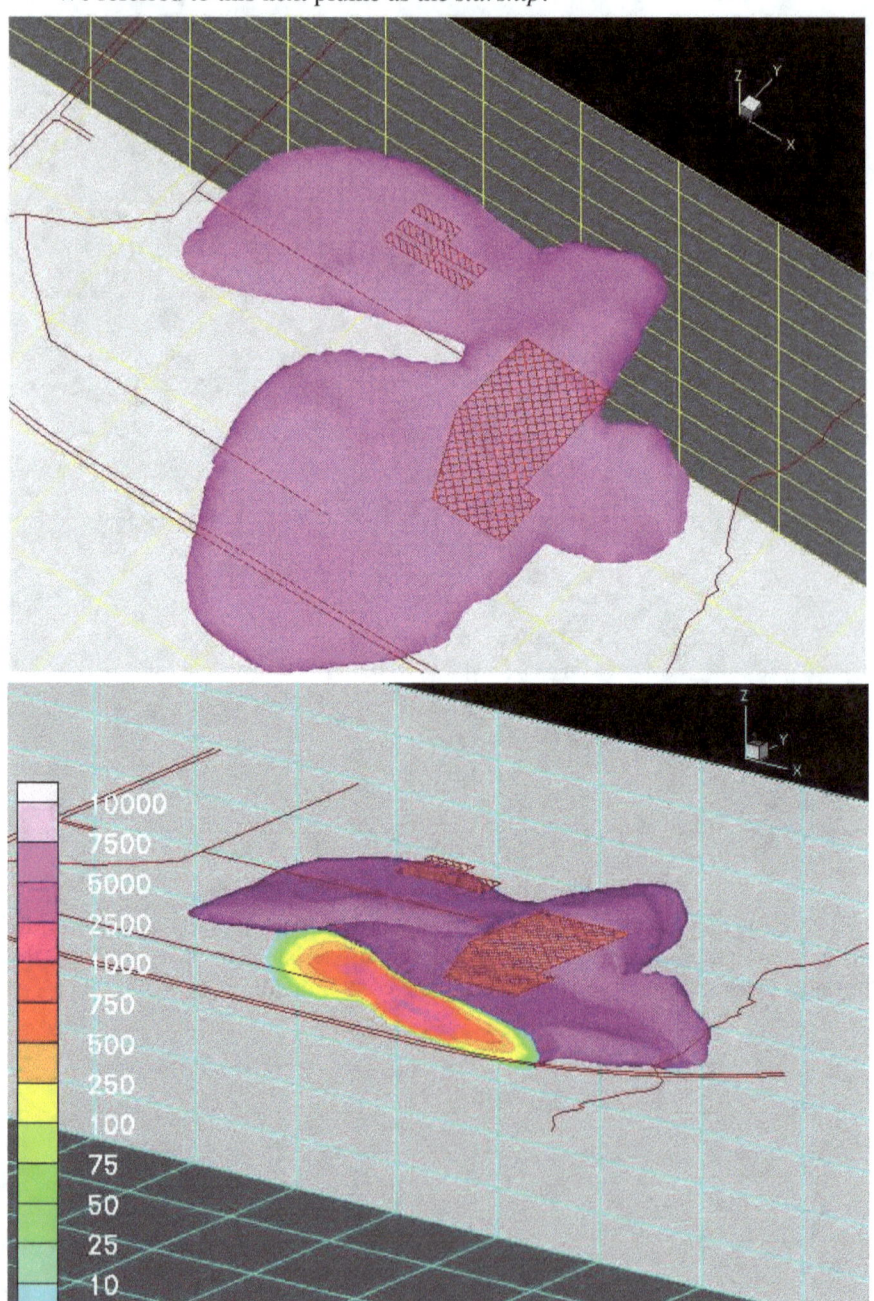

The plume was approximated by 706,042 seeds, which were sprinkled:

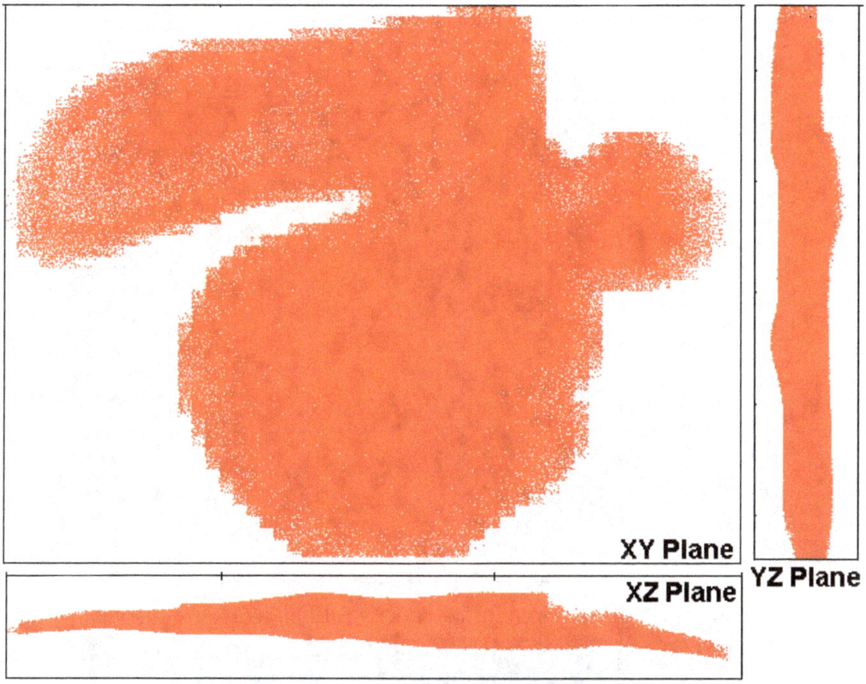

Figure 245. Seeds Representing Plume

Inside the particle tracker at initialization (time=zero):

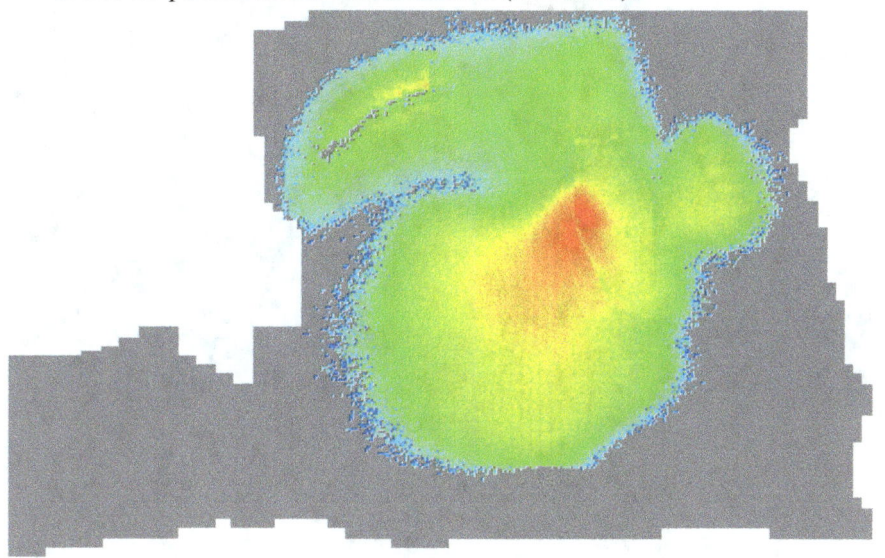

Figure 246. Initial Particles

After 30 years without remediation, the expected concentrations are almost unchanged:

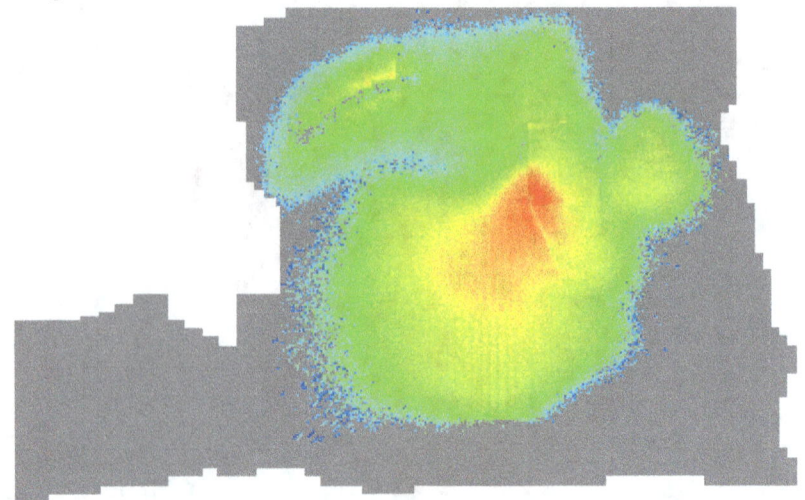

Figure 247. Expected Particles after 30 Years

A different view of the same initial concentrations:

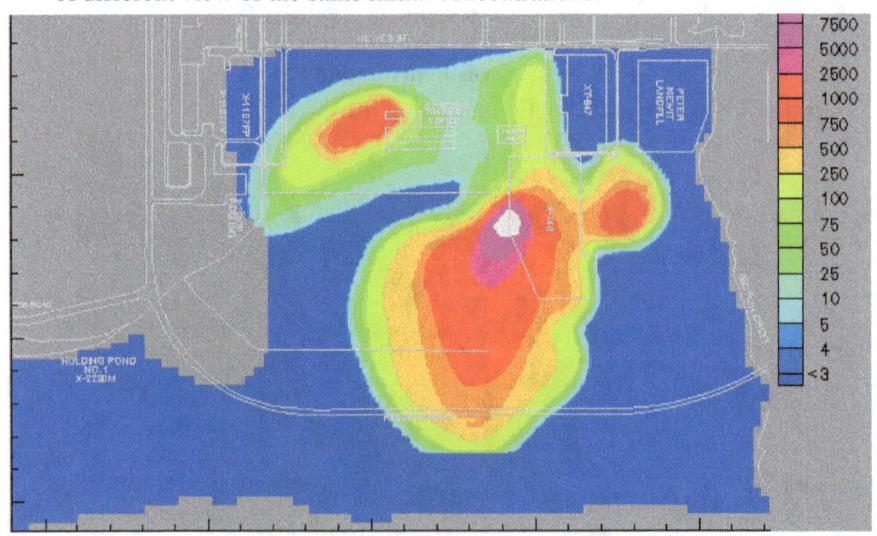

Figure 248. Initial Concentrations

And after 30 years of remediation (first proposed approach):

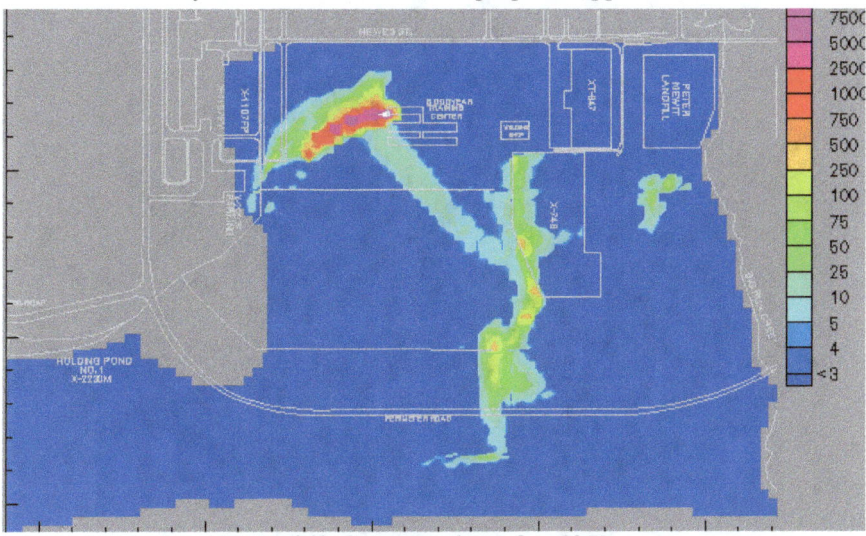

Figure 249. Concentrations after 30 Years

Concentration data (along with white control points) for this last plume are shown in the figure below:

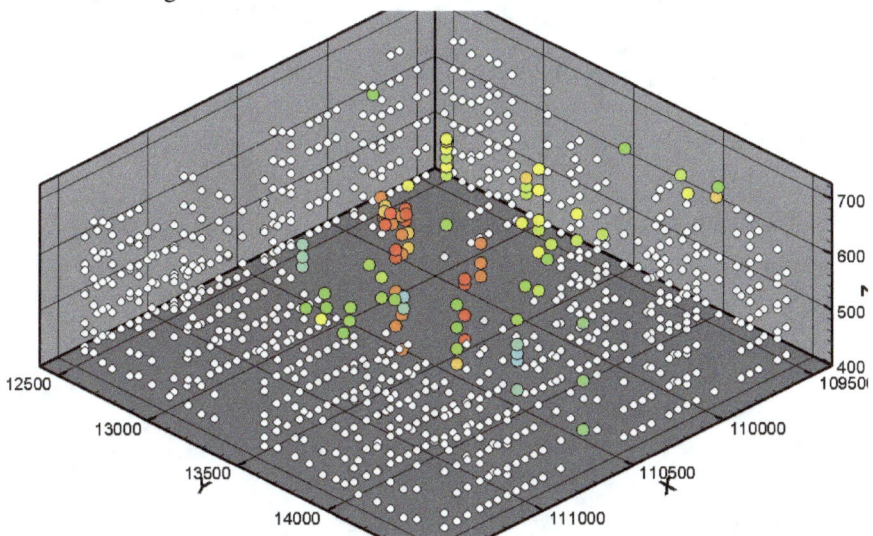

Figure 250. Measured Data + Control Points

Concentration contours for this field are shown below:

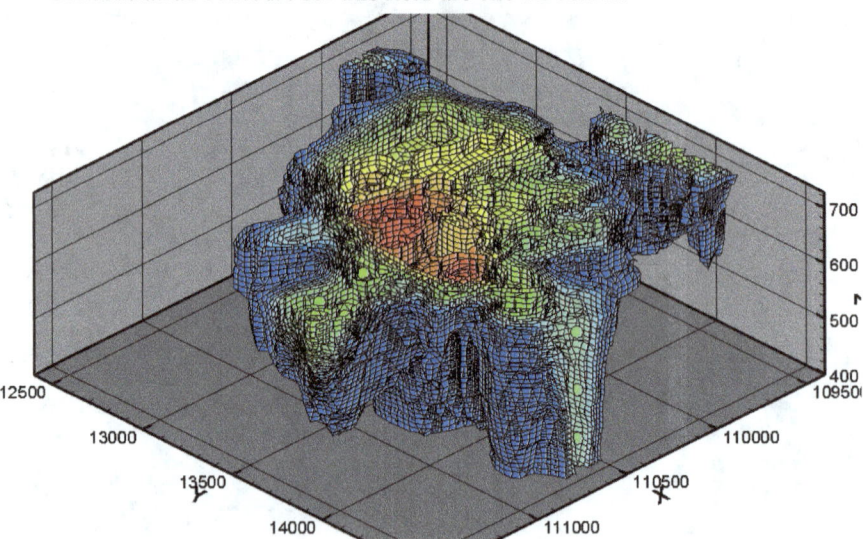

Figure 251. Initial Concentrations

A plan view (XY plane) of the seeds:

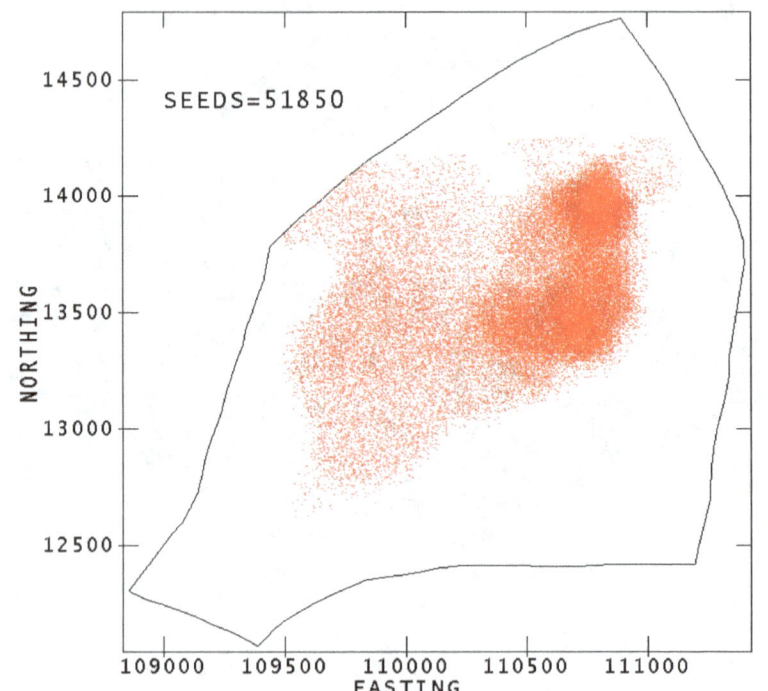

Figure 252. Initial Seed Locations Plan View

A perpendicular view of the seeds (XZ plane):

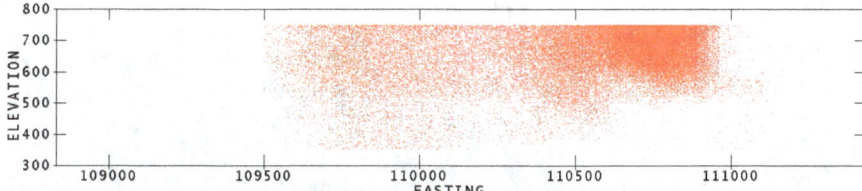

Figure 253. Initial Seed Locations Side View

The first scenario for this plume at time t=0 is:

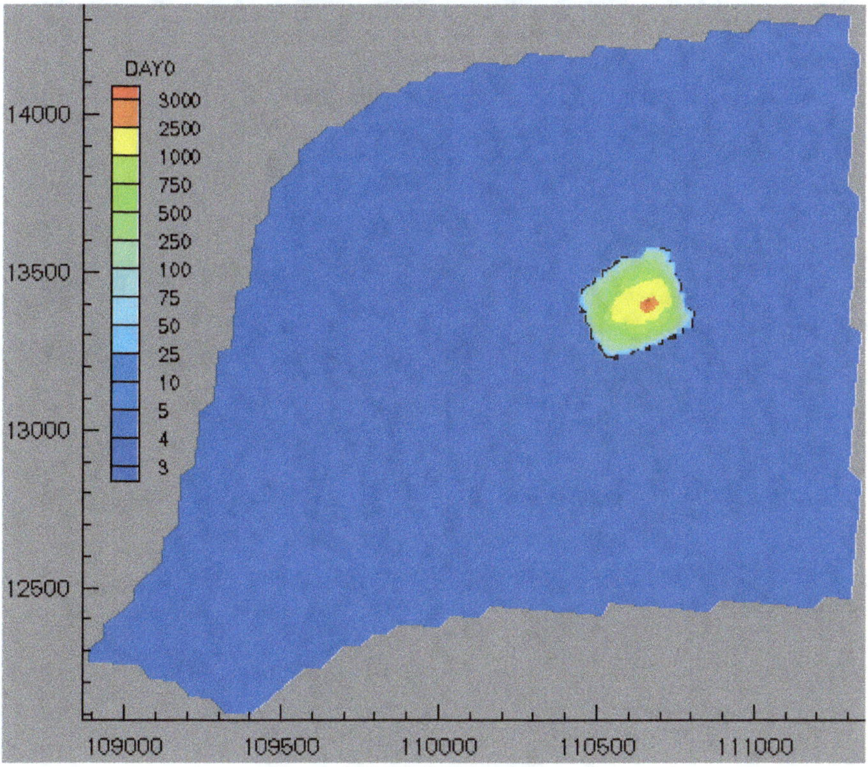

Figure 254. Model Results at t=0

At day 6083 (16.6 years) of the original remediation plan the predicted location and concentration of the plume is:

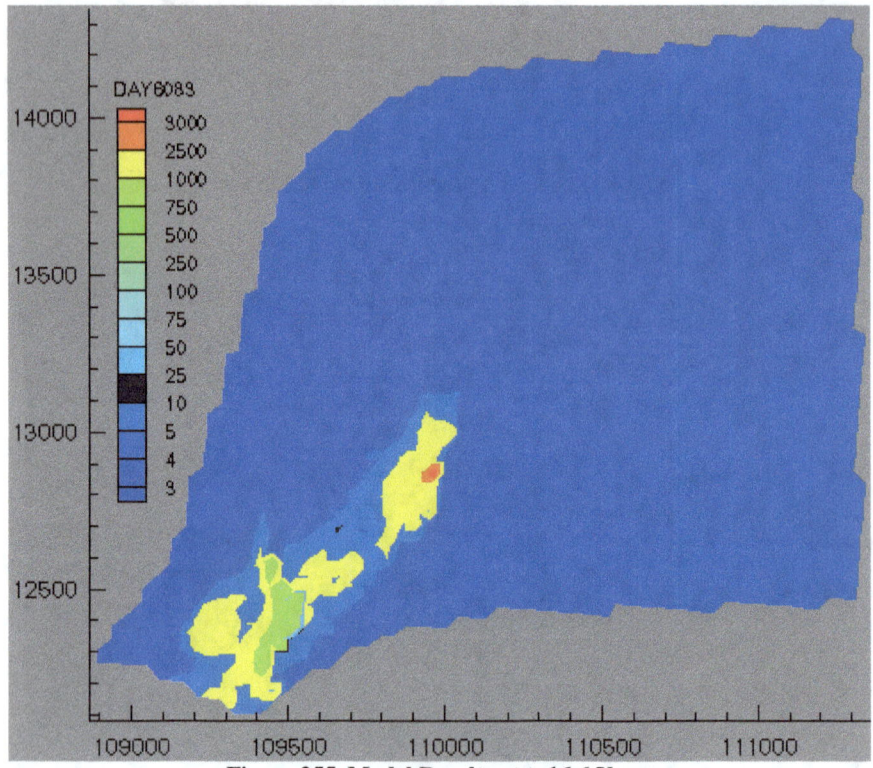

Figure 255. Model Results at t=16.6 Years

Appendix A. Displaying Data in 3⁺D

Most any model spanning three spatial dimensions plus time—actually 4D—is capable of generating an enormous amount of data, requiring specialized software to display. In previous texts, I have discussed two such programs: TP2 and Tecplot™. I developed TP2 beginning in 1980 as TPLOT. TP2 (the second generation of TPLOT) is available free on the Web and can handle many types of data. Slices of a 3D field are shown in the figure below. The central view is from the top, the right view is looking in the side from the X direction, and the lower view is looking in the side toward the Y direction. The black crosshairs indicate the respective cutting planes, which are also listed in the lower right corner. TP2 can handle such data as text and also in binary, which decreases file size. We will generate both types of files for the 3D examples (see Appendices B and C for details).

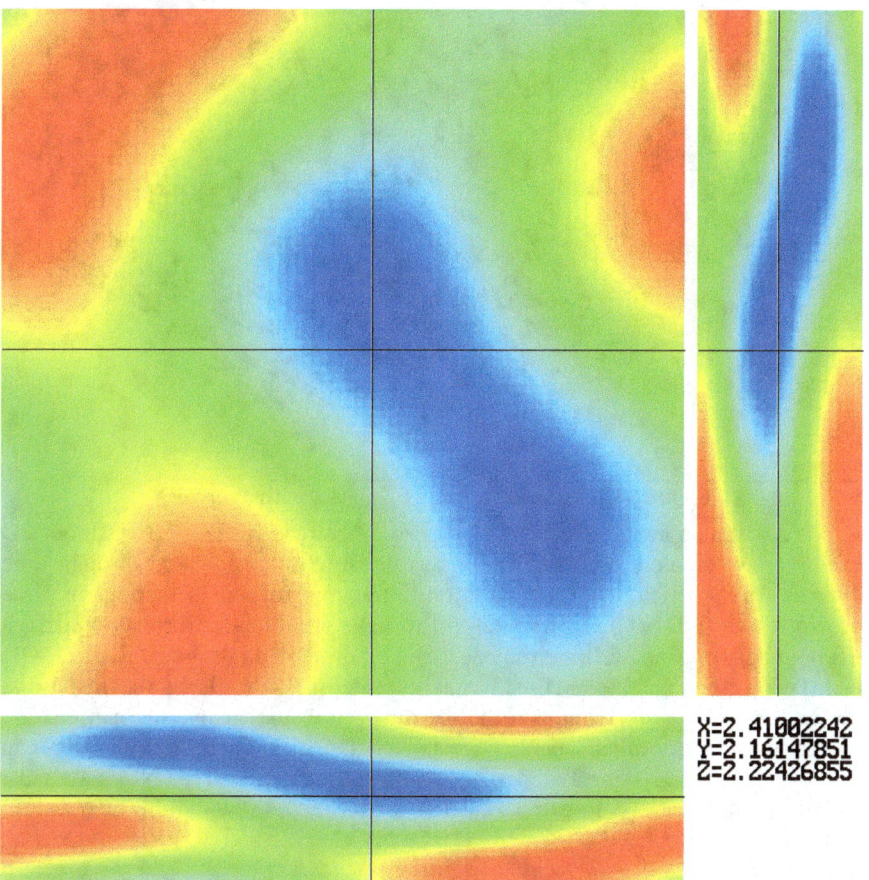

Figure 256. Slices of 3D Results in TP2

Tecplot™ is an excellent commercial tool[54] and can also handle many types of data. Besides the issues of cost (TP2 is free) and user support (TP2 has none), perhaps the biggest difference between TP2 and Tecplot™ is that the file extension (e.g., the ".dat" in myfile.dat) tells TP2 what type of data the file contains while headers within the file (whose extension is immaterial) provide this information to Tecplot™. This same data set viewed in Tecplot™ is shown below:

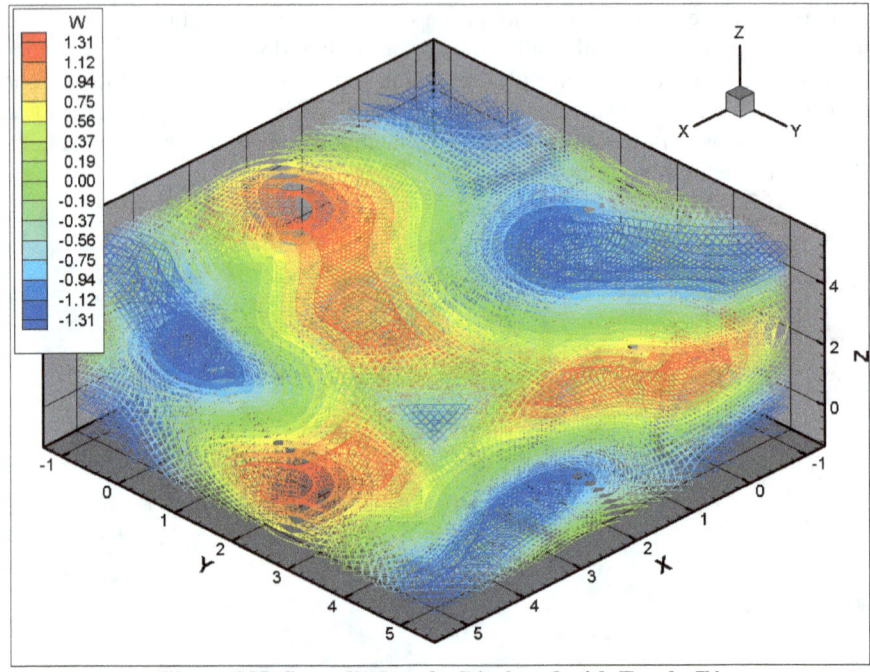

Figure 257. Same 3D Results Displayed with Tecplot™

We will also generate output files that can be read by Tecplot™. In our 2D examples we created the visualizations as the solution progressed through time but this isn't practical in 3D. In addition to the colorful representations in these previous two figures, it is often helpful to have additional information such as the boundary outline, air base runways, and cities shown in Figure 21. Both TP2 and Tecplot™ have the facility to incorporate such details in the respective layout files. This information in the form of a transparent overlay can also be embedded as an optional layer in the binary TP2 file (type MAP), which can also be compressed using Lempel-Ziv arithmetic encoding (type NM8) to make it even smaller. An example of a transparent overlay is shown in this next figure:

---

[54] This excellent product can be found at their web site https://www.tecplot.com/

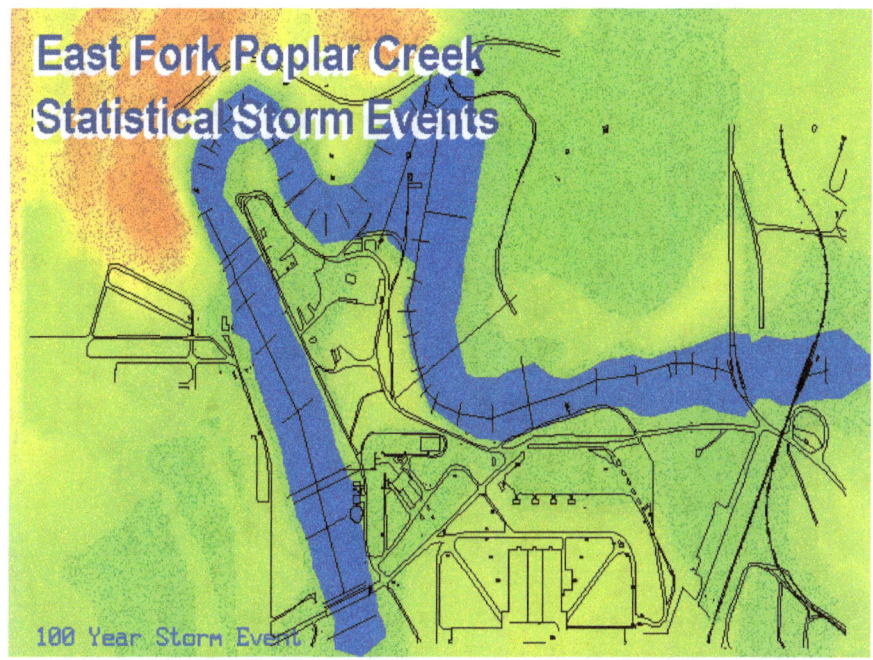

Figure 258. Example of Transparent Overlay Using TP2

Tecplot™ also provides transparent overlays. These are accomplished using what are called "geometries" within the layout file (name.LAY), an excerpt of which is listed below:

```
$!ATTACHGEOM
  ATTACHTOZONE = YES
  COLOR = RED
  FILLCOLOR = BLACK
  GEOMTYPE = LINESEGS3D
  PATTERNLENGTH = 4
  ARROWHEADANGLE = 20
  SCOPE = GLOBAL
  RAWDATA
1
14
870277 237001.25 100
870274.75 237014.25 100
870280.25 237152.75 100
870275.5 237170 100
870268.5 237178.75 100
870259 237184.5 100
870244.5 237186.75 100
870225.25 237186 100
870199.25 237041.5 100
```

A Tecplot™ transparent overlay is illustrated in this next figure:

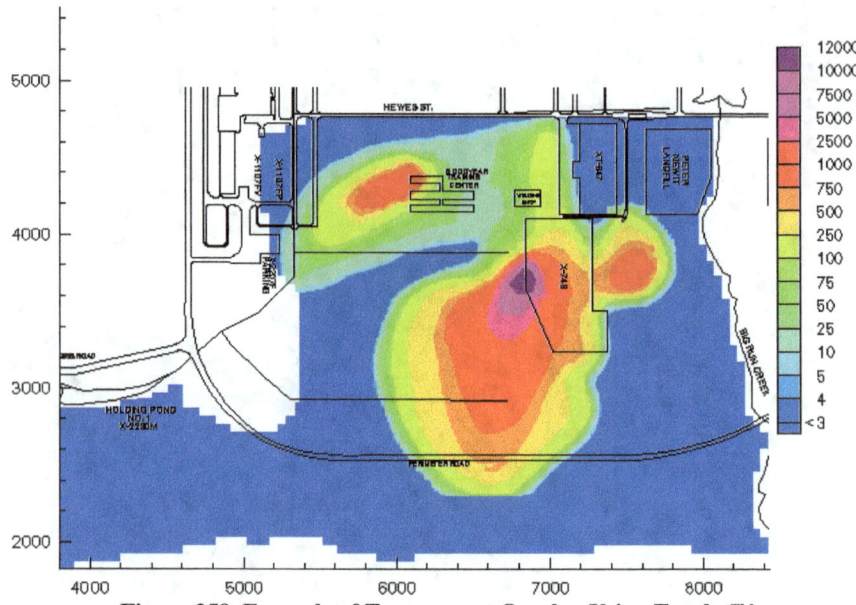
Figure 259. Example of Transparent Overlay Using Tecplot™

Appendix B. 3D Data files for Tecplot™

We will produce file directly for Tecplot™ but it may be useful to know that TP2 will convert a 3D file from its format to that of Tecplot™. Simply launch TP2, go to the menu, select "Convert", and then "volume => raw data". The resulting file can be loaded directly into Tecplot™ by using the menu to select "File" and then "Load DataFile(s)". The resulting file looks like this:

```
# converted 3D tabular data
# Nx=201, Ny=201, Nz=21
# -2500≤X≤2500
# -2500≤Y≤2500
# -25≤Z≤25
# -5.80914≤W≤4.82807
#   X    Y    Z    W
# this file can be read "as-is" by Tecplot
VARIABLES="X", "Y", "Z", "W"
ZONE I=201, J=201, K=21
-2500 -2500 -25 -5.80914
-2475 -2500 -25 -5.80914
-2450 -2500 -25 -5.80914
```

There is one "zone" or group of data, which has three dimensions, indicated by I, J, and K, having size 201x201x21. The headers are followed by 848,412 lines of X Y Z W. This particular file is 20,511,296 bytes in length. The original file in TP2 format was only 7,596,336 bytes in length, almost a factor of 3. For this particular format (IJK) the X, Y, and Z values are repetitious, whether evenly spaced or not. Filing the same values of X, Y, and Z over and over again is wasteful. Note that the file extension is immaterial.

This first format consists of regularly-spaced point (i.e., nodes). Tecplot™ will also accept several types of finite element data, including hexahedra (bricks). TP2 will also convert files to that format with menu convert volume => elements. This format looks like:

```
VARIABLES="X", "Y", "Z", "C"
ZONE N=99372 E=91575 F=FEPOINT ET=BRICK
109369 12051 359.8 0
109397 12071 360.6 0
109350 12079 360.3 0
etc.
110862 14745 701 0
110815 14753 701.2 0
110890 14764 700.9 0
 1 2  6 3 3823 3824 3828 3825
 2 5 10 6 3824 3827 3832 3828
 3 6 12 7 3825 3828 3834 3829
etc.
```

Appendix C: 3D Data Files for TP2

TP2 reads 3D regularly spaced data from a file having the extension TB3 (i.e., a 3D table of values). The format is quite simple: the number of X values, followed by the Xs, the number of Y values, followed by the Ys, the number of Z values, followed by the Zs, and the number of W values, followed by the Ws. For example:

```
201
-2500
-2475
-2450
...
201
-2500
-2475
-2450
...
21
-25
-22.5
-20
...
848421
-1.05735
0.0649466
2.53878
...
```

Appendix D. Inverse Distance Interpolation

The inverse distance interpolation method is one of the most useful techniques for interpolating data, especially spatially disperse and even more so irregularly spaced. In short, the closer a known point is, the more influence it should have on the local (i.e., interpolated) value. Often a power of 2.5 is applied to the distance. This can be expressed by Equation 25.1 and also the following code, which can be found in the online archive in the examples\ozone folder:

```
for(S=Z=j=0;j<Nd+Ni;j++)
  {
  D=hypot(Xd[j]-X,Yd[j]-Y);
  if(D<DBL_EPSILON)
     D=DBL_EPSILON;
  D=pow(D,2.5);
  S+=1./D;
  Z+=Zd[j]/D;
  }
Z/=S;
```

It is necessary to limit the distance so as to not divide by zero. If you're that close, it doesn't matter what the other points are. Depending on how the known points are scattered, unwanted artifacts can arise. For example, if the known points are all clumped together on one side or corner of the domain, this will produce shadows, over-emphasizing the clumped data. These unwanted artifacts can often be eliminated by considering only the closest point in each of four quadrants (or eight octants).

A mistake often made when implementing a quadrant or octant search is to use the a=arctan2(y,x) function and then a series of if(a<M_PI/4.) statements. The arctan function takes far longer to execute than the multiplications, divisions, and even raising distance to a non-integral power. Do not use arctan for this purpose. A far easier and vastly faster process is:

```
for(i=0;i<n;i++)
   {
   dX=X-Xd[i];
   dY=Y-Yd[i];
   D=hypot(dX,dY);
   q=0;
   if(dX>0.)
      q|=1;
   if(dY>0.)
      q|=2;
   if(fabs(dX)>fabs(dY))
      q|=4;
   if(D<Dq[q])
      {
      Dq[q]=D;
      iQ[q]=i;
```

}
}

The three simple (and fast) comparisons (i.e., dX>0, dY>0, and |dX|>|dY|) uniquely determine the octant (0-7) by conditionally adding 3 bits (1, 2, and 4), which can only have values 0 through 7. Using the arctan takes at least eight times as long and provides no advantage whatsoever. The code above saves the (double) distance in Dq[8] and (integer) index in iQ[8]. The summation and application of pow(D,2.5) is applied as before. Of course, depending on where you are in the field, some of the octants may be empty, which is why I first fill the array iQ with minus ones.

Inverse distance code for 2D and 3D can be found in the online archive in folder examples\invdist and also several of the other examples, including ozone. Use of octants can be selected by changing the corresponding conditional compilation statements. Typical 2D results are shown in the following figure:

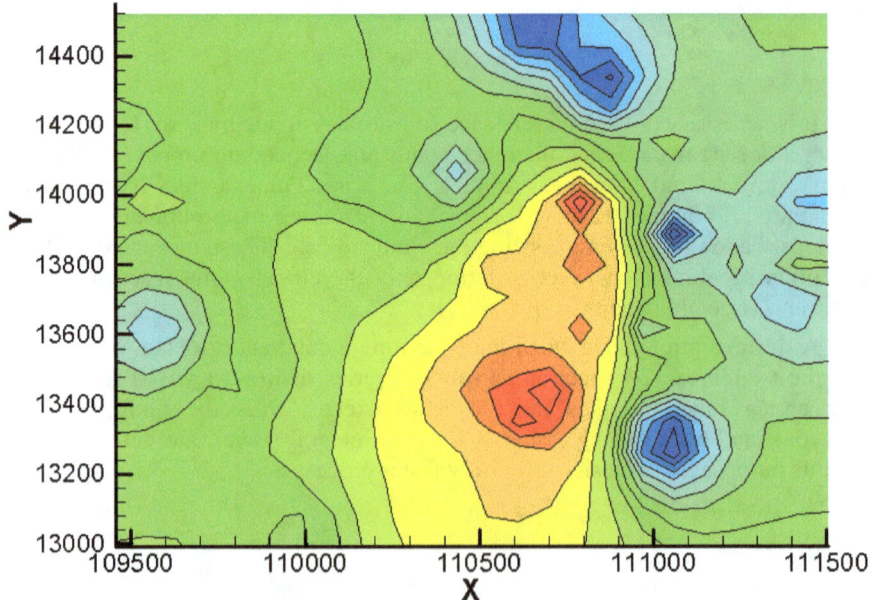

Figure 260. Results of Inverse Distance Interpolation

<u>Inverse Distance in Three Dimensions</u>

While there is sometimes disparity in transport or spreading in the two horizontal directions (X and Y), more often there is disparity between the horizontal and vertical. When this is the case, inverse distance interpolation must be modified to account for this. Most of the time this can be simply adjusted with a scaling factor, as in:

```
D=sqrt(dX*dX+dY*dY+A*dZ*dZ);
```

The code (invdist.c) will read in the data and determine if it is 2D or 3D and interpolate accordingly. The domain is optionally expanded by some percentage, typically 5% or 10%. Typical 3D data are shown in the following figure:

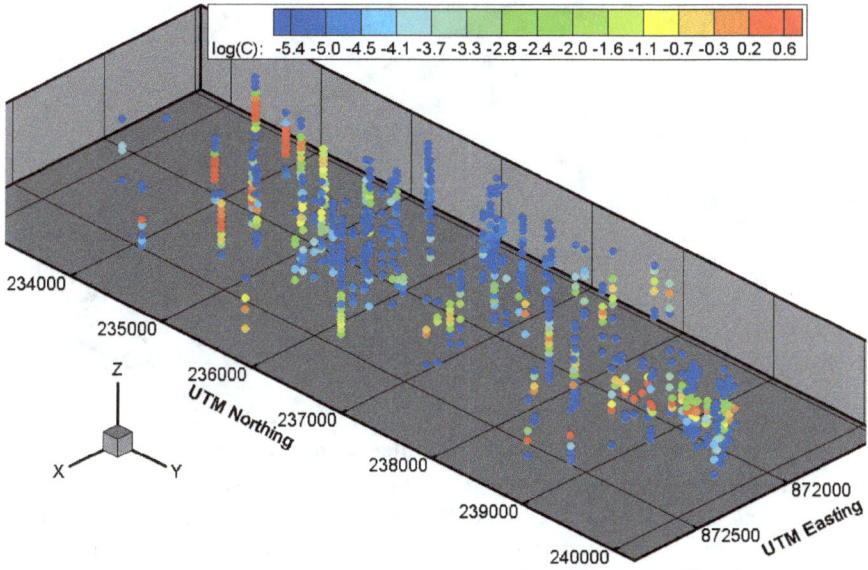

Figure 261. Typical 3D Data

Three-dimensional data (i.e., X,Y,Z,C) is a volume and must be sliced or converted to iso-surfaces in order to visualize it. Both Tecplot™ and TP2 can perform this task. One slice of contours is shown in this next figure:

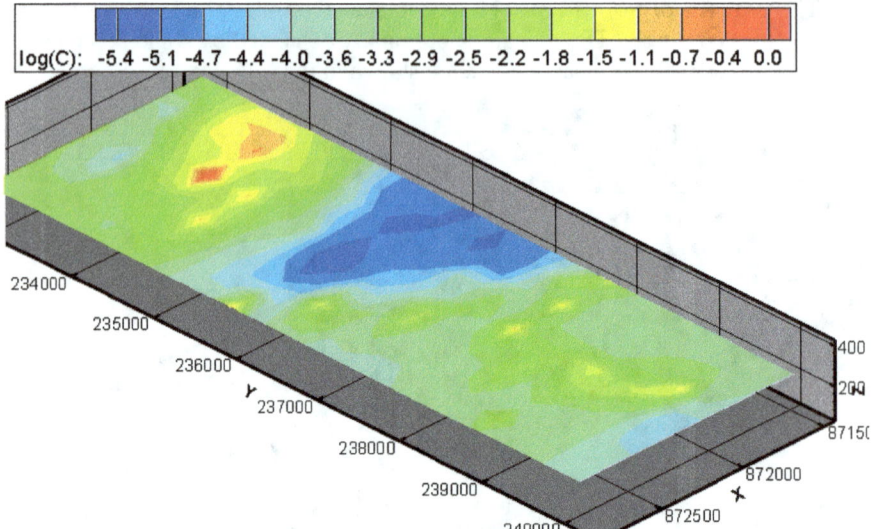

Figure 262. Typical 3D Interpolation Results

Note that both Tecplot™ and TP2 will perform the inverse distance method for you and also display the data. Several options are available. TP2 also performs the relaxation method described in Appendix E. Tecplot™ will relax a 2D or 3D field for you, but not with the control point strategy, which is often essential in obtaining acceptable results. Tecplot™ is an excellent commercial product. TP2 is available free online at the address provided in the Forward.

Appendix E. Relaxation Method

The relaxation method is sometimes called *smoothing*, for it has this effect, especially if viewed sequentially, were one to create an animation of the process. The basis for this method is actually solving Laplace's equation. In two dimension, this is expressed:

$$\frac{\partial^2 \varphi}{\partial x^2} + \frac{\partial^2 \varphi}{\partial y^2} = 0 \qquad (E.1)$$

A graphical representation for the finite difference approximation of this partial differential equation is:

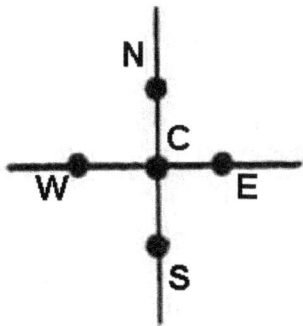

Figure 263. Finite Difference Points

The corresponding finite difference equation for uniformly spaced points is:

$$\frac{(\varphi_N + \varphi_S + \varphi_E + \varphi_W - 4\varphi_C)}{\Delta^2} = 0 \qquad (E.2)$$

This can be rearranged to form:

$$\varphi_C \frac{(\varphi_N + \varphi_S + \varphi_E + \varphi_W)}{4} \qquad (E.3)$$

This is the same as averaging the four surrounding points. The same thing extends to three dimensions:

$$\varphi_C \frac{(\varphi_N + \varphi_S + \varphi_E + \varphi_W + \varphi_{UP} + \varphi_{DOWN})}{6} \qquad (E.4)$$

This process is very simple to implement. You typically have control points or locations where you know the value, for instance, field data or boundaries. This is handled by a Boolean index for each point (0 or 1). The relaxation method requires a lot more memory than the inverse distance method described in Appendix D. Depending on the number of data points, one method may take considerably more computational time than the other. The relaxation method performs much better with more calculation points (i.e., finer granularity) than the inverse distance method, which result is independent of the number of calculation points. The code (relax.c) can be found in the online archive in folder examples\relax. The same data files from folder examples\invdist can be used. The results for the 2D example are:

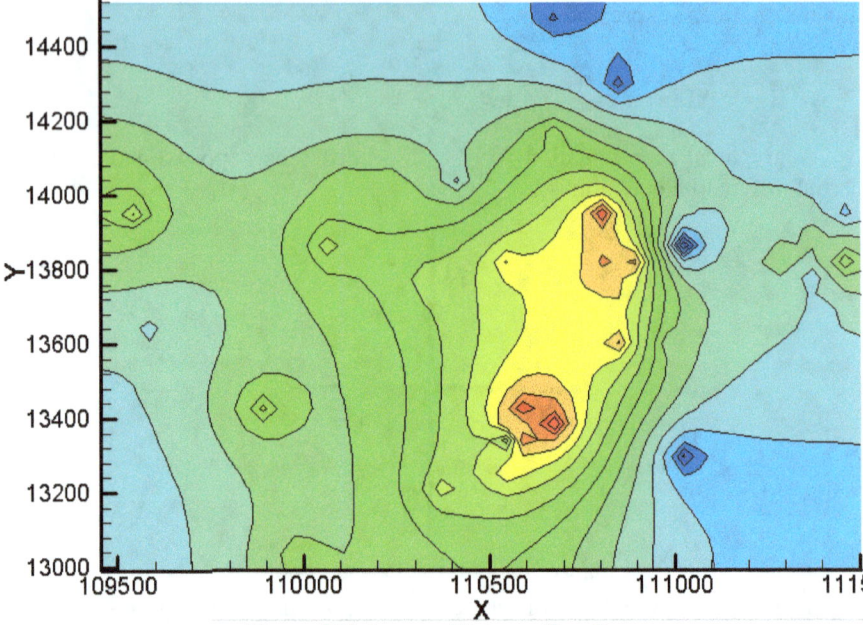

Figure 264. Inverse Distance Results

You can easily add scale disparity to the calculation, as this is simply a weighting (more or less) of the points in one direction (East/West, North/South, or Up/Down). The relaxation loop is quite simple:

```
for(w=z=0;z<Nz;z++)
  {
  for(y=0;y<Ny;y++)
    {
    for(x=0;x<Nx;x++,w++)
      {
      if(fix[w])
        continue;
      Cr[w]=n=0;
      if(x>0)
          {
          Cr[w]+=Cr[w-1];
          n++;
          }
      if(x<Nx-1)
          {
          Cr[w]+=Cr[w+1];
          n++;
```

```
            }
          if(y>0)
            {
            Cr[w]+=Cr[w-Nx];
            n++;
            }
          if(y<Ny-1)
            {
            Cr[w]+=Cr[w+Nx];
            n++;
            }
          Cr[w]/=n;
          }
      }
  }
```

The 3D example results are shown in the figure below:

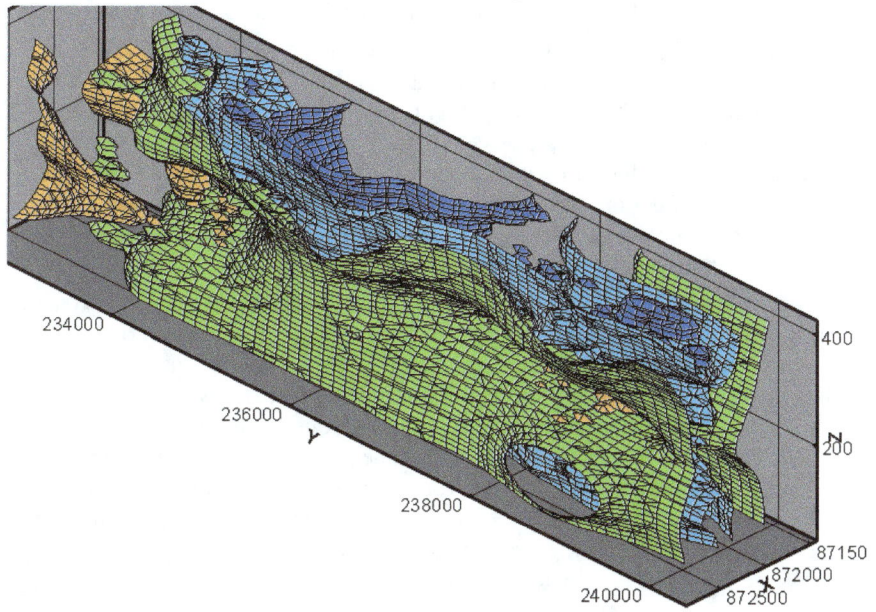

Figure 265. Shells of Constant Value

Appendix F. Kriging

Kriging is a form of Gaussian regression or interpolation used primarily in geostatistics. The original method was developed by Matheron[55] and based on the work of Krige.[56] There are actually a variety of formulas and techniques associated with the term *kriging*, making this designation rather ambiguous. Even a cursory search of the Internet will produce a dozen formulas, many of which (e.g., cos) bear no resemblance to the concept of Gaussian. Illustrations in one dimension show each data point blending into the next, but this is far removed from two- and three-dimensional hydrogeological data such as we have been discussing in this text.

Tecplot™ contains an implementation of this method that works well for some data but not for others. The same could be said for every other application I have used, so this is by no means a criticism of Tecplot™. There is a comparison of inverse distance and kriging in Chapter 30. In that particular case, kriging is a complete flop. Still, we won't discard the concept altogether.

Recall from Appendix E that the relaxation method is like solving Laplace's equation, which is steady-state conduction or diffusion. The control points (i.e., locations where the value is known) might be viewed as point sources (locations of fixed temperature, potential, or concentration). The 2D/3D field arises from the point sources. In fact, this is the same solution we would get if we were to solve a heat conduction or diffusion problem with the data points as fixed values. There is no particular function associated with the intervening spaces, except that it must satisfy Laplace's equation.

If we want kriging to be Gaussian and we consider these two approaches (kriging and relaxation) together, this leads to the question: Why not solve the problem of point sources having exponential decay? This gets us back to some of the equations you will find online, in particular $\Sigma C_I/\exp(-r^2/\lambda^2)$. We must first select a length scale, λ, and then determine the contribution of each data point (C_I). We could build a set of NxN simultaneous linear equations and simply solve for the values. That might work for a few dozen data points, perhaps even 100. Given 1000 data points, the chance of obtaining meaningful results from direct solution of 1000x1000 linear equations is nil.

Noting that the current point will always have more influence than any other (i.e., $\exp(0)=1$, $\exp(<0)<1$), the diagonal elements of the resulting matrix will always be the largest terms, if not strictly dominating. Also, we note that concentrations and conductivities might be zero, but are not negative, which

---

[55] Georges François Paul Marie Matheron (1930–2000) was French mathematician and civil engineer of mines, known as the founder of geostatistics and a co-founder (together with Jean Serra) of mathematical morphology.

[56] Danie Gerhardus Krige (1919–2013) South African statistician and mining engineer who pioneered the field of geostatistics and was professor at the University of the Witwatersrand.

forces each contribution to be positive. These combine to make iterative solution of the resulting system of simultaneous equations tractable. Because the diagonal elements are not strictly dominant, iterations tend to over-shoot. Rather than using Successive Over Relaxation (SOR), we use under-relaxation (not related to Appendix E). The iterations also tend to oscillate; so we force them to converge with progressive dampening. The inner loop is quite small:

```
for(w=0;w<64;w++)
  {
  for(d=0;d<Nd;d++)
    {
    C=Cd[d];
    for(c=0;c<Nd;c++)
      if(c-d)
        C-=Ck[d]*Bd[Nd*d+c];
    Ck[d]=fmax(Ck[d]/2.,fmin(2.*Cd[d],
(w*Ck[d]+C)/(w+1)));
    }
  }
```

Progressive dampening is implemented by the term $w/(w+1)$. The number of iterations may require adjustment, here 64 are taken. The influence factors (Bd) need be calculated only once, as these are the $\exp(-r^2/\lambda^2)$ terms, which don't change with iteration. The code (kriging.c) is in the examples\kriging folder.

This modified kriging process yields very smooth results—too smooth in some cases. The 2D data set from the examples\invdist is shown below:

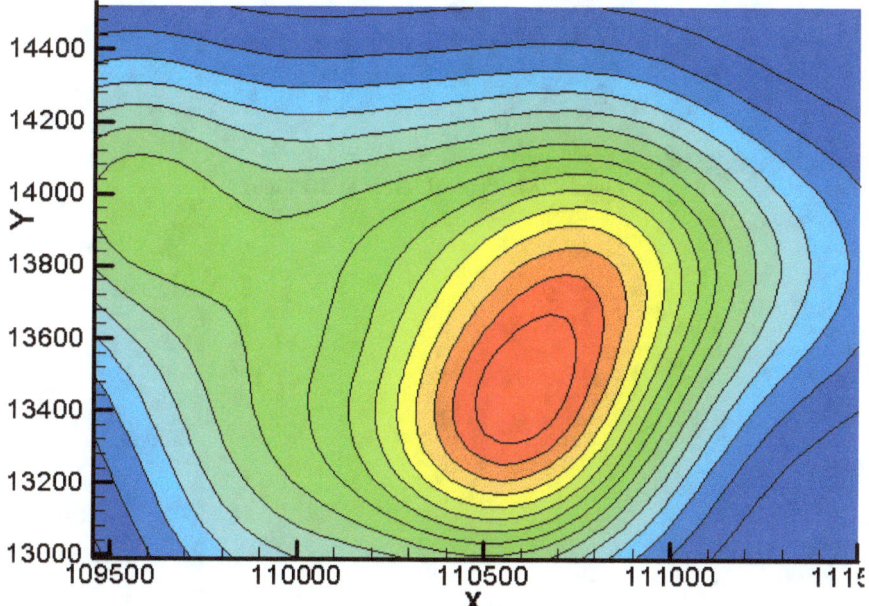

Figure 266. Typical Contours

The 3D data set from the examples\invdist is shown in this next figure:

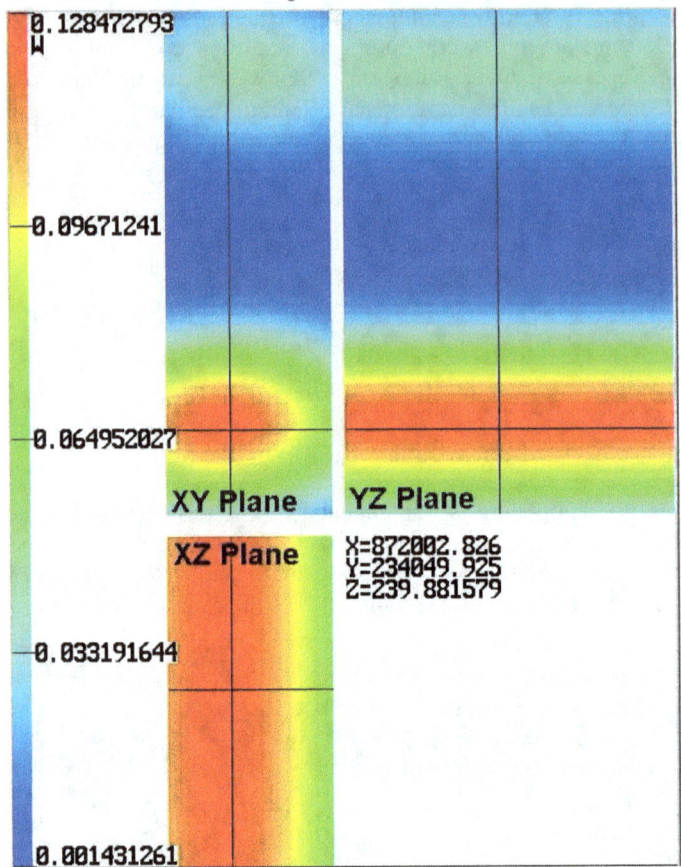

Figure 267. Typical Slices of 3D Field

Appendix G. BMP to GIF Conversion

The code required to create a single frame and save it as a BMP file is rather small (see for instance, ozone.c in examples\ozone). Creating a GIF is considerably more complex (but much less complex than a JPEG). I have provided a utility with source code (bmp2gif) that uses wild card file matching (e.g., ozone*.bmp) to combine multiple images into a single animated GIF. You can download it any time from my web site. It is included in the examples for *Compression & Encryption*:

http://dudleybenton.altervista.org/software/Compression.zip

and also for *Orthogonal Functions*:

http://dudleybenton.altervista.org/software/Orthogonal Functions/orthogonal_function_examples.zip

Appendix H. Potential Fields

Potential fields satisfy Laplace's partial differential equation. In Cartesian coordinates this is:

$$\frac{\partial^2 \varphi}{dx^2} + \frac{\partial^2 \varphi}{dy^2} + \frac{\partial^2 \varphi}{dz^2} = 0 \tag{H.1}$$

where φ is the potential. In cylindrical coordinates, this can be written:

$$\frac{1}{r}\frac{\partial}{\partial r}\left(r\frac{\partial \varphi}{\partial r}\right) + \frac{1}{r^2}\frac{\partial^2 \varphi}{\partial \theta^2} + \frac{\partial^2 \varphi}{dz^2} = 0 \tag{H.2}$$

Inviscid flow (i.e., flow of a fluid with zero or negligible viscosity or having an effectively infinite Reynolds number) satisfies this partial differential equation. Electrical and magnetic fields can often be described by this same partial differential equation. Heat conduction and molecular diffusion when linear in nature can be described by this same partial differential equation. This means that analytical solutions developed for these otherwise diverse disciplines are equally applicable to any other potential field.

Two-dimensional potential solutions are often expressed using complex variables (x+iy), where i=√-1. These complex solutions can be surprisingly simple formulas. Several examples can be found in the online archive in folder examples\potflow. Many more examples can be found in the online archive accompanying my text, *Complex Variables*, which can be freely downloaded from the same site.

For example, a source in polar coordinates is simply:

$$w = Q(\ln r + i\theta) \tag{H.3}$$

An irrotational (i.e., free) vortex is:

$$w = \frac{Q}{2\pi}(\theta - i\ln r) \tag{H.4}$$

Flow over a corner, step, or wedge is:

$$w = z^n \tag{H.5}$$

Stream Function

In two-dimensional flow, we can often—though not always—derive a stream function, whose curves of constant value are called streamlines. Streamlines are the paths that a particle would take if traveling through the field. The stream function is most often given the symbol ψ (Greek psi). The stream function is perpendicular to the potential function. In Cartesian coordinates this relationship can be expressed:

$$\frac{\partial \psi}{\partial x} = -\frac{\partial \phi}{\partial y}$$
$$\frac{\partial \psi}{\partial y} = \frac{\partial \phi}{\partial x}$$
(H.6)

It can be shown through the chain rule and combining A.6 with A.1 that the stream function also satisfies Laplace's Equation. Solutions exist (especially in 3D) where there is a stream function but no velocity potential and vice versa.

Irrotational Flow

When reading about potential flow, you may come across a statement that it is *irrotational*. You may also come across an example (such as in potflow.c and track1.c) that mentions *circulation* (as in flow over a cylinder with circulation, see page 5). How can this be? It depends on what you mean by *irrotational* and *circulation*. These have very specific mathematical definitions. To say that a flow (or any field) is *irrotational*, is to say that the *curl* of the velocity vector (i.e., the *vorticity*, which is also a vector) is zero. This is written:

$$\vec{\omega} = \nabla \times \vec{V} = 0 \quad (H.7)$$

where ∇ is the del operator and V is the vector velocity:

$$\vec{V} = u\hat{i} + v\hat{j} + w\hat{k} \quad (H.8)$$

In Cartesian coordinates, the curl of V is:

$$\nabla \times \vec{V} = \left(\frac{\partial w}{\partial y} - \frac{\partial v}{\partial z}\right)\hat{i} + \left(\frac{\partial u}{\partial z} - \frac{\partial w}{\partial x}\right)\hat{j} + \left(\frac{\partial v}{\partial x} - \frac{\partial u}{\partial y}\right)\hat{k} \quad (H.9)$$

Appendix I. Boundary Element Method

The boundary element method arises from Green's Lemma[57]. This remarkable relationship between the integral of a function over an area and the integral of a corresponding function around the perimeter of the same domain is usually covered in advanced calculus. Green's Lemma can be expressed by the following integral:

$$\iint \nabla^2 \varphi \, dA = \int \frac{\partial \varphi}{\partial n} dS \quad (I.1)$$

In Equation I.1 φ is the potential, dA is the differential area within the domain, dS is a differential distance along the boundary, and n is the normal (perpendicular) at each location along the boundary. The potential could be simply that (i.e., electrostatic potential or invicid flow), temperature, concentration, stress, strain, or anything else that satisfies Laplace's equation.

We must also have a *fundamental* solution to Laplace's equation—in this case, a general, homogeneous solution (i.e., works for any case and is zero on the right-hand side). The derivation is a bit circuitous, except in polar coordinates.

$$\varphi = \ln\left(\frac{1}{r}\right) \quad (I.2)$$

We can at least show by substitution that Equation I.2 satisfies Laplace's equation in polar coordinates:

$$\frac{1}{r}\frac{\partial}{\partial r}\left(r\frac{\partial \varphi}{\partial r}\right) + \frac{1}{r^2}\frac{\partial^2 \varphi}{\partial \theta^2} = 0 \quad (I.3)$$

For Cartesian (x,y) coordinates, r in the above equation is the distance from some as yet unspecified location $(x=a, y=b)$, or $r^2=(x-a)^2+(y-b)^2$. We further propose that the potential, φ, and the derivative with respect to the normal, $\partial \varphi / \partial n$, have some specific, finite, non-trivial value at each of the points along the boundary. Consider two points along the boundary (1 and 2). We can write Equation I.2 for these two points:

$$\varphi(r) = \left\{ \frac{\varphi_1(r - r_2)}{\ln\left(\frac{1}{r_1}\right)} - \frac{\varphi_2(r - r_1)}{\ln\left(\frac{1}{r_2}\right)} \right\} \frac{\ln\left(\frac{1}{r}\right)}{(r_1 - r_2)} \quad (I.4)$$

---

[57] George Green (1729-1841): British mathematical physicist best known for work with electric fields and magnetism.

At x_1,y_1 $r=r_1$ and at x_2,y_2 $r=r_2$ so that $\varphi(r_1)= \varphi_1$ and $\varphi(r_2)= \varphi_2$; therefore, Equation I.4 is the particular form of the fundamental equation that satisfies Laplace's equation and matches at these two points along the boundary. We can construct a similar equation for the right side of Equation I.1. When integrate Equation I.4 from point 1 to point 2, we will get some constant times φ_1 plus some other constant times φ_2. We will use $H_{i,j}$ to represent these constants on the left side and $G_{i,j}$ to represent the corresponding constants on the right side of Equation I.1. The indices i and j indicate each segment along the boundary and each pair corresponding to each segment. The resulting set of equations can be written:

$$\sum_{i=}^{n}\sum_{j=1}^{n} H_{i,j}\varphi_j = \sum_{i=}^{n}\sum_{j=1}^{n} G_{i,j} \frac{\partial \varphi_j}{\partial n} \qquad (I.5)$$

Equation I.5 constitutes a set of simultaneous linear equations for the potential function and its derivative at the n points along the boundary. We will need three different formulas for the integrals. We can readily integrate this equation around the boundary except at the two points (here, 1 and 2). At those points $r=r_1$ or $r=r_2$ and the standard result is indeterminate. We use a different formula for that one segment. We use these two formulas for points along the boundary. For points inside the boundary we use a third formula. We need this third integral to evaluate the results of the solution inside the boundary.

I have been vague up until this point on exactly what formulas are integrated and how, sparing you the gory details. In most applications, numerical integration (e.g., Gauss Quadrature) is used, because the analytical integral is unknown to the programmer. In fact, that's the way I present this as an example in my book *Numerical Calculus*. Here, you have the benefit of the analytical solution, which is precise, instead of a numerical solution that is approximate. Of course, I didn't figure this out the hard way. I used Maple® to do that for me. The resulting formulas are indeed tedious. You will find everything you need (source code and examples) in the online archive in folder examples\bem.

Appendix J. Explicit Runge-Kutta Methods

Runge-Kutta is a type of marching method in that we start with some initial values and then step along through time (or space). We will first consider explicit methods. The seminal reference on the Runge-Kutta and related methods was published by Butcher.[58] There are countless articles on the Web dealing with Runge-Kutta. In marching methods, we consider differential equations of the following form:

$$\frac{dy}{dx} = f(x, y(x)) \qquad (J.1)$$

As we shall see, higher order differentials are easily handled by extending this formula. The initial position is represented by x and the time step, Δx, is represented by h. The symbol k is used to represent some particular value of $f(x,y(x))$. The simplest procedure is known as Euler's explicit method, which is implemented:

$$\begin{aligned} k_1 &= f(x, y(x)) \\ y(x+h) &= y(x) + hk_1 \end{aligned} \qquad (J.2)$$

This is exactly the same as:

$$y_{x+\Delta x} = y_x + \Delta x \left(\frac{dy}{dx}\right)_x \qquad (J.3)$$

Euler's explicit method is sometimes called 1st order Runge-Kutta. In general, these and similar methods can be expressed by the following formula, where n is the number of steps, which is not necessarily the same as the order:

$$\begin{aligned} k_1 &= f(x, y) \\ k_2 &= f(x + c_2 h, y + h(a_{21} k_1)) \\ k_3 &= f(x + c_3 h, y + h(a_{31} k_1 + a_{32} k_2)) \end{aligned} \qquad (J.4a)$$

$$k_i = f\left(x + c_i h, y + h \sum_{j=1}^{i-1} a_{ij} k_j \right) \qquad (J.4b)$$

$$y = y + h \sum_{i=1}^{n} b_i k$$

---

[58] Butcher, J. C., The Numerical Analysis of Ordinary Differential Equations: Runge-Kutta and General Linear Methods, John Wiley & Sons Ltd., New York, 1987.

Butcher Tableaus

Butcher expressed the preceding set of equations in tabular form, called a tableau, having the following form:

c_1	a_{11}	a_{12}	a_{13}	...	a_{1n}
c_2	a_{21}	a_{22}	a_{23}	...	a_{2n}
...	...	...	...	...	...
c_n	a_{n1}	a_{n2}	a_{n3}	...	a_{nn}
	b_1	b_2	b_3	...	b_n

(J.5)

The Butcher tableau for Euler's explicit method (Equation 11.2) is:

0	0
	1

(J.6)

We will present all of these methods in this way and then implement them in a code that can handle any formula in this form. There are three common variants of 2nd order Runge-Kutta. The first variant is:

0	0	0
1/2	1/2	0
	0	1

(J.7)

The second variant is called Huen's method:

0	0	0
1	1	0
	1/2	1/2

(J.8)

The third variant is called Ralston's method:

0	0	0
2/3	2/3	0
	1/4	3/4

(J.9)

There are also two common variants of 3rd order Runge-Kutta. The first is:

0	0	0	0
1/2	1/2	0	0
1	-1	2	0
	1/6	2/3	1/6

(J.10)

The second variant of 3rd order Runge-Kutta is:

0	0	0	0
1/3	1/3	0	0
2/3	0	2/3	0
	1/4	0	3/4

(J.11)

There are also two common variants of 4th order Runge-Kutta. The first is:

0	0	0	0	0
1/2	1/2	0	0	0
1/2	0	1/2	0	0
1	0	0	1	0
	1/6	1/3	1/3	1/6

(J.12)

This formula is reminiscent of Simpson's method for numerical integration. The second variant is:

0	0	0	0	0
1/3	1/3	0	0	0
2/3	-1/3	1	0	0
1	1	-1	1	0
	1/8	3/8	3/8	1/8

(J.13)

This formula is reminiscent of Simpson's 3/8ths rule. Implementation of C.12 is quite simple:

```
void RungKutta4(void dydx(double,double*,double*),
    double*x,double dx,double*y,double*dy,int n)
    {
    int i,j;
    static double a[4]={0.,0.5,0.5,1.};
    static double b[4]={1./6.,1/3.,1./3.,1./6.};
```

```
double*w,*v;
w=calloc(4*n,sizeof(double));
v=calloc( n,sizeof(double));
dydx(x[0],y,w);
for(j=1;j<4;j++)
   {
   for(i=0;i<n;i++)
      {
      dy[i]=a[j]*w[n*(j-1)+i];
      v[i]=y[i]+dx*dy[i];
      }
   dydx(x[0]+dx*a[j],v,w+n*j);
   }
for(i=0;i<n;i++)
   {
   dy[i]=0;
   for(j=0;j<4;j++)
      dy[i]+=b[j]*w[n*j+i];
   y[i]+=dx*dy[i];
   }
x[0]+=dx;
free(w);
free(v);
}
```

Implementation of C.13 just requires changing the preceding a[] and b[] data statements. User-defined function dydx returns the differential:

```
void dydx(double x,double*y,double*dy)
   {
   dy[0]=f1(x,y);
   dy[1]=f2(x,y);
   etc.
   }
```

The stepping process is done inside a loop:

```
for(i=0;i<steps;i++)
   RungKutta4(dydx,&x,dx,y,dy,n);
```

where n is the number of variables.

Appendix K. Build3D Model Builder

Build3D was developed to gather diverse information and build the corresponding input files to facilitate complex three-dimensional modeling, particularly for groundwater problems but also for other types of applications. Build3D reads in a variety of data, including topography, geological formations, soil types and depth, rivers and streams, water table, and recharge (rainfall). Each of these data sets must be in a different file and have a specific three-letter file extension (i.e., filename.ext).

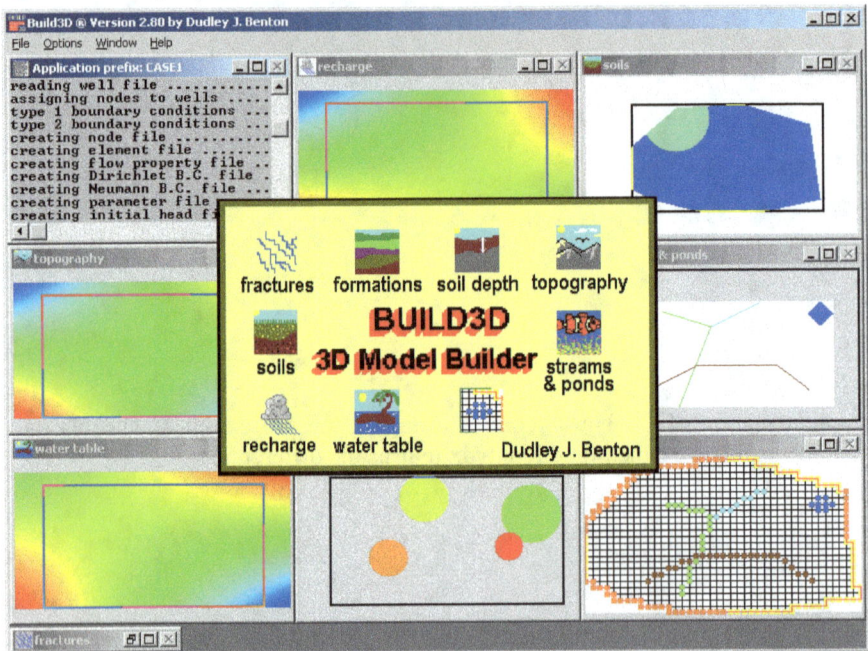

Figure 268. Typical Build3D Display

Build3D can create regular grids or optionally read in a prepared finite element grid. Build3D can handle finite elements (FEM) and finite differences (FDM). Three types of elements are recognized: hexahedra (bricks), pentahedra (prisms), and tetrahedra (three-sided pyramids).

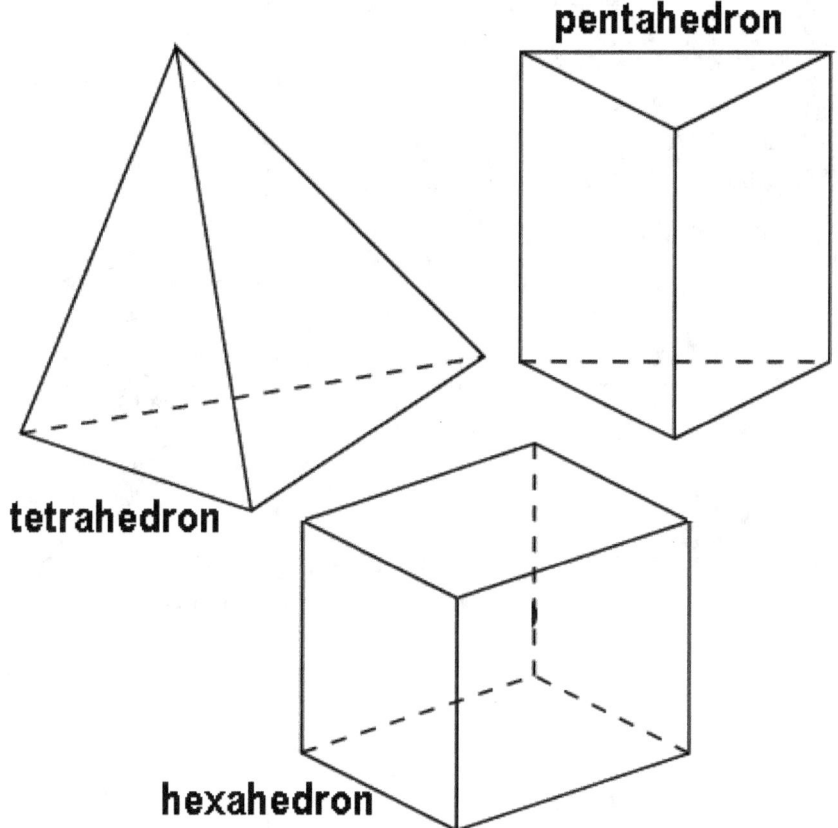

Figure 269. Three Types of Elements Recognized by Build3D

Hexahedral elements are shown in the lower right corner of the first figure in this appendix. Pentahedral elements are shown in the following figure:

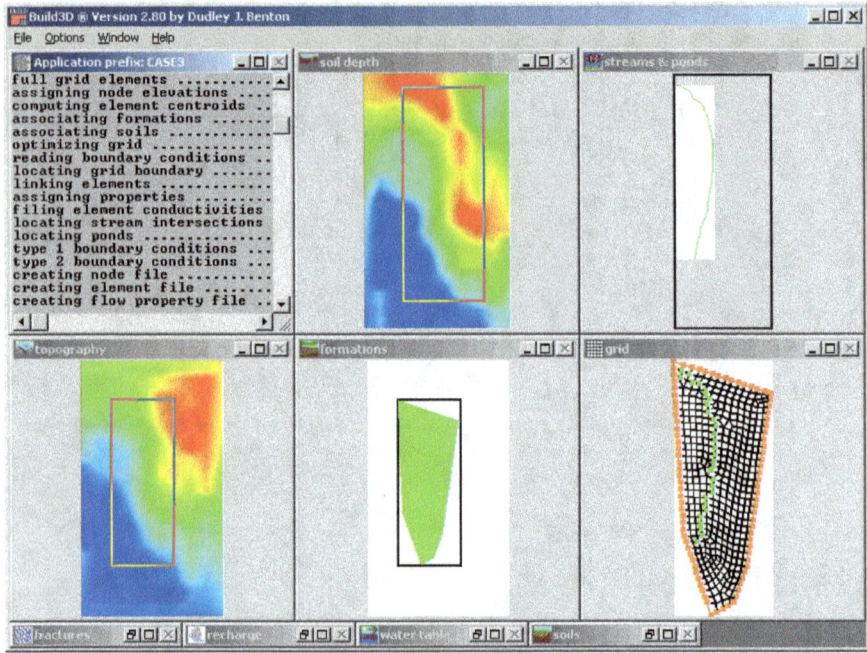

Figure 270. Pentahedral Elements (lower right corner)

Tetrahedral elements are shown in the "grid" box of this next figure:

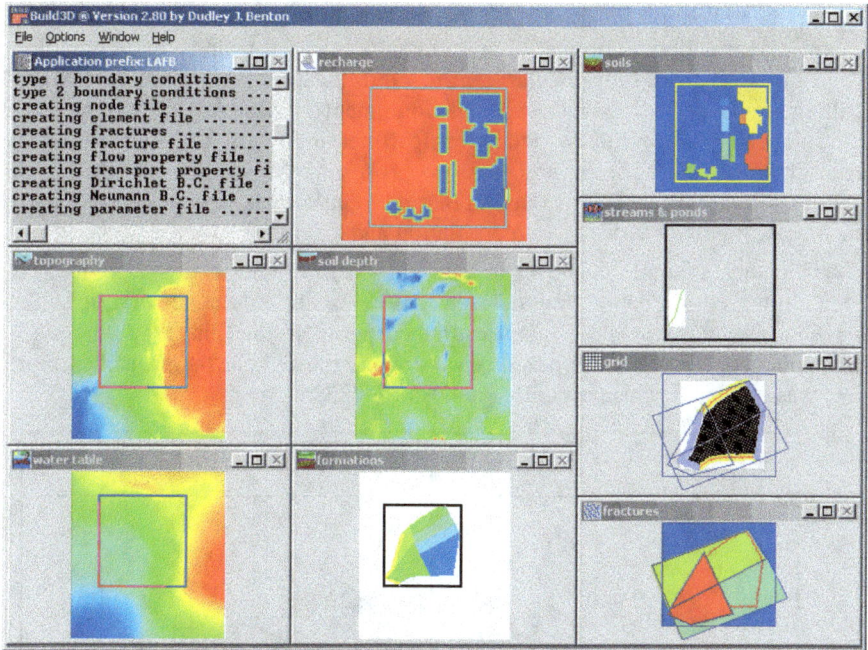

Figure 271. Data Combined to Create a Model

Build3D will create input files for MODFLOW and also FRAC3D. Build3D can be downloaded and used free of charge at the address below. Along with the executable and help file, three complete examples are included in the archive. Build3D also prepares input files for the particle tracker, PTRAX.

http://dudleybenton.altervista.org/projects/Build3D/index.html

Appendix L. Initial Concentrations

In the preceding examples we have used rather simple initial concentrations. In real like remediation projects, the initial concentrations can be quite complex. Most often data are only available at discrete locations, as sampling requires drilling wells and extraction. Special equipment is used to sample wells at controlled depth intervals to obtain vertical measurements. Once this data has been obtained, there are several ways to process it. Most often we need a 3D field of initial concentrations. Both TP2 and Tecplot™ have featured to handle this, including inverse distance interpolation and kriging.

One such example is shown in plan view below. The scattered points have been collapsed in depth and colored based on the log(concentration). The magenta polygon was the original estimated perimeter and the gray polygon is the expanded one. Changes like this were made throughout the project as more data was collected. This particular data were for TCE.

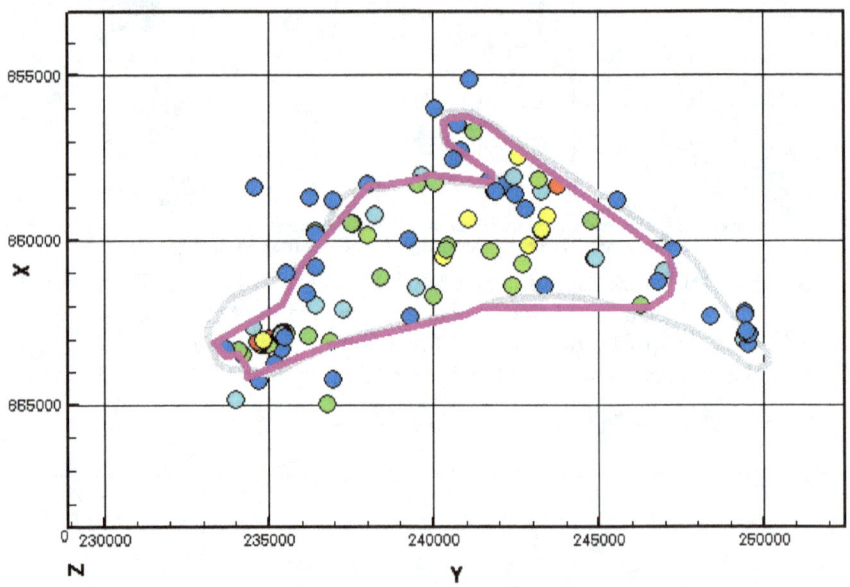

Figure 272. TCE Concentration Data from Multiple wells

This same data in 3D are shown in the next figure:

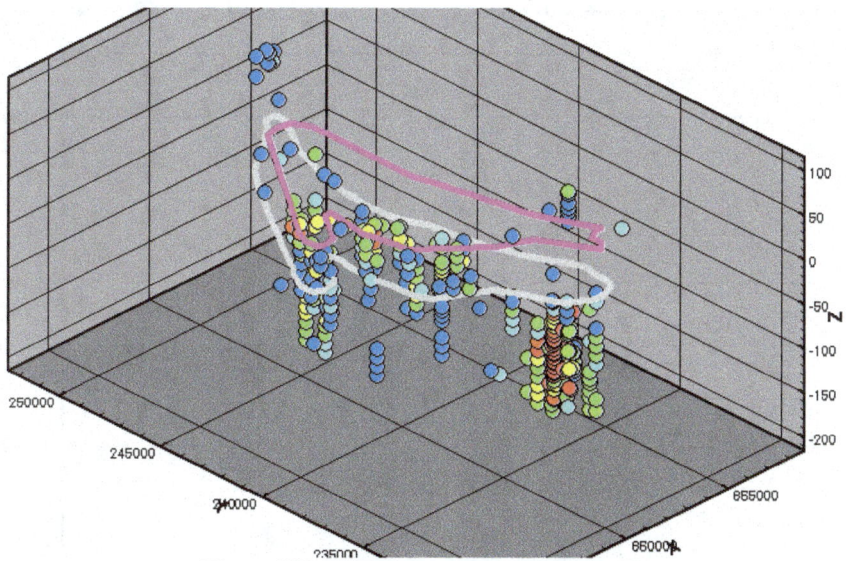

Figure 273. TCE Concentration Data in 3D

This same data shown with an interpolated plane slicing through the domain.

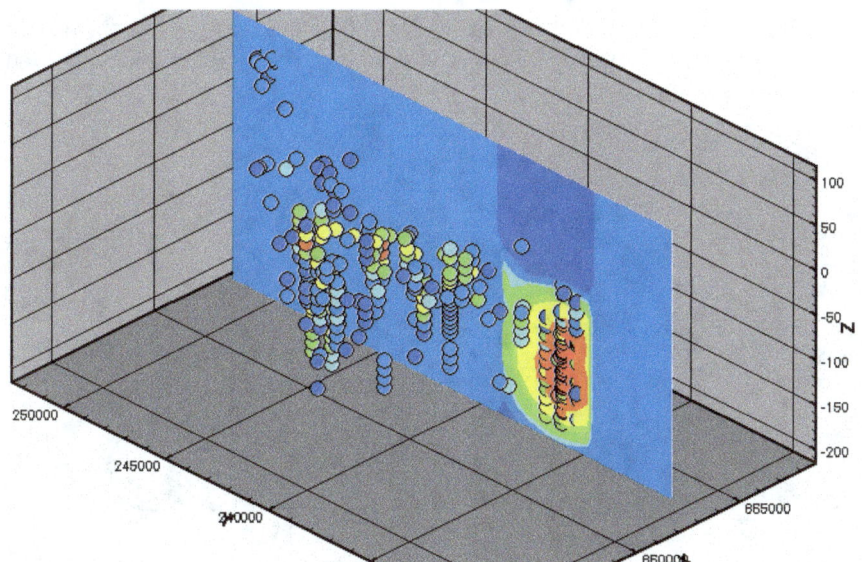

Figure 274. TCE Data with Slicing Plane

Data for a different substance (PCE) at the same location is shown in this next figure along with some initial contours.

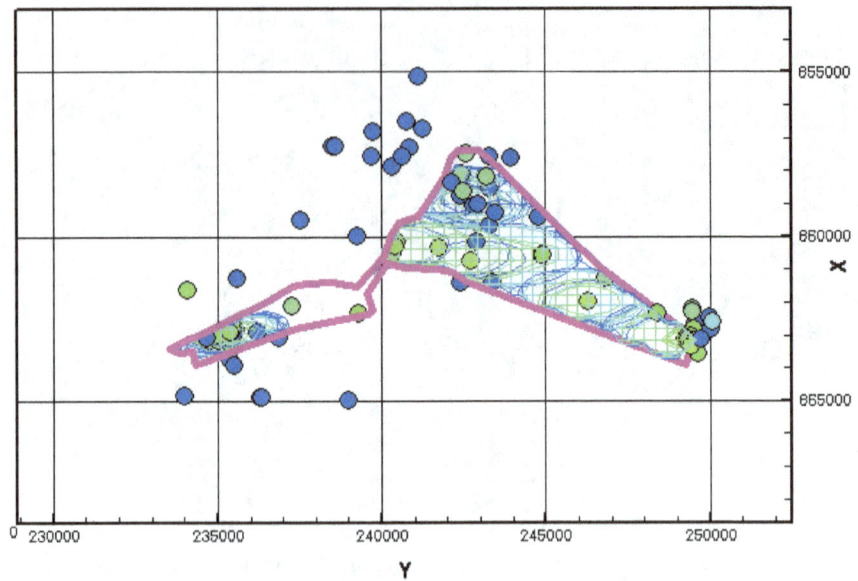

Figure 275. PCE Concentrations at the Same Wells

Both TP2 and Tecplot™ can interpolate and paint slices through the data in any of the three principle directions, which can be quite useful in evaluating and quantifying the initial conditions.

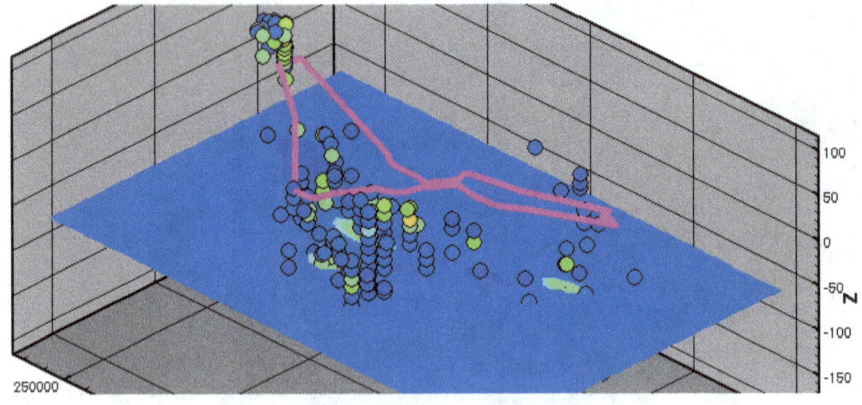

Figure 276. PCE Concentrations in 3D

Appendix M. Validation of PTRAX

PTRAX was developed in 1995 by a team at Environmental Consulting Engineers, Inc., under contract to Martin Marietta Energy Systems, Inc., for the U. S. Department of Energy and later for the U. S. Department of Defense. All of the associated documents are available through the Freedom of Information Act (FIOA). The development team consisted of Steve Young, Nick Williams, and myself. Steve was the geohydrologist, Nick built all of the test cases and ran all of the validations, and I wrote the PTRAX software in the C programming language, targeting the Windows® NT operating system. The complete validation report can be downloaded from this link:

http://dudleybenton.altervista.org/publications/Description_and_Verification_of_PTRAX.pdf

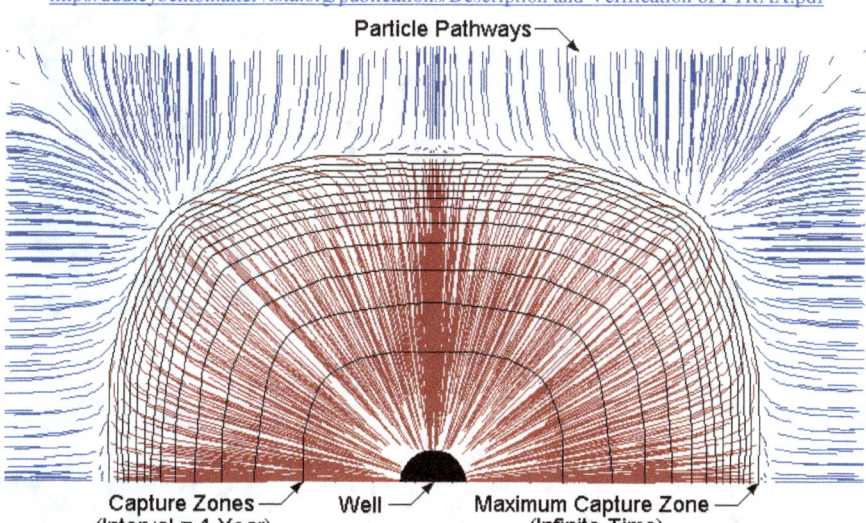

Figure 277. Computed Particle Tracks

Test Case	α_X	α_Y	α_Z	Model	15 Years						30 Years					
					X	Y	Z	σ_X	σ_Y	σ_Z	X	Y	Z	σ_X	σ_Y	σ_Z
1	3	3	0.3	Numerical	109.4	0.4	0	25.6	25.8	8.0	218.7	0.8	0	37.1	36.9	11.3
				Analytical	109.5	0	0	25.6	25.6	8.1	219.0	0	0	36.2	36.2	11.4
2	3	0.3	0.3	Numerical	109.3	0.1	0	25.9	8.7	8.1	218.7	0.3	0.1	37.1	12.0	11.4
				Analytical	109.5	0	0	25.6	8.1	8.1	219.0	0	0	36.2	11.4	11.4
3	12	12	0.3	Numerical	109.1	0.5	0	51.1	51.1	8.1	217.7	1.4	0.1	73.5	72.8	11.4
				Analytical	109.5	0	0	51.3	51.3	8.1	219.0	0	0	72.4	72.4	11.4
4	12	1.2	0.3	Numerical	109.5	0.3	0.2	50.6	16.4	8.2	218.2	0.5	0.1	72.7	23.3	11.4
				Analytical	109.5	0	0	51.3	16.2	8.1	219.0	0	0	72.4	22.9	11.4

The original validation consisted of well capture and comparison to an analytical model, AT123D. Both tests were successful. Over the next several years, I added some features and the model was validated with considerable field data with surprisingly good results. A selection of figures is included here. These simulations were performed with 800,000 particles.

Analytical Solution - 15 Years

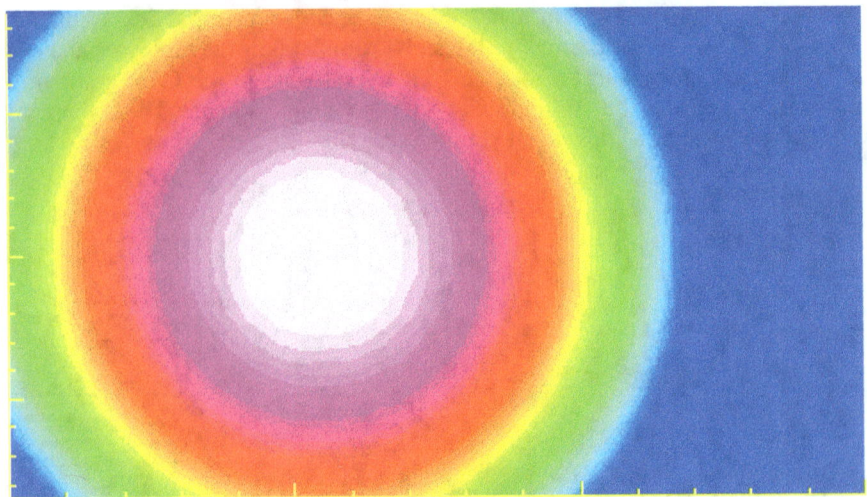

PTRAX - 15 Years

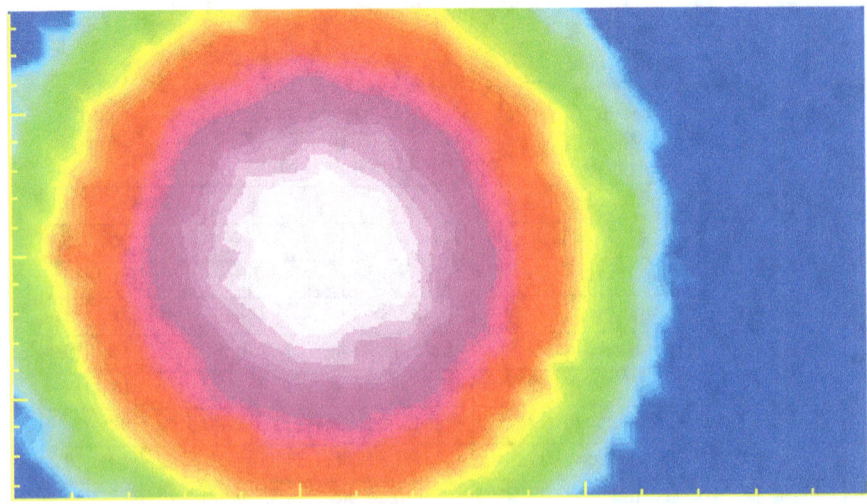

Figure 278. Dispersion Resulting from ax=ay=12m

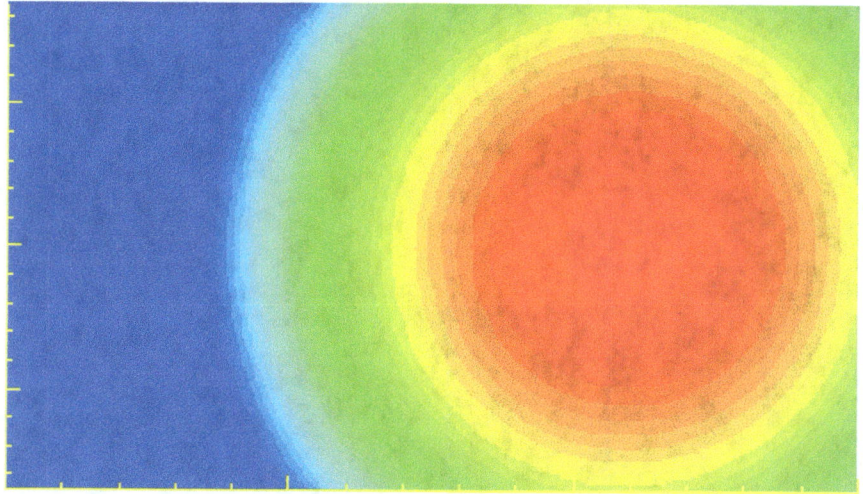
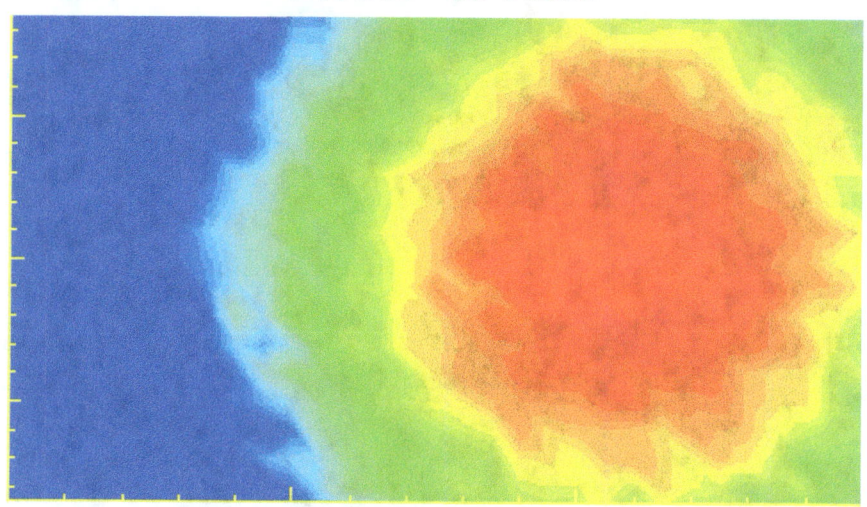

Figure 279. Dispersion Resulting from ax=ay=12m

Analytical Solution - 15 Years

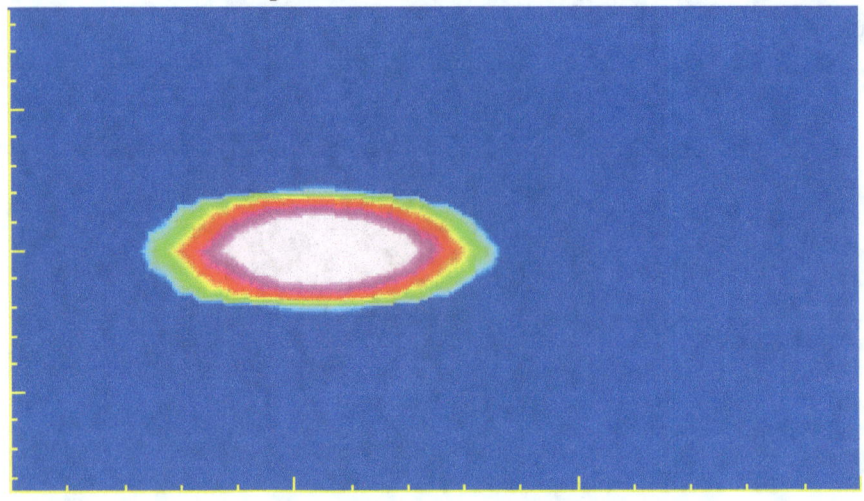

PTRAX - 15 Years

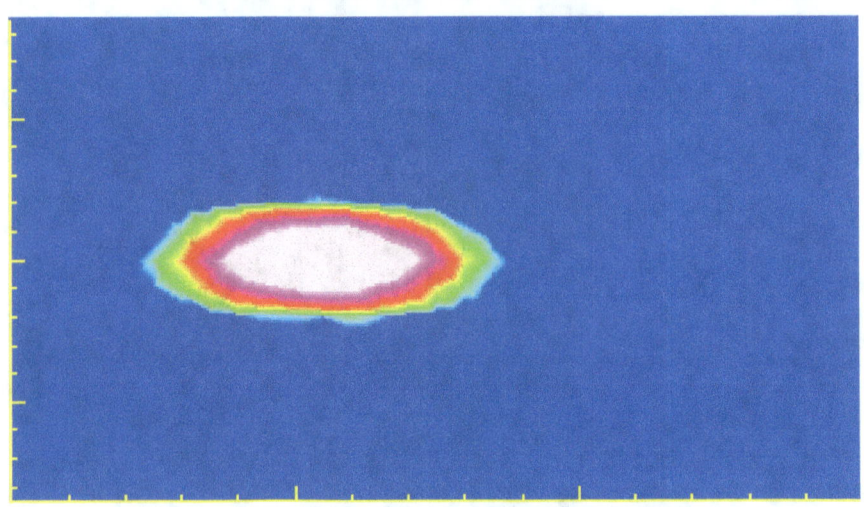

Figure 280. Dispersion Resulting from ax=3m, ay=0.3m

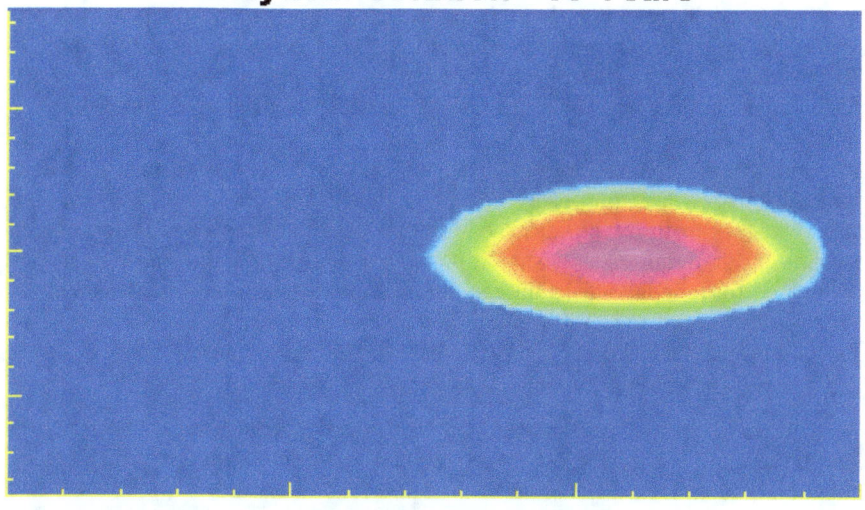

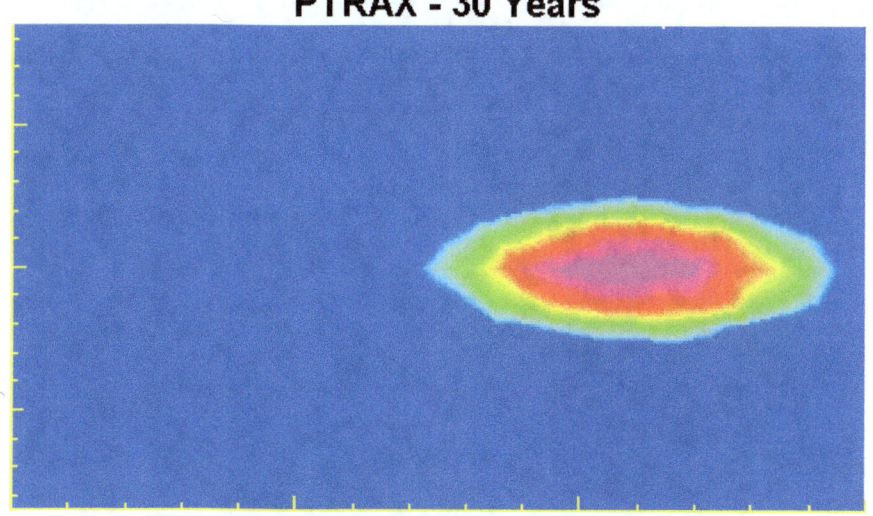

Figure 281. Dispersion Resulting from ax=3m, ay=0.3m

Additional publications are also available, including:

http://dudleybenton.altervista.org/publications/Development of the Fast 3D Particle Tracker PTRAX.pdf

Appendix N. PTRAX Coding

The particle tracking code is written entirely in the C programming language. The code is highly optimized, as evidenced by the illustration in Chapter 24 of tracking three million particles in fourteen minutes on a single 3GHz Intel® processor. There are no elements of object-oriented programming in PTRAX. While such might be efficient for development, object-oriented programming is anything but efficient for execution. This also means that all dynamic arrays are purposely allocated and initialized, not leaving this important task to throwing and catching exceptions. Exceptions (in the object-oriented programming sense) should be reserved for fatal application errors and never used for mundane tasks, like uninitialized variables. Just because the pizza place is next door to the firehouse doesn't mean that pulling the fire alarm and dispatching the big red truck is an efficient way to deliver pizza.

Geometric Functions

One of the most time-consuming aspects of particle tracking through a discrete domain is figuring out which element a particle is in and which ones it might enter next. Below are some of the basic functions used:

```c
void EquationLine(double*Xp,double*Yp,double*C) /*
    general equation of a line */
    { /* equation: C0*X+C1*Y=C2 */
    C[0]=Yp[1]-Yp[0];
    C[1]=Xp[0]-Xp[1];
    C[2]=Xp[0]*Yp[1]-Yp[0]*Xp[1];
    }
void EquationPlane(double*Xp,double*Yp,
       double*Zp,double*C) /* general equation of a plane */
    { /* equation: C0*X+C1*Y+C2*Z=C3 */
    double Y01,Y02,Y12,Z01,Z02,Z12;
    Z01=Zp[1]-Zp[0];
    Z02=Zp[2]-Zp[0];
    Z12=Zp[2]-Zp[1];
    Y01=Yp[1]-Yp[0];
    Y02=Yp[2]-Yp[0];
    Y12=Yp[2]-Yp[1];
    C[0]= Yp[0]*Z12-Yp[1]*Z02+Yp[2]*Z01;
    C[1]=-Xp[0]*Z12+Xp[1]*Z02-Xp[2]*Z01;
    C[2]= Xp[0]*Y12-Xp[1]*Y02+Xp[2]*Y01;
    C[3]=C[0]*Xp[0]+C[1]*Yp[0]+C[2]*Zp[0];
    }
double Line2Point(double*Xp,double*Yp,double X,double Y)
    {
    double C[3],H;
    EquationLine(Xp,Yp,C);
    H=_hypot(C[0],C[1]);
    if(H>DBL_EPSILON)
        return((C[0]*X+C[1]*Y-C[2])/H);
    else
```

```
    return(0);
  }
double Point2Line(double*Xp,double*Yp,double X,double Y)
  {
  double A,B,C,S,R;
  A=_hypot(X-Xp[0],Y-Yp[0]);
  B=_hypot(X-Xp[1],Y-Yp[1]);
  C=_hypot(Xp[0]-Xp[1],Yp[0]-Yp[1]);
  S=min(A,B);
  R=max(A,B);
  if(R*R-S*S>C*C)
    return(S);
  else
    return(Line2Point(Xp,Yp,X,Y));
  }
double Plane2Point(double*Xp,double*Yp,double*Zp,double
    X,double Y,double Z)
  {
  double C[4],H;
  EquationPlane(Xp,Yp,Zp,C);
  H=hypot3D(C[0],C[1],C[2]);
  if(H>DBL_EPSILON)
    return((C[0]*X+C[1]*Y+C[2]*Z-C[3])/H);
  else
    return(0);
  }
double Point2Plane(double*Xp,double*Yp,double*Zp,double
    X,double Y,double Z)
  {
  double D1,D2,D3,Dm,Dx,S1,S2,S3,Sm,Sx;
  D1=hypot3D(X-Xp[0],Y-Yp[0],Z-Zp[0]);
  D2=hypot3D(X-Xp[1],Y-Yp[1],Z-Zp[1]);
  D3=hypot3D(X-Xp[2],Y-Yp[2],Z-Zp[2]);
  Dm=min(D1,min(D2,D3));
  Dx=max(D1,max(D2,D3));
  S1=hypot3D(Xp[1]-Xp[0],Yp[1]-Yp[0],Zp[1]-Zp[0]);
  S2=hypot3D(Xp[2]-Xp[1],Yp[2]-Yp[1],Zp[2]-Zp[1]);
  S3=hypot3D(Xp[0]-Xp[2],Yp[0]-Yp[2],Zp[0]-Zp[2]);
  Sm=min(S1,min(S2,S3));
  Sx=max(S1,max(S2,S3));
  if(Dx*Dx-Dm*Dm>Sx*Sx)
    return(Dm);
  else
    return(Plane2Point(Xp,Yp,Zp,X,Y,Z));
  }
double Point2Edge(double*X,double*Y,double*Z)
  {
  double D1,D2,D3;
  D1=hypot3D(X[1]-X[0],Y[1]-Y[0],Z[1]-Z[0]);
  if(D1<Ltiny)
```

```
     return(D1);
  D2=hypot3D(X[2]-X[0],Y[2]-Y[0],Z[2]-Z[0]);
  if(D2<Ltiny)
     return(D2);
  if((X[1]-X[0])*(X[2]-X[0])>0)
     return(min(D1,D2));
  if((Y[1]-Y[0])*(Y[2]-Y[0])>0)
     return(min(D1,D2));
  if((Z[1]-Z[0])*(Z[2]-Z[0])>0)
     return(min(D1,D2));
  D3=hypot3D(X[2]-X[1],Y[2]-Y[1],Z[2]-Z[1]);
  if(D3<Ltiny)
     return(min(D1,D2));
  return(2*AreaPlane(X,Y,Z)/D3);
  }
double IntersectCircle(double Xp,double Yp,double
   U,double V,double Xw,double Yw,double Rw)
  {
  double A,B,C,D,D2,dX,dY,P,R2,T; /* compute the closest
     */
 /* intersection between      */
  A=U*U+V*V; /* a ray and a circle         */
  dX=Xp-Xw; /* the ray begins at Xp,Yp   */
  dY=Yp-Yw; /* and extends along U,V     */
  R2=Rw*Rw; /* the circle has radius Rw */
  D2=dX*dX+dY*dY; /* and center Xw,Yw         */
  if(D2<=R2) /* if the velocity is zero    */
     return(0); /* then an intersection can  */
  else if(A<DBL_EPSILON) /* only occur if the particle
     */
     return(-1); /* is already within the     */
 /* capture radius              */
 /* for an intersection to occur */
  P=V*dX-U*dY; /* the minimum distance between */
  if(P*P>A*R2) /* the ray and the well center   */
     return(-1); /* must not be greater than    */
 /* the capture radius             */
  B=2*(U*dX+V*dY);
  C=D2-R2; /* the point of intersection */
  D=B*B-4*A*C; /* results in a quadratic    */
  if(D<0) /* equation which must have */
     return(-1); /* real roots (i.e., DòO)     */
  A*=2;
  if(D>0) /* avoid as many operations */
     D=sqrt(D)/A; /* as possible, especially û */
  B/=A;
  T=-B-D;
  if(T>=0) /* return the smallest */
     return(T); /* non-negative root    */
  else
```

```
      return(-B+D);
   }
   double IntersectCylinder(double Xp,double Yp,double
     Zp,double U,double V,double W,double Xw,double
     Yw,double B,double T,double Rw)
   {
   double D2,dX,dY,R2,T1,T2,Tc;
   dX=Xp-Xw;
   dY=Yp-Yw;
   D2=dX*dX+dY*dY;
   R2=Rw*Rw;
   if(D2<=R2)  /* test for particle */
     if(B<=Zp&&Zp<=T) /* already in cylinder */
        return(0);
   if(fabs(W)>DBL_EPSILON) /* compute time to intersect
     */
     { /* upper and lower ends */
     T1=(B-Zp)/W;
     T2=(T-Zp)/W; /* if W÷0 then no intersection */
     } /* is possible unless Z lies */
   else if(Zp<B||Zp>T) /* within the cylinder already */
     return(-1);
   /* compute time to */
   Tc=IntersectCircle(Xp,Yp,U,V,Xw,Yw,Rw);  /* intersect
     circle */
   if(Tc>=0)
     if(min(T1,T2)<=Tc&&Tc<=max(T1,T2))
        return(Tc);
   return(-1);
   }
```

Animations and Concentrations

Updating the animations and concentration maps as each particle is tracked so as to create the ensemble impact is perhaps the second most time-consuming process.

```
   void UpdateParticles(int why,int steps)
     {
     BYTE b;
     int c,d,m,r,step,t,x,y,z;
     double T,T1,T2,X,Y,Z;
     if((!keep_trap)&&(why==TRAPPED))
        return;
     if(seed_fils<2)
        track_color=brand(blue,red);
     for(t=step=0,T=Tmin;t<Show.Nt;t++,T+=Show.dS)
        {
        if(T<Time[0]) /* skip snapshots before */
           continue; /* particle is seeded    */
        if(T>Time[steps-1]) /* test for snapshot later */
           break; /* than particle lifetime  */
```

```
   while(step<steps-1&&T>Time[step+1]) /* locate
particle time        */
   step++; /* step containing snapshot */
T1=Time[step]; /* time at beginning of step */
T2=max(T1+Ttiny,Time[step+1]); /* time at end of
step */
X=((T2-T)*Xloc[step]+(T-T1)*Xloc[step+1])/(T2-T1);
x=(int)((X-Xmin)/Show.dX);
if(EorF[step]<0&&x>0&&x<Nx-1)
   x+=(rand()%3)-1;
Y=((T2-T)*Yloc[step]+(T-T1)*Yloc[step+1])/(T2-T1);
y=(int)((Y-Ymin)/Show.dY);
if(EorF[step]<0&&y>0&&y<Ny-1)
   y+=(rand()%3)-1;
if(Nd>2)
   {
   Z=((T2-T)*Zloc[step]+(T-T1)*Zloc[step+1])/(T2-T1);
   z=(int)((Z-Zmin)/Show.dZ);
   if(EorF[step]<0&&z>0&&z<Nz-1)
      z+=(rand()%3)-1;
   }
d=(int)((T-Tmin)/Show.dT);
if(file_part&4)
   {
   c=y;
   r=z;
   }
else if(file_part&2)
   {
   c=x;
   r=z;
   }
else
   {
   c=x;
   r=y;
   }
if(c<0||c>=Show.Nc)
   continue;
if(r<0||r>=Show.Nr)
   continue;
if(d<0||d>=Show.Nd)
   continue;
m=(c*Show.Nr+r)*Show.Nd+d;
b=Show.trk[m];
if(b>black&&b<white)
   Show.trk[m]=track_color;
   }
}
void UpdateConcentration(int why,int steps)
```

```
{
int c,d,m,r,step,x,y,z;
double M,M1,M2,T,T1,T2,X,Y,Z;
if((!keep_trap)&&(why==TRAPPED))
  return;
for(d=step=0,T=Tmin;d<Show.Nd;d++,T+=Show.dT)
  {
  if(T<Time[0]) /* skip snapshots before */
    continue; /* particle is seeded    */
  if(T>Time[steps-1]) /* test for snapshot later */
    break; /* than particle lifetime */
  while(step<steps-1&&T>Time[step+1]) /* locate
 particle time       */
    step++; /* step containing snapshot */
  T1=Time[step]; /* time at beginning of step */
  T2=max(T1+Ttiny,Time[step+1]); /* time at end of
 step */
  M1=Mass[step];
  M2=Mass[step+1];
  if(M1>0&&M2>0)
    M=M1*Exp(log(M2/M1)*(T-T1)/(T2-T1));
  else
    M=((T2-T)*M1+(T-T1)*M2)/(T2-T1);
  X=((T2-T)*Xloc[step]+(T-T1)*Xloc[step+1])/(T2-T1);
  x=(int)((X-Xmin)/Show.dX);
  if(EorF[step]<0&&x>0&&x<Nx-1)
    x+=(rand()%3)-1;
  Y=((T2-T)*Yloc[step]+(T-T1)*Yloc[step+1])/(T2-T1);
  y=(int)((Y-Ymin)/Show.dY);
  if(EorF[step]<0&&y>0&&y<Ny-1)
    y+=(rand()%3)-1;
  if(Nd>2)
    {
    Z=((T2-T)*Zloc[step]+(T-T1)*Zloc[step+1])/(T2-T1);
    z=(int)((Z-Zmin)/Show.dZ);
    if(EorF[step]<0&&z>0&&z<Nz-1)
      z+=(rand()%3)-1;
    }
  if(file_conc&4)
    {
    c=y;
    r=z;
    }
  else if(file_conc&2)
    {
    c=x;
    r=z;
    }
  else
    {
```

```
      c=x;
      r=y;
      }
   if(c<0||c>=Show.Nc)
      continue;
   if(r<0||r>=Show.Nr)
      continue;
   m=(c*Show.Nr+r)*Show.Nd+d;
   if(Show.con[m]>black&&Show.con[m]<white)
      Conc[m]+=(float)(M/eV);
   }
if(why==TRAPPED||why==CIRCULATION)
   {
   if(d<Show.Nd)
      {
      step=steps-1;
      M=Mass[step];
      x=(int)((Xloc[step]-Xmin)/Show.dX);
      y=(int)((Yloc[step]-Ymin)/Show.dY);
      if(Nd>2)
         z=(int)((Zloc[step]-Zmin)/Show.dZ);
      if(file_conc&4)
         {
         c=y;
         r=z;
         }
      else if(file_conc&2)
         {
         c=x;
         r=z;
         }
      else
         {
         c=x;
         r=y;
         }
      if(c>=0&&c<Show.Nc)
         {
         if(r>=0&&r<Show.Nr)
            {
            m=(c*Show.Nr+r)*Show.Nd+d;
            while(d<Show.Nd)
               {
               if(Show.con[m]>black&&Show.con[m]<white)
                  Conc[m]+=(float)(M/eV);
               d++;
               m++;
               }
            }
```

also by D. James Benton

3D Articulation: Using OpenGL, ISBN-9798596362480, Amazon, 2021 (book 3 in the 3D series).

3D Models in Motion Using OpenGL, ISBN-9798652987701, Amazon, 2020 (book 2 in the 3D series).

3D Rendering in Windows: How to display three-dimensional objects in Windows with and without OpenGL, ISBN-9781520339610, Amazon, 2016 (book 1 in the 3D series).

A Synergy of Short Stories: The whole may be greater than the sum of the parts, ISBN-9781520340319, Amazon, 2016.

Azeotropes: Behavior and Application, ISBN-9798609748997, Amazon, 2020.

bat-Elohim: Book 3 in the Little Star Trilogy, ISBN-9781686148682, Amazon, 2019.

Boilers: Performance and Testing, ISBN: 9798789062517, Amazon 2021.

Combined 3D Rendering Series: 3D Rendering in Windows®, 3D Models in Motion, and 3D Articulation, ISBN-9798484417032, Amazon, 2021.

Complex Variables: Practical Applications, ISBN-9781794250437, Amazon, 2019.

Compression & Encryption: Algorithms & Software, ISBN-9781081008826, Amazon, 2019.

Computational Fluid Dynamics: an Overview of Methods, ISBN-9781672393775, Amazon, 2019.

Computer Simulation of Power Systems: Programming Strategies and Practical Examples, ISBN-9781696218184, Amazon, 2019.

Contaminant Transport: A Numerical Approach, ISBN-9798461733216, Amazon, 2021.

CPUnleashed! Tapping Processor Speed, ISBN-9798421420361, Amazon, 2022.

Curve-Fitting: The Science and Art of Approximation, ISBN-9781520339542, Amazon, 2016.

Death by Tie: It was the best of ties. It was the worst of ties. It's what got him killed., ISBN-9798398745931, Amazon, 2023.

Differential Equations: Numerical Methods for Solving, ISBN-9781983004162, Amazon, 2018.

Equations of State: A Graphical Comparison, ISBN-9798843139520, Amazon, 2022.

Evaporative Cooling: The Science of Beating the Heat, ISBN-9781520913346, Amazon, 2017.

Forecasting: Extrapolation and Projection, ISBN-9798394019494, Amazon 2023.

Heat Engines: Thermodynamics, Cycles, & Performance Curves, ISBN-9798486886836, Amazon, 2021.

Heat Exchangers: Performance Prediction & Evaluation, ISBN-9781973589327, Amazon, 2017.

Heat Recovery Steam Generators: Thermal Design and Testing, ISBN-9781691029365, Amazon, 2019.
Heat Transfer: Heat Exchangers, Heat Recovery Steam Generators, & Cooling Towers, ISBN-9798487417831, Amazon, 2021.
Heat Transfer Examples: Practical Problems Solved, ISBN-9798390610763, Amazon, 2023.
The Kick-Start Murders: Visualize revenge, ISBN-9798759083375, Amazon, 2021.
Jamie2: Innocence is easily lost and cannot be restored, ISBN-9781520339375, Amazon, 2016-18.
Kyle Cooper Mysteries: Kick Start, Monte Carlo, and Waterfront Murders, ISBN-9798829365943, Amazon, 2022.
The Last Seraph: Sequel to Little Star, ISBN-9781726802253, Amazon, 2018.
Little Star: God doesn't do things the way we expect Him to. He's better than that! ISBN-9781520338903, Amazon, 2015-17.
Living Math: Seeing mathematics in every day life (and appreciating it more too), ISBN-9781520336992, Amazon, 2016.
Lost Cause: If only history could be changed..., ISBN-9781521173770, Amazon, 2017.
Mass Transfer: Diffusion & Convection, ISBN-9798702403106, Amazon, 2021.
Mill Town Destiny: The Hand of Providence brought them together to rescue the mill, the town, and each other, ISBN-9781520864679, Amazon, 2017.
Monte Carlo Murders: Who Killed Who and Why, ISBN-9798829341848, Amazon, 2022.
Monte Carlo Simulation: The Art of Random Process Characterization, ISBN-9781980577874, Amazon, 2018.
Nonlinear Equations: Numerical Methods for Solving, ISBN-9781717767318, Amazon, 2018.
Numerical Calculus: Differentiation and Integration, ISBN-9781980680901, Amazon, 2018.
Numerical Methods: Nonlinear Equations, Numerical Calculus, & Differential Equations, ISBN-9798486246845, Amazon, 2021.
Orthogonal Functions: The Many Uses of, ISBN-9781719876162, Amazon, 2018.
Overwhelming Evidence: A Pilgrimage, ISBN-9798515642211, Amazon, 2021.
Particle Tracking: Computational Strategies and Diverse Examples, ISBN-9781692512651, Amazon, 2019.
Plumes: Delineation & Transport, ISBN-9781702292771, Amazon, 2019.
Power Plant Performance Curves: for Testing and Dispatch, ISBN-9798640192698, Amazon, 2020.
Practical Linear Algebra: Principles & Software, ISBN-9798860910584, Amazon, 2023.
Props, Fans, & Pumps: Design & Performance, ISBN-9798645391195, Amazon, 2020.

ROFL: Rolling on the Floor Laughing, ISBN-9781973300007, Amazon, 2017.
Seminole Rain: You don't choose destiny. It chooses you, ISBN-9798668502196, Amazon, 2020.
Septillionth: 1 in 10^{24}, ISBN-9798410762472, Amazon, 2022.
Software Development: Targeted Applications, ISBN-9798850653989, Amazon, 2023.
Software Recipes: Proven Tools, ISBN-9798815229556, Amazon, 2022.
Steam 2020: to 150 GPa and 6000 K, ISBN-9798634643830, Amazon, 2020.
Thermochemical Reactions: Numerical Solutions, ISBN-9781073417872, Amazon, 2019.
Thermodynamic and Transport Properties of Fluids, ISBN-9781092120845, Amazon, 2019.
Thermodynamic Cycles: Effective Modeling Strategies for Software Development, ISBN-9781070934372, Amazon, 2019.
Thermodynamics - Theory & Practice: The science of energy and power, ISBN-9781520339795, Amazon, 2016.
Version-Independent Programming: Code Development Guidelines for the Windows® Operating System, ISBN-9781520339146, Amazon, 2016.
The Waterfront Murders: As you sow, so shall you reap, ISBN-9798611314500, Amazon, 2020.
Weather Data: Where To Get It and How To Process It, ISBN-9798868037894, Amazon, 2023.

www.ingramcontent.com/pod-product-compliance
Lightning Source LLC
Chambersburg PA
CBHW071445220526
45472CB00003B/681